D0843456

*Around the World
with Chemistry*

Around the World with Chemistry

Kurt Lanz

McGraw–Hill Book Company

New York   St. Louis   San Francisco   Auckland   Bogotá
Hamburg   Johannesburg   London   Madrid   Mexico
Montreal   New Delhi   Panama   Paris   São Paulo
Singapore   Sydney   Tokyo   Toronto

Library of Congress Cataloging in Publication Data

Lanz, Kurt, date.
  Around the world with chemistry.

  Translation of Weltreisender in Chemie.
  Includes index.
  1. Farbwerke Hoechst AG.   2.   Chemical industries.
3. Voyages and travels.   4.   Chemists—Germany, West—
Biography.   I. Title.
HD9654.9.F33L3613   338.4'7'6600904   79-21209
ISBN 0-07-036356-0

German text first published by Econ Verlag GmbH, Düsseldorf
and Wien, under the title *Weltreisender in Chemie*.

Copyright © 1978 by Econ Verlag GmbH.

Translated by David Goodman.
Edited by Edward van Vlaanderen

Copyright © 1980 by McGraw Hill, Inc. All rights reserved.
Printed in the United States of America. No part of this
publication may be reproduced, stored in a retrieval system,
or transmitted, in any form or by any means, electronic,
mechanical, photocopying, recording, or otherwise, without
the prior written permission of the publisher.

1234567890   DODO   89876543210

*The editors for this book were Robert L. Davidson and
Geraldine Fahey, the designer was Naomi Auerbach, and the production
supervisor was Paul Malchow. It was set in Palatino by Haddon Craftsmen.*

Printed and bound by R. R. Donnelley & Sons Company.

Contents

Preface

Long before Kurt Lanz was elected President of CEFIC, the European chemical industries' association, we had met and over the years we had become friends.

It was quite some time ago that he called on me at the Ministry for Foreign Affairs where, as State Secretary, I was responsible for European economic integration. My return to Paris for a second term as German Ambassador had just been announced.

Lanz had a delicate problem involving Franco-German relations: Hoechst where Kurt Lanz was Vice Chairman of the Managing Board was about to acquire majority control of a highly reputed French company—with the owner's consent incidently—but in a situation where a certain reserve could be expected in French political circles traditionally adverse to foreign investment.

But first our conversation concentrated on new trends in German, French, American and English literature, the theatre, and music. We soon discovered that although our professional lives diverged—he a leading executive in a big chemical concern and I a diplomat—our personal approaches and our methods were not so different: it is always important to have people behind you 100% in whatever you want to achieve.

When we got down to the actual reason for his visit I realized that Kurt Lanz had a great deal of experience, was able and well-versed in the economic and political problems of many countries, and was a connoisseur of the fine arts. He is fluent in several languages which he perfected during extensive travels.

Lanz began his career in the same company of which he became a Member of the Managing Board at the early age of 38 and Vice Chairman of the Board 10 years later.

Since then he has been responsible for all commercial activities of an enterprise which employs 185,000 people. It is one of the largest companies in Germany and leads the chemical industry with subsidiaries throughout the world, including the United Kingdom and the United States where alone about 10,000 people are employed.

The book takes us on a world tour where Lanz re-establishes former contacts lost during the war and creates new ones through exports, investments, and joint ventures that touch many countries and all continents. We witness the growth of an industry from humble beginnings in a three-room office to plants that now employ many people. We learn that thousands of chemical and pharmaceutical substances have to be tested before a product can be introduced into the market and that it may take up to ten years and cost millions in research and development before the end product can be sold and earn money.

Lanz puts all such experiments and business ventures into the wider perspective of their political environment and we are often reminded of the different personalities in power at the time and their effect on the policies of a given country. We travel with him to dictatorships and democracies, to countries with all kinds of legal or economic systems only find that the secret of success lies in free enterprise and in good human relations.

At the United Nations where I was Ambassador of the Federal Republic for almost six years, a Japanese colleague once told me that the art of diplomacy was to read what is *behind* a paper to discover its motives. Kurt Lanz has been able to do this and it is fully demonstrated in this account of his travels.

While at the United Nations I also learned about macroeconomics and the problems of the underdeveloped and undernourished parts of the world. It was only Kurt Lanz' book which made me realize that the entrepreneur of today not only has to deal with industrial and commercial problems but must also concern himself with a positive and practical response to the many questions arising from hunger, under-development, and disease.

The so-called multinationals, which are now quite unjustly the target of blame and criticism, were in those days not as prominent in the political limelight as they are today. Hoechst with its Tower and Bridge logo visible in about 130 countries belongs in that category. Lanz refers to the multiplicity of their functions: to be a catalyst and a carrier for growing international trade and to provide safe jobs for many.

I am not a chemist and the names of chemical products and medical specialties mean little to me. However, I realize that the importance of this industry lies in the assistance it provides mankind, whether in the areas of food, health, the environment, or in the whole process of our civilization. The importance of the chemical industry thus becomes apparent to all readers far beyond the mere figures of sales and production.

What struck me reading the book is Lanz' expertise in commerce and industry, his eye for political realities, and his sheer humaneness.

We can only wish that, after this book, Kurt Lanz will sit down again pen in hand and that other highly responsible personalities will find the time to pass on their knowledge and experience with such clarity and vision.

Sigismund Freiherr von Braun

1
The Rise and Fall
of the I. G. Farbenindustrie

Surrounded by well-tended parks, the mighty I. G. Farben building
towered above the distinguished villas along Frankfurt's Grüneburg-
weg. Berlin architect Hans Poelzig, designer of Berlin's main theater
and exhibition grounds, had created an appropriate setting for Germany's largest concern. The six seven-story wings that fanned out
from the eight-story center somewhat presaged the later Pentagon in
Washington, seat of the U.S. Department of Defense.

The I. G. Farben board management had its offices on the vast first
floor of the building. The board was not a "council of the gods,"
hovering far above the earthly scene—a description popular among its
critics; not a monolithic body of industrial princes with absolute
power nor a secret alliance of capitalists bent on exploitation. Rather,
it was a motley group of scientists and commercial experts of varying
strengths of personality. They owed their positions to ability, loyal
service, and in some cases, to a little luck or patronage. Each of them
had experienced triumphs and defeats in his career, had successfully
maneuvered his plans through one of the governing technical or commercial committees, or had seen them rejected. Each had encountered

1

discord and envy, fairness and decency, and had learned that outstanding ability is not necessarily coincident with character of similar
distinction, or indeed that at times the reverse applies.

In short, the I.G. board was a group of people who, although undoubtedly more powerful than some other contemporary captains of
industry, were not very different from present-day board members of
large companies.

Of course, I knew little of these things when, on April 1, 1937, just
before 8 A.M., I entered the Grüneburg. I was one of two or three dozen
young men who had succeeded in finding an opening as a trainee with
the mighty company.

Born in Mannheim in 1919, I lived first in Kehl on the Rhine and
later in Lahr, a little town of about 16,000 inhabitants in the state of
Baden, with some attractive medieval monuments and half-timbered
houses. It was in Lahr that I passed my university entrance examination in 1937.

My father came from northern Baden. However, he had lived for
many years in Russia. It was there that he married my mother, who
was born in Riga on the Baltic, but had spent her youth in Russia.
They were interned in Russia until the end of the First World War. My
father's last job in Russia was with the Mannheim company of Heinrich Lanz who, despite his name, was not related to my family. Afterward, he became manager of the Kehl branch of the Deutsche Bank
and, finally, manager of a French confectionary company. My father
died in 1937 at the rather young age of fifty-three. His death compelled me to enter professional life as soon as possible. University
study, probably in law or industrial economics, was out of the question.

The I.G. Replies First

From my early youth, I had enjoyed learning foreign languages, especially English and French. This inclination increasingly helped to
shape my decision to seek employment abroad. As a result, when my
high school training was finished, I applied to three companies that
were particularly active overseas: Bosch, Zeiss, and I. G. Farbenindustrie. Since I.G. was the first to reply to my application, I ended up in
Frankfurt. My involvement with the chemical industry was thus
somewhat accidental.

A young man entering the reception lobby of the I.G. for the first time on April 1, 1937, had to be vastly impressed by this building, "elevated," to him, in every way. Standing in the proud Grüneburg, as the site of the I.G. building was called, it was rather easy to forget that the I.G. Farbenindustrie was a combine not only formed in response to economic stresses following World War I, but also one which had been hit for a second time by the worldwide depression in the early thirties. Of course, this experience was not confined to the I.G., which in spite of its size accounted for only about a quarter of total German chemical sales. The depression had similarly affected the many small and medium-sized companies so characteristic of the German chemical industry. Nevertheless, between 1925 and 1937, the year I joined the company, Germany's share in world chemical trade had again steadily increased to over 28 percent, the 1913 level. Germany's share in world chemical production, which had dropped considerably between 1924 and 1932, also rose to 19 percent during 1936 and 1937.

I'm with I.G.

A young man in his early twenties, coming from the comparative tranquility of a small provincial town like Lahr to a commercial metropolis like Frankfurt, was bound to be fascinated by the magnificent way in which city and company complemented each other. Here, on the one hand, was a city, a commercial center with international tradition. In spite of the political violence of the thirties, it had remained a tiny oasis of liberal life. On the other hand, there was an internationally oriented industrial complex whose members had helped make German products famous and were leaders of modern chemistry. All this deeply impressed a young man in search of a career. To be able to say, "I am with I.G.," was a badge of prestige at that time.

In the months before joining the company, I had read some of the history of the I.G. Farben. Certainly I knew of Carl Duisberg and Carl Bosch who, against great obstacles, had forged the company during the crisis years after World War I. The concept of closer cooperation among German chemical companies had already emerged shortly after the turn of the century. The head of Bayer in Leverkusen, Carl Duisberg, had brought back new insights from a trip to America. There the era of the industrial magnates, such as Hill, Harriman, Rockefeller, Morgan, and Mellon, was in its heyday. Giant industrial trusts were being formed, some of which attained optimum efficiency and productivity. Nonetheless, the markets for dyestuffs and pharmaceuticals were dominated by German imports or the products of German subsidiaries in the United States.

Thus, each of the German chemical companies had some justification at the time to feel confident of its own strength. The impressionable Duisberg, on the other hand, seemed to have a sixth sense for the need to greatly strengthen German chemical industry against growing competition in the American market.

First Mergers before the 1914–1918 War

Indeed, the first move had been quietly made in 1904. Hoechst formed a partnership with Leopold Cassella & Co. in Frankfurt, much to the surprise of Carl Duisberg, on his honeymoon in Italy at the time. Duisberg at once applied his outstanding organizational talent. It was not long before he matched the Hoechst duet with a trio made up of Bayer (Leverkusen), BASF (Ludwigshafen), and Agfa (Berlin).

At first, Hoechst's and Duisberg's groups faced each other with some misgivings. But both soon realized that they would not be able to afford an unduly narrow competitive philosophy, especially with regard to the foreign markets. In the end, however, it was probably the shock of the war and the loss of world markets, rather than considerations of company policy, that induced the two groups to sit down together round the table.

The rupture in Germany's international trade, caused by the war, soon led to quite severe shortages of raw materials. Simultaneously, Germany's prewar customers throughout the world keenly felt the severance of commercial relations with German industry. Above all, they increasingly missed the products of the German chemical industry. After all, at the outbreak of war, German factories were supplying some 80 percent of the world requirements of synthetic dyestuffs. Drugs from Germany, the "world's pharmacy," accounted for an almost equal share of the world market.

The problem was probably greatest for the United States. Cut off from German exports, the country had to meet not only domestic needs from its own resources but also the Allies' demands for materials. A large and rapidly expanding nation suddenly faced the need to replace, as quickly as possible and from its own resources, the interrupted supplies from Germany, 5000 kilometers away.

It was a need that called for both financial potential and creative initiative. The challenge was met with American efficiency and enthusiasm. It was clear that this massive mobilization of resources would greatly advance both chemical research and production in the United States. It also meant future powerful competition for the German companies, especially in overseas markets.

These prospects provided a fresh impetus for Germany's dyestuffs chiefs to begin talks. By August 1916, they decided to form a "little I.G." composed of the two groups formed in 1904, together with Griesheim Elektron (near Frankfurt) and Chemische Fabriken Weiler-ter Meer (at Uerdingen, near Cologne). The governing body of this small I.G. was a council on which the leading men of each constituent company were represented.

From the Little to the Big I.G.

When I started my training with I.G. in Frankfurt, this chapter of its history had already lasted twenty years. Many upheavals had occurred during this period before the giant I.G. organization assumed its final form.

At first, there was the hopelessness that lay oppressively over Germany as the result of the lost war. In the chemical industry, the losses in production and profitability had exceeded even the most pessimistic forecasts. Germany's plants in the allied countries had been seized, thousands of patents had been lost, the entire German industry was subjected to exaggerated demands and unrealistic reparation claims. The country was convulsed by successive political and economic crises. How could the German chemical industry regain its former world prominence under those circumstances?

Once again, it was Carl Duisberg who took the initiative. When the German currency had finally been stabilized, merger talks were renewed. But this time, to Duisberg's surprise, the objective was no longer just a loose association. Halfway solutions were not discussed. Even more forcefully than Duisberg, Carl Bosch, chief of BASF, now threw all the weight of his personality behind a "major solution" of the problem.

As a first step in the merger, BASF changed its name to I.G. Farbenindustrie Aktiengesellschaft. It moved its headquarters to Frankfurt and increased its capital by adding the funds introduced by the other companies that joined the new organization. These were Bayer, Hoechst, Agfa, Griesheim Elektron, and Weiler-ter Meer.

When the I.G. Farbenindustrie was founded on December 2, 1925, the *Frankfurter Zeitung* commented that a transaction of these dimensions had never before taken place in the history of European finance. Carl Bosch was appointed chairman of the board of management. He conducted the affairs of the new concern for almost a dozen years through the greatest difficulties, especially after Hitler came to power and the internationally oriented German chemical industry was confronted with many new problems.

Carl Bosch was born in Cologne in 1874. The famous industrialist Robert Bosch was a member of his family. Carl Bosch joined BASF shortly before the turn of the century and was soon immersed in attempting the synthesis of ammonia. His success in translating a laboratory discovery by a well-known chemist, Fritz Haber, into an industrial process is among the outstanding achievements of chemistry. Bosch was awarded the Nobel prize in 1931 for his development of chemical high-pressure methods which made possible the production of fertilizers from nitrogen.

Carl Duisberg became the first chairman of the I.G. supervisory board. He too had been trained as a chemist. Born in Barmen in 1861, Duisberg devoted his Bayer years to basic research and the development of new dyestuffs. The close collaboration between science and industry in Germany is in no small measure due to his efforts.

When I.G. was founded, Germany had largely recovered from the effects of World War I. And in the company's first years, although numerous demanding investments were undertaken, sales increased from slightly over 1 billion Reichsmarks (RM) in 1926 to RM 1.4 billion in 1928. Of course, the subsequent world economic crisis took its toll of I.G. sales. By 1932, they had dropped to an all-time low of RM 871 million (equal to approximately US$207 million at that time).

In 1930, the upright but dull Heinrich Brüning became Chancellor of Germany. However, like Hermann Müller (1926–1930) before him, and Franz von Papen (May to November 1932) and General Kurt von Schleicher (December 1932 to Hitler's accession in January 1933) after him, he was unable to master the crisis. Unemployment had reached more than 6 million by that time. Bosch had great hopes in Brüning, of whom he was very fond. At Brüning's personal request, he agreed to Professor Hermann Warmhold's leaving the I.G. board to become minister of economics in Brüning's cabinet.

When I joined I.G. in 1937, the business had greatly recovered. Sales had reached RM1.5 billion, at a time when there were still a million unemployed in Germany. Hermann Göring, who was in charge of the then-current four-year plan, had coined the fateful phrase "guns instead of butter." Through autobahn construction and increasingly through rearmament, the number of jobless soon dropped significantly.

For Carl Bosch, the years after 1935 were filled with internal conflicts. He was fond of telling his colleagues, "If you have to choose between a genius and a man of character, forget the genius." Doubtless he was deeply affected by the "characters" who had now floated to the top.

At the same time, he resolved not to let go of the reins, come what might. At a ceremony in Leverkusen in 1933 celebrating the completion of fifty years' service by Carl Duisberg and Arthur von Weinberg, he gave an address that revealed his warm, humanistic attitude toward life.

Carl Duisberg greatly esteemed Bosch, whose lively temperament contrasted sharply with his own comparative ponderousness. But their first human encounter, in the true sense of the word, did not take place until Duisberg was near death. Bosch was summoned, and the two men spoke their final words to one another. Duisberg said afterward: "In the end, I won the friendship of Bosch after all." Bosch confirmed later that he "came to a complete understanding with Duisberg during this discussion."

That was in 1935. In the meantime, Bosch had become chairman of the supervisory board. Privy councillor Dr. Hermann Schmitz had been appointed as chairman of the board of management. During his career with BASF, he was regarded as a talented financial expert. Above all, he had the confidence of the major divisions of I.G., but he was hardly more than a first among equals.

One reason why Carl Bosch withdrew to the chairmanship of the supervisory board was his deepening feeling of depression over political developments in Germany, especially after having met Adolf Hitler. At the end of 1933, the new Chancellor was planning to establish a General Economic Council and wanted to meet Bosch, who had been selected as the representative of the chemical industry. After their meeting, Bosch knew what to think of the nation's new rulers. He warned Hitler about the danger of isolating Germany. He pointed out that considerable impairment would result from the reduction in technological performance as a consequence of forcing outstanding Jewish scientists to leave the country without regard to their services. Bosch suggested that such unhappy consequences should be avoided. Eventually, Hitler interrupted him, saying curtly, "You do not understand anything of these matters."

Bosch was distraught when Hitler, knowing little of science and nothing of economics, rejected his pleas with the words: "Never mind, then we'll work for a hundred years without physics or chemistry." The experience plunged Bosch into a serious depression, from which he sought refuge in scientific preoccupations or in travel.

Bosch was personally distressed by the increasing elimination of Jewish scientists and industrialists, many of them his friends. The Nuremburg Laws, decreed in 1935, relegated all Jewish citizens to second-class citizenship. Two years later, the Jewish members of I.G.'s supervisory board had to leave, including both Carl and Arthur von Weinberg as well as the highly nationalistic *grand seigneur,* Richard Merton, son of the founder of the Metallgesellschaft, Germany's largest nonferrous-metals producer.

Arthur von Weinberg had been owner and head of Cassella (located in an industrial suburb of Frankfurt). He had initiated cooperation with Hoechst and had played a decisive part in the I.G. merger. A great chemist, industrialist, patriot, and personality, he had been made an honorary citizen of Frankfurt, which had even named a street after him. Unlike Merton, who yielded to the urging of his friends and emigrated to England at the last moment, von Weinberg remained in Germany. He could not believe he was in personal danger and paid for this error with his life. In 1942 the Gestapo took him to the Theresienstadt concentration camp. I have often wondered whether the I.G. Farben company might not have been able to rescue this well-deserving founding member from this fatal sentence.

The political influence of the I.G. has been greatly overestimated. It is in any case an ideological distortion of history to claim that industry, the "ominous capital," had in reality dominated the Führer and his party from the beginning. No one could rule Hitler. Although the National Socialists (the Nazis) moved more cautiously in industry than in other fields, they betrayed their intentions in the autumn of 1933, when they began stripping the former Association of German Industry of its powers. Krupp von Bohlen Halbach replaced Duisberg as chairman of this organization in 1932. Furthermore, Economics Minister Hermann Göring was granted almost unlimited powers as administrator of the four-year plan to convert large sectors of industry to arms production.

Unlike some other companies, I.G. had not supported the Nazis before they took power. On the contrary, I.G. was internationally oriented and had concluded numerous licensing agreements, especially with American companies. Many of these agreements were of a long-term nature.

Nevertheless, the management of I.G., in order to survive, had to come to terms with the new rulers so as to avoid state intervention, such as total nationalization of the firm or management by party or army bosses. Perhaps this danger was overestimated; perhaps some of the leaders of the I.G. were excessively obsequious; perhaps some were overly impressed by the economic progress of the Nazi government, which had been aided by the incipient worldwide economic recovery.

How rigorously the "supreme judge of the nation" proceeded against those with different views was amply proved in the aftermath of the Reichstag fire and the murders of Ernst Röhm and General Schleicher. One needs to bear these facts in mind when questioning why Germany's largest company was so accommodating to the new rulers. Carl Bosch's sincere attitude becomes even more admirable in this light.

In the beginning, the Nazi rulers themselves treated the I.G. extremely politely, given the circumstances. They recognized its worldwide reputation; indeed, the company name had become synonymous with German research and product quality. Of course, in the end they made use of these capabilities, for example, by transferring new investments to central Germany or the eastern border areas to promote a policy of self-sufficiency. Within these limits, however, the company enjoyed a degree of autonomy, at least during the first Nazi years.

When I joined the I.G. in 1937, the five-day week had just been abolished. Things were indeed improving. For a number of years there had not been a market economy. In almost every important field there were cartels, syndicates, and conventions. They represented the magic formula for manipulating an economy that had lost confidence in the working of the free market. Many countries had erected high tariff barriers in order to protect their own industries.

The high tariff barriers affected not only the German dyestuffs manufacturers, which enjoyed no such state protection, but also the Swiss companies, highly dependent on exports. For this reason, the Swiss dyestuffs manufacturers had formed a pool as early as 1918. Ten years later, I.G. established an international cartel with this group and with French manufacturers. The Tripartite Cartel laid down worldwide manufacturing and selling quotas. The I.G. was allocated 71.7 percent, the Swiss group 19 percent, and the French group, which had joined the German-Swiss association, 9.3 percent. Britain's Imperial Chemical Industries joined this cartel in 1932.

Cartels were encountered not only in the chemical industry. In the late 1920s, European industry appeared to be a web of cartels and syndicates. Fortunately, this inflexible system, restricting if not suffocating true competition, was permanently banished after World War II.

First Training Station: Hoechst

To the outside world, the I.G. presented itself as a monolithic empire. But even a young insider soon realized that the companies and plants which had risked the great merger of 1925 had—even by 1937—not nearly reached the level of integration envisaged by I.G.'s founding fathers. To some degree, this lack of integration also applied to the top people. In their hearts, Carl Bosch had always remained "the man from Ludwigshafen" and Duisberg "the man from Leverkusen." Most of the directors, staff, and workers regarded themselves as employees not so much of the I.G. as of Bayer, BASF, or Hoechst. The slow pace of integration probably was least felt in the Grüneburg headquarters. It was much more apparent in the works outside.

I spent my first year of training at Hoechst. Since the founding of the I.G., this site has been part of the Central Rhine group. The other groups were the Upper Rhine, centered at BASF, and the Lower Rhine, led by Bayer, with the ever-growing group in central Germany (Bitterfeld, Schkopau, and Leuna).

Apart from this regional arrangement, there was also a vertical organization. It consisted of three divisions, each responsible for certain products or product groups. Division I was responsible for nitrogen-based products, methanol, synthetic fuel, oil, and lubricants; Division II handled heavy chemicals, organic intermediates, dyestuffs, plastics, synthetic rubber, detergents and pharmaceuticals, and plant protection. Division III was concerned with photographic items (Agfa), rayon, and other synthetic fibers.

The chiefs of these divisions, the "I.G. princes," played a special role in the board of management, since they were members of a central committee, the top body of the I.G., headed first by Bosch and later by Schmitz. So the board of management consisted of equals, but some were more equal than others. This appears to be rather customary with big companies even today. Anyhow, when I joined Hoechst, it was not adequately represented on the board.

Hoechst: A "Sleeping Factory"

I need not give a detailed account of the history of Hoechst. It has already appeared in the centenary volume *A Century of Chemistry* by Ernst Baümler, and, above all, in *Challenging Years,* the reminiscences of Professor Karl Winnacker, who so decisively and successfully molded the development of Hoechst after the Second World War. Winnacker had started as a young scientist in Hoechst in 1933. At first sight, he wrote in his memoirs, Hoechst struck him as a "sleeping factory." I had the same impression on first working with Hoechst four years later.

True entrepreneurial spirit hardly had room to breathe here. Major decisions could not be made at Hoechst. Where dyestuffs and chemicals were concerned, the appropriate division and joint sales office in Frankfurt had jurisdiction; pharmaceuticals and agrochemicals were controlled by the divisional headquarters and the joint sales office in Leverkusen. The finance and economic divisions, central raw material purchasing, and public relations were centralized in Berlin.

Hoechst was an extreme example of a merger-handicapped organization. Hoechst managers had fought tenaciously in the protracted tugs-of-war when big decisions were made on allocations of products to the various sectors. Ultimately, however, they could not succeed against the dominant personalities on BASF (Ludwigshafen) and Bayer (Leverkusen).

Both these companies had, in a sense, "outstripped" Hoechst technologically in the twenties and early thirties. Ludwigshafen had high-pressure synthesis of nitrogen, methanol, and fuel; Leverkusen pioneered the development of synthetic rubber. Eventually, even Hoechst pharmaceuticals were taken over by Bayer. And yet, Hoechst had been the cradle of chemotherapy, the birthplace of Ehrlich's world-famous Salvarsan and Behring's vaccines.

Hoechst Pharmaceuticals under the Bayer Cross

These and other great achievements were never adequately acknowledged either at the merger of I.G. or thereafter. The many Hoechst pharmaceuticals available to doctors throughout the world had been transferred to the Bayer organization. Henceforth, the Bayer identifying cross appeared also on Hoechst pharmaceutical packages. In fact, I.G. focused its entire pharmaceutical promotion on this cross. It was only a question of time before it would drive the symbol of Meister Lucius & Brüning into oblivion.

Bayer's worldwide organization—especially strong in the pharmaceutical sector—was massively reinforced during the I.G. era. After World War II and Bayer's dismemberment, much effort was needed to convince even Hoechst's own people that an independent Hoechst needed its own trademark and, of course, its own organization, both at home and abroad. Subsequent history has proved how right this conviction was.

On their first working day, I.G. trainees in Frankfurt were assigned to either the dyestuffs or the chemicals division. As I was sent to the dyestuffs division, my practical experience began with dyeing wool and cotton in the Hoechst colorants department. Later, I served in the shipping department, color store, plant accounting, stock control, purchasing, costing, and finally pharmaceutical packaging control.

The year in Hoechst passed quickly and I returned to the Grüneburg headquarters. My last post there was in dyestuffs exports, in the Scandinavian section. The I.G. trainee program was excellent. Still, the job changed rather frequently; there was at times little opportunity for practical work and, instead, one learned the sequence of operations. On the other hand, for someone eager to look and listen, this variety was very instructive. I often attended the evening lectures arranged by leading people of I.G. almost weekly. For studying alone, a business correspondence course proved very helpful.

An Island of Culture

Having come to Frankfurt from the tranquil provinces, I tried, of course, to enjoy life in the atmosphere of this city. The cultural scene impressed me especially deeply.

The *Frankfurter Zeitung,* regularly read even in my father's house, helped prepare me for my encounter with the city and its economic and intellectual life. In Frankfurt I continued to read this newspaper everyday. I was much impressed when I heard that Carl Bosch had ensured that the *Frankfurter,* which had a difficult time under the Nazis, would remain financially secure. This support had been neither expected nor requested.

Even in the late thirties, Frankfurt appeared to be a kind of island where some vestige of its once-free spirit was preserved. For example,

Frankfurt has a university which owes its existence entirely to the initiative of its citizens. Opened in October 1914 without any help from the state, it was financed solely by private donations.

Glimmerings of Former Times

Foundations have always been symbols of the open-mindedness, the sense of social responsibility, of this city; among them are the Senckenberg Natural Research Society, the Physical Society, and particularly the Georg-Speyer Foundation for the promotion of chemotherapy. This foundation was established by Franziska Speyer and gained worldwide note through its support of Paul Ehrlich, the famed biologist. Until the early thirties, Frankfurt was truly a center of scientific and cultural vitality. And even in the years after Hitler came to power in 1933, a few glimmers of this former glory could be detected in the city.

I was eager to absorb all this glory. I often visited the Städel Institute of Art and its famous art gallery. Here, the masters of Dutch and Italian painting impressed me profoundly. This municipal gallery had also opened its doors to modern art, although most of it was, of course, regarded as decadent by the Nazis. The monumental sculptures of the late Middle Ages, exhibited in the Gothic Dominican church, were unforgettable. And the Römerberg open-air festival with Gerhart Hauptmann's *Florian Geyer* was tremendously impressive.

The opera, a landmark of Frankfurt even as a ruin, had a talented company at the time. With few practicable contemporary works available, the playhouse concentrated on classical dramas. One can understand why Frankfurt audiences broke into applause when, in Schiller's *Don Carlos,* Marquis Posa retorted to the despot, "Give us freedom of thought, Sire." And the Schumann theater, a popular variety theater, often featured Claire Waldoff. She became famous for her politically risqué chansons, poking fun at some of the political figures of the day.

In the book world, Nazi literature, including blood-and-soil romanticism, seemed to dominate. Many people therefore turned to light fiction. Among the widely read authors of the time were Hans Carossa, Ernst Wiechert, Manfred Hausmann, and the English writer A. J. Cronin. I confess that Reinhold Conrad Muschler's sentimental novel *The Unknown Woman of the Seine* affected me quite deeply in those days.

Racial Barriers, Then Racial Persecution

Frankfurt in 1937 and 1938 had another dimension. Since taking power, the Nazis had made it clear that they were in deadly earnest about their racial theories, however muddled their ideas. The emigration of Jewish scientists and artists, the unknown number of arrests, and the increasing violence had transformed the practice of racial legislation into latent terror, alarming even those Germans who had been ready to give the new rulers a chance.

Scarcely a day passed without some news in the *Frankfurter* about new measures against civil servants, lawyers, scientists, merchants, and even artists. Writing much more between the lines became a specialty of the *Frankfurter*. Like many other Germans, I soon learned to spot these hidden messages.

The "crystal night" of November 9, 1938, was the momentary climax of terror, later to be escalated beyond belief. In one single night, almost 300 synagogues were burned down, a thousand Jewish shop windows shattered (hence the term "crystal night") 20,000 Jews arrested, and dozens of them killed.

Frankfurt once had the largest Jewish community in Germany. All the more reason, then, that this pogrom was bound to strengthen the doubts in the conscience of a young person like myself. All these events reinforced my desire to leave Germany, perhaps with the help of the company. But I was not to succeed in this goal during the I.G. era.

Nine Grands Prix for the I.G.—or the Reverse Side of the Picture

Of course, in 1938 most of us did not know how close we were to war. The I.G. Farben was immersed in fascinating work, as was confirmed by the nine Grands Prix won at the Paris World Exhibition in 1937. The awards were first of all for Indanthrene dyestuffs, and then for the new chemotherapeutic agent Prontosil. This was the first sulfonamide, whose efficacy had been discovered by Gerhard Domagk. The third honor was for the successful liquefaction of coal. Among the other products that gained awards were Buna, the synthetic rubber, Vistra fiber, and Kalle's cellophane.

The growth potential of the divisions making these products was shown in subsequent sales. Oil and synthetic gasoline sales rose from RM162 million (US$65 million) in 1939 to RM352 million (US$141 million) in 1943. Buna and plastics rose from RM90 million (US$36 million) to RM406 million (US$161 million), rayon staple and artificial silk from RM163 million (US$65 million) to RM219 million (US$87.5 million), and pharmaceuticals and pesticides from RM152 million (US$61 million) to RM294 million (US$118 million). These were increases averaging 55 percent.

These years strengthened the realization that with the rapid progress of chemistry and its technological foundations, a broadly based, steadily deepening research program was needed to achieve the economic targets. It also became clear that none of the companies alone could provide on its own the large-scale syntheses required, for example, for nitrogen production, coal liquefaction, Buna rubber, or manmade fibers. Projects of such magnitude could be handled only by really strong groups that could provide both the necessary funds and the brains of their large scientific and technical staffs.

One of the most interesting developments of those I.G. years was rubber synthesis. Earlier research work was resumed in Leverkusen in 1926. Further developments centered on synthesizing types of rubber with certain characteristics superior to those of natural rubber. Both Buna and Perbunan were successful: aging and heat resistance and greatly improved abrasion resistance (particularly important for car tires), as well as swelling resistance to gasoline and oil, were properties in which the synthetic products surpassed the natural substance.

Further experiments up to 1935 encouraged I.G. to construct an experimental plant at Schkopau near Halle with an initial capacity of 2400 tons. The lignite deposits of I.G. in this area assured cheap electricity for the carbide furnaces. Hydrogen for the third stage of the butadiene process was piped in from the nearby Leuna works through a high-pressure line.

As BASF had been during coal hydrogenation, I.G. was confronted for the first time with considerable financing difficulties. Schkopau was to cost RM200 million (US$80 million), beyond the means of even I.G. Only after long negotiations with the government was a solution found. The state guaranteed Buna sales of up to 30,000 tons a year and provided a loan for construction. Production problems also had to be overcome through collaboration with the German rubber processors and the tire industry.

Nevertheless, early experience in Schkopau was so favorable that a second plant was erected in 1938, this time in Hüls near Recklinghausen. A third Buna plant was added in 1942 through 1944 in Ludwigshafen. In 1943, total output of these three plants reached 120,000 tons of Buna a year.

Such developments were grist to the self-sufficiency mill of the Nazi government. Coal liquefaction, and thus the manufacture of synthetic fuel, would become a critical need when connections with the world market were severed. By 1933, high-pressure hydrogenation of lignite and lignite tar had developed to such an extent that translation of the processes to an industrial scale had become feasible. These plans were in the interests of I.G., but they also provided aid to the job-creation program, and they were helpful because of the ever-increasing shortage of foreign currency.

In the bold and comprehensive job-creation program of the Nazi rulers, Hjalmar Schacht, president of the Reichsbank and, later, minister of economics, played a vital role with his ingenious advance-financing plan. This experienced banker was indispensable to the rulers of Germany, at least for some time. Like Alfred Hugenberg, the nationalistic Berlin newspaper baron, Schacht had been a protector of the "Harzburg Front." This was an association of the German nationalists, the Stahlhelm (union of ex-servicemen), and the SA, the Nazi stormtroopers. The Harzburg Front was the major agitational force of National Socialism. With it Hitler launched his final attack on the crisis-ridden Weimar Republic. Once Schacht had gained full financial control, he created the often-cited "Mefo" notes which were drawn, as a substitute for nonexistent cash, on the Metallforschungs-GmbH (Metal Research, Ltd.), founded specifically for the purpose. The notes could be extended for a period of five years. This transaction assisted the take-off of the government's economic program. However, the government later had no qualms over sending the by-then irksome Schacht first into the wilderness and then into a concentration camp.

At the end of 1933, a long-term agreement was concluded with the Nazi Ministry of Economics requiring the I. G. Farben to raise its production of gasoline from lignite to 350,000 tons a year. The total contribution of all its projects is best illustrated by noting that the Leuna plant, near Magdeburg, achieved an output of 600,000 tons of gasoline and diesel fuel in 1943.

I. G. Farben and the War

The preparation for war was naturally accompanied by an expansion of all industries needed for the production of armaments and the supply of the armed forces. But, however important the role of I.G. in Germany's self-sufficiency and rearmament, however attractive the enormous technological possibilities, none of the leading personalities of the company could have wanted the war. Nor, despite the later Nuremberg charges, could they have played an active part in bringing it about.

Such accusations are based on basic misconceptions. Surely the experience of World War I argues overwhelmingly against the attractions of war for large-scale chemical manufacturers. They had worldwide organizations which once supplied up to 80 percent of global requirements of dyestuffs and pharmaceuticals. A war—whether won or lost—could result only in the destruction of patiently built-up international sales networks and, more important, the loss of internationally recognized patents and manufacturing facilities abroad.

These are not tempting prospects for industrialists. That I.G. did not expect war is indicated by its close cooperation with American companies like Standard Oil or E. I. du Pont de Nemours. As late as 1938, I.G. and Standard Oil had begun negotiations over the mutual exchange of synthetic rubber technology. Earlier, I.G. had granted foreign licenses for its coal hydrogenation processes. Just before the war, I.G. and Du Pont had agreed on the exchange of licenses for the production of Perlon and nylon, respectively. Fritz ter Meer, board member and head of Division 2 of I.G., describes this incident in detail in his book *The I. G. Farben* (published in 1953).

Nobody knows how many top officials of I. G. Farben sympathized wholeheartedly with the new rulers. Many had to cooperate, for better or for worse, in one instance or another. Some may have doubted the enthusiasm with which economic problems were tackled, others may have tried to anticipate state supervision or even nationalization. In either case, the dangers inherent in the ideological background against which all these events took place were soon realized by many individuals. No internationally renowned scientist or industrialist was prepared to identify himself with the ideological beliefs of the Nazi state. Of course, there were exceptions, but the attitude of most people was clearly shown by deliberate detachment from the Nazi regime.

Individualists Are Unsuitable as Ideologists

So far as I, a young businessman working in a large company just before the war, was concerned, my heritage and upbringing militated against the Nazi regime. A healthy degree of inborn individualism and strong skepticism regarding any loss of freedom kept me from losing myself in utopian fancies of whatever kind. Individualists just are poor material for collectivist doctrines.

During my traineeship in Hoechst, any youngster was, of course, required to become involved in Nazi organizations. The state was anxious to gain total control of the young at the earliest opportunity. In I. G. Farben as in all other industrial units, there was an offshoot of the Labor Front in which all former trade unions and staff associations had been unified in October 1934. Membership was obligatory. But in contrast with other party organizations, the Hoechst group did not have a pronounced political profile. Membership was more of an excuse for avoiding other compulsory duties. Although, on joining one had to acquire the corresponding blue uniform by installments, I never wore it.

My training was completed in the spring of 1940. As already mentioned, my last job was in the sales department for the Scandinavian countries to which Hoechst was still exporting. This may have been one reason why I did not don my army uniform until October 1940. I had no more liking for the armed forces then I had for the new rulers. I had heard enough from my contemporaries about what went on behind the front, especially in the eastern area.

After basic military training, I succeeded in joining an interpreters' company. In 1941, I passed my interpreter's exams in French, English, and Spanish, and thereafter was assigned to the language service. Most of my time was spent in field hospitals for French and British prisoners of war. I returned to Frankfurt as "Corporal, retired" in the spring of 1945—with proper discharge papers, of course. Although bombed out and hungry, I had the feeling of being a free man at long last.

2
Globe-trotter on a Bicycle

May 1945, Frankfurt am Main: Like many other Germans, I experienced conflicting emotions on my return. Frankfurt had been widely devastated by aerial raids. Whole streets, whole quarters, had disappeared. The beautiful city center had been completely wiped out. A cross dangled symbolically from the dome of the church of St. Paul. The bridges across the Main had been destroyed and the river could be crossed only by ferry. Traffic between the main and eastern railway stations was maintained by overburdened, horse-drawn carts.

The city had been peacefully occupied by the Americans on March 29. The beginnings of a civil administration were already in evidence. A mayor, Wilhelm Hollbach, and a citizens' committee had been appointed by the city commandant, Lieutenant Colonel Howard C. Chiswell. The American flag fluttered from the former I.G. headquarters. Separated from the German population by a high wire fence, the military government had resided there since May 8, the day of capitulation. The military government was the highest authority in the zone occupied by the Americans. The commander-in-chief, General Dwight D. Eisenhower, had his headquarters in Kronberg Castle, built by Empress Victoria, wife of Frederik III, before the turn of the century and used as her retirement abode after the death of her son, the Kaiser.

But those days in 1945 also had some positive aspects. The nightmare that had haunted us in the past was beginning to recede. Of course, we were ruled by a military government that imposed upon us a stream of new regulations and restrictions. There was, for example, the nightly curfew. "As from today and until further notice," the decree said, "no one is permitted to be in a public street within the city limits of Frankfurt between 20:00 hours and 07:00 hours without the express permission of the military government." The citizens were warned that "the military police have orders to fire at all persons who are encountered outside their homes during the hours of curfew and who attempt to hide or to escape." Nobody knew what was in store, how long hunger, misery, and devastation would last. But there was one thing that we all hoped for and anticipated: a more peaceful and more liberal life with fewer uniforms.

My civilian wardrobe was meager. I had a suit, two shirts, and a pair of shoes. Although I had invested some of my trainee remuneration and my modest I.G. salary in a wardrobe appropriate to my position, it had been lost when my small apartment in the Fichardstrasse in Frankfurt was bombed. But what matter! I was twenty-six years old and healthy. I completed the indispensable formalities, reported to the police station and to the ration-card office. Then I started to restore a bombed-out apartment in the Schleidenstrasse, near the Eschenheim Gate. However, I cannot lay claim to any particular mastery of construction skills. In any case, lack of talent hardly mattered in those days of devastation.

A precious possession then, when there were neither trams nor trains, was an old bicycle. It was the only souvenir from my soldiering days. The bike took me everywhere in Frankfurt, and I also used it for the indispensable "hoarding trips" to the countryside. But, since I had little to barter and no particular gift for haggling with the farmers, I was lucky to bring home a few kilos of potatoes in my rucksack.

Griesheim, the First Stop

On September 14, 1945, an especially beautiful, hot summer's day, I cycled down to Frankfurt to the office of the I.G., or more precisely, to the remains of what was once Germany's largest concern. The familiar drabness of the streets was reflected in the scene inside the building: frightened, emaciated people, dressed in drab grey, some still in shabby uniforms. A notice in the building said that those who wished could have their uniforms re-dyed. Farbwerke Hoechst supplied the dyestuff. It was, however, necessary to submit a proper "re-dyeing application."

Yes, the Griesheim plant had a job for me. Since I had not been a member of the Nazi Party, I could submit an unsullied questionnaire. Moreover, I was able to refer to my interpreter's examination. I was therefore sent to Frankfurt-Griesheim to work as translator in the administration of the American Control Office of I. G. Farben, located inside the half-destroyed Griesheim factory.

Naturally, I knew the Griesheim plant from my first year at Hoechst. I was familiar with some aspects of its interesting history: for example, the process had been pioneered there for the electrical decomposition of common salt through electrolysis to produce caustic soda and chlorine gas. I was, of course, also familiar with the name of Fritz Klatte, the Griesheim chemist whose name is closely linked with the beginning of the plastics age. Although not the first to polymerize a synthetic resin with the aid of sunlight, Klatte was the first to recognize the industrial potential of the plastics thus produced. His process was patented on July 4, 1913, as "a method for the production of technologically valuable products from organic vinyl esters." Polyvinyl acetate began its victorious career in world markets under the trade name of Mowilith. It is still being sold today in every corner of the earth.

500 Marks and a Hot Meal

In those days, Griesheim looked more like an American army camp than an industrial plant. Nevertheless, I had not done badly. My first salary was RM500. And every day I had a hot meal without having to surrender any coupons! In the middle of 1945, that was some attraction. At that time, rations for the "normal user" in the Frankfurt area were close to the minimum subsistence level.

The weekly ration card provided for 100 grams of margarine, 600 grams of cereals, 62.5 grams of cheese, 62.5 grams of butter, 200 grams of meat, 1600 grams of bread, 200 grams of coffee substitute, and 250 grams of legumes. But even these rations were not always available in shops whose "windows" usually were made of plywood.

For many contemporaries, the only way out was recourse to the black market or the Frankfurt horse butchers, where long queues of hungry people formed whenever there was a rumor that meat was being sold. There were also the Frankfurt soup kitchens which had provided mass feeding since December 1945. The "invitations" to these meals requested the participants to provide their own crockery.

Insofar as food was concerned, I lost out when I switched over in May 1946 to the economic department of the Hoechst unit. Although the staff restaurant in Hoechst served warm meals, meat was a rarity. In addition, one had to surrender one's food coupons. The best day was Thursday, because fish was then on the menu. Whenever possible, I invited my secretary to lunch on a Thursday. She didn't like fish.

Hungry for Information

In those first postwar years, I was hungry also for quite a different fare: newspapers and magazines of every kind that would inform me openly and clearly and without the influence of the former Ministry of Propaganda. Initially, the Americans met this need with their German-language army papers. Later on, they granted licenses for the publication of German papers. As a result, on Wednesday, August 1, 1945, the citizens of the ruined city were able to buy their first copy of the *Frankfurter Rundschau,* the first postwar local newspaper. About six months later, the *Frankfurter Neue Presse* began to publish.

It took another three years, until November 1949, before the *Frankfurter Allgemeine Zeitung* rounded off Frankfurt's journalistic scene. I have been a reader of that paper since its first issue. In Freiburg, within the French zone of occupation, a small group of former editors of the old *Frankfurter* published *Die Gegenwart,* a weekly held in particular esteem because of its excellent commentaries. Occasionally, if one was lucky, it was possible to get a copy of the Swiss *Weltwoche* or the *Neue Zürcher Zeitung,* or even a paper from France or England. The business sections of the German papers, required reading for me nowadays, at that time were very scanty. After all, what was there to report about German companies? Many firms were trying to repair their bomb-damaged plants as best they could, always assuming that there was a prospect of getting a permit to produce. In the city center of Frankfurt, the new mayor, Dr. Walter Kolb, launched a campaign to clear the streets of rubble and debris. Every citizen was called upon to help on weekends. The portly Dr. Kolb frequently joined the squads, gaining unrivaled popularity in those years.

Hoechst under American Control

Although not destroyed, Hoechst had been occupied on March 28, 1945, two days after the plant management had halted production. Factories, offices, and apartments were requisitioned by American soldiers. On July 5, 1945, the Americans decreed the official requisition of the entire assets of the I.G. and therefore also those of the Hoechst plants. A unit manager was appointed and the manufacture of certain products, especially pharmaceuticals, was authorized. The plant manager, and later the trustee, had to travel every few days to the I. G. Farben Control Office in Frankfurt to give a detailed account of what was happening in the plant. At the time, the Americans deliberately attempted to dismember the I.G. and to replace the large plant complexes by many small, self-supporting units. The economic policy toward the vanquished was based on a "First Industrial Plan," announced in March 1946. It provided for an output of 50 to 55 percent of the volume of 1938. All other capacities were to be either destroyed or sent abroad by way of reparations. The manufacture of certain products, including numerous chemicals, was prohibited. The capacities for base chemicals were fixed at 50 percent of those in 1936. These quotas were nothing to cheer about. On the other hand, it seemed clear that the Americans had finally abandoned the Morgenthau plan in whatever form it may have existed. To create a largely agrarian state in the heart of Europe had proved to be an absurd idea.

A First Turning Point in the Policy toward Germany

Soon afterward, on September 6, 1946, U.S. Secretary of State James F. Byrnes gave a speech in Stuttgart announcing a new, more positive phase in the United States policy toward vanquished Germany. Byrnes, as I was to learn later, came from Spartanburg in South Carolina, a state he served as governor. Today, Hoechst operates an extensive Trevira production facility in Spartanburg. Another development in American policy affecting Germany was the Truman Doctrine. Harry S. Truman, who had become President of the United States in March 1945 after the death of Franklin D. Roosevelt, regarded Stalin as "good old Joe." But whatever his feelings with regard to Stalin—whom he considered a pawn of the Soviet's Politboro—Truman had few illusions about Soviet actions in Europe. On March 12, 1947, he proposed military and economic aid to Greece and Turkey, proclaiming that "it must be the policy of the United States to support free peoples who are resisting attempted subjugation by armed minorities or by outside pressures." It was the first step toward a policy of containment of Soviet communism.

The winds of change were blowing through world politics. Every page of the American press showed this clearly, even if some of the unswerving "reeducators" in Germany lagged behind these currents. The only surprising feature, especially if we look back on those events today, was the speed with which the changes took place. Immediately after the war, Germany was a politically proscribed, economically drained country close to total collapse; five years later, it was a remarkable convalescent; and ten years later, it was a respected state whose astonishing economic renewal was widely admired. Equally surprising was the speed with which the Allies granted the Germans a seat at the negotiating table, in total contrast to the delay after World War I.

How to Trap a Secretary

My first office was a former experimental laboratory and my desk a laboratory bench covered with lead sheeting. My secretary, as well as the young woman working for my one colleague, was obviously very friendly with American officers from the I.G. control office. They were nice girls, but it was clear that in their employment interviews no undue importance had been attached to their secretarial skills. So one day I cycled from Hoechst to Griesheim to waylay a certain young lady as she left her office and to entice her to come to work at Hoechst. Frau Melitta Radtke, until she retired in 1978, was thereafter the undisputed ruler of my office and especially of my appointment calendar.

It was not easy at that time to recruit qualified people for the company. For many, salaries and wages in Reichsmarks offered little attraction in a period when a cigarette served as a currency guide. A pack of cigarettes fetched up to RM100 marks, and a pound of butter as much as RM200 in the black market.

Every plant management was therefore anxious to offer its employees additional benefits. During the course of 1946, I received several packets of sweetener, a few pounds of molasses, vinegar essence, washing powder, and for Christmas, two candles at a price of 5 reichspfennigs per candle. Shoelaces, razor blades, shoe polish, and laces were also among the highly welcome salary supplements distributed from time to time. Employees with two or more children under ten years of age received a doll or a school bag on September 3, 1947.

How the Company Symbol Evolved

Since Hoechst was no longer part of the I.G. and had, in fact, been operating since August 26, 1946, as Farbwerke Hoechst, U.S. Administration, we needed a new company symbol.

All employees were invited to submit their proposals. The three best designs would be awarded a first prize of 20 boxes of washing powder, 50 packets of saccharin, and 3 bottles of vinegar; a second prize consisting of 10 boxes of washing powder, 35 packets of saccharin, and 2 bottles of vinegar; and a third prize of 5 boxes of washing powder, 12 packets of saccharin, and 1 bottle of vinegar. The bait of those precious commodities—and that they really were in those days—stimulated the 200 employees to submit a total of 500 designs. But they were motivated more by zeal than by talent. The competition failed to yield a single worthwhile idea in graphically acceptable form.

The very first Hoechst company symbol, introduced by Meister Lucius & Brüning, was a recumbent lion, probably in imitation of the heraldic animal of Nassau. The king of the beasts held the intertwined initials MLB in its right paw. This symbol was used throughout the world for thirty-three years, from 1877 until around 1910. Not that it was without its problems. Especially after Hoechst's introduction of Paul Ehrlich's Salvarsan in 1910, pharmaceutical traders abroad sensed an opportunity of sharing in the considerable commercial success of this product by offering an imitation Salvarsan. Especially in China, there were various imitations, not only of Salvarsan itself but also of the Hoechst company seal. A fascinating chapter of its own could be written about these imitations.

The old company symbol was used until 1910, when the lion was retired in favor of a simple seal composed solely of the initials MLB encircled by a ring within a square. In 1926, the I.G. created its own symbol, "I.G. in the retort." The Hoechst symbol became topical again only after the dismemberment in the postwar years. In 1947, Professor Liesker of Frankfurt created the first version of the tower-and-bridge symbol.

Then, at the beginning of the fifties, the Frankfurt graphic artist Robert Smargo was invited to create a variation of the tower-and-bridge motif, initially as a symbol for Hoechst's pharmaceuticals. This version was soon adopted for all the Hoechst products. But it proved very difficult to spread its acceptance, especially among the long-service employees in the pharmaceutical division. Furthermore, the somewhat unorthodox design and the difficulty of pronouncing the name "Hoechst" in almost every foreign language rendered its introduction abroad rather challenging. But perhaps the name and seal are all the more memorable because of their complexity.

Hoechst's trustee since 1946 was Dr. Michael Erlenbach, born in 1902 in Nuremburg and a man of resolute character. Partly of Jewish ancestry, he was sheltered by good friends during the Nazi era, and thus survived as a chemist in the crop-protection research department of Hoechst. Erlenbach was benevolently skeptical about the company's modest attempts to get a small export trade going.

Unanswered Applications

Indeed, it seemed almost lunacy to think in terms of large-scale exports. Each export project had to be approved by the American authorities through a special agency, the Joint Export-Import Agency (JEIA). The many JEIA forms called for information about the products available for export, the quantities, the prices, and the countries involved. The appropriate application had to be sent to the central JEIA office in Berlin. In most cases nothing more was ever heard. The economic council for the American and British zones, established in Frankfurt in May 1947, with Ludwig Erhard becoming economic director in 1948, was unable to help us.

However, we were not deterred from submitting a stream of new applications, each one with the requisite number of copies. This activity continued for about two years until, gradually, chemical traders from neighboring countries came to see us of their own accord. (This development will be described in greater detail later.) They asked whether we had any raw materials or intermediates not available to them. Could they supply us with finished products, or could they buy them from us?

Before the currency reform in 1948, such business was conducted on a barter basis. In one instance, for example, we supplied an American firm with organic chemicals. In return, we received CARE parcels to supplement the somewhat spartan menus of our "Russian Court." On Sundays, I usually had my lunch at an automatic restaurant in Frankfurt's Kaiserstrasse. Seventy-five Reichspfennigs bought a plain meal, usually a chicken stew. The remainder of my Sunday budget of 5 Marks was spent on a theater ticket. Incredibly, the theaters of Frankfurt had been performing again since the end of 1945. Sometimes it appeared as though people thought nothing was more important than going to the theater. They stoically endured freezing cold indoors, watching performances with overcoat collars upturned.

My first "foreign" trip led me to Saarbrücken. The Saar was at that time under French control, and it was by no means certain whether Germany would ever regain this territory. The Villeroy & Boch Company in Saarbrücken had the ceramic tiles we needed for our electrolysis process. We arranged a barter deal that lasted until the currency reform.

X Day

Eventually, on that memorable June 20, 1948—X Day—the long-awaited currency reform was announced in the three Western zones. I was spending a short vacation in the Black Forest when, on Saturday, June 19, I heard this radio news bulletin:

> The first law for the reorganization of the German monetary system has been announced by the military governments of Great Britain, the United States, and France, and comes into force on June 20. All German currency valid hitherto is withdrawn under this law. The new currency is called Deutsche Mark. Each Deutsche Mark has 100 Deutsche Pfennige. The old money, the Reichsmark, the Rentenmark, and the Allied Military Mark become invalid as of June 21, 1948.

Returning to Frankfurt by train the next day, I collected my quota of 40 new Deutsche Marks. Three years after Germany's unconditional surrender, Ludwig Erhard's reform of the currency and of the economy laid the economic foundation for the liberal development of the Federal Republic of Germany, even before that state was formed.

Like many other people, I found it difficult to believe that the new currency would remain stable. In spite of the sudden temptation to buy a wide range of goods, I decided to manage very carefully with my allocation. In any case, in its Circular 218 of June 23, 1948, the Hoechst plant management advised prudence in managing the allocation. According to radio and press reports, it was considered unlikely that salaries and wages would be paid on time.

Erhard had also canceled controls on many commodities. When

General Lucius D. Clay, the American military governor, accused him of having amended the regulations of his own accord, Erhard is alleged to have replied: "I've not amended them, I have canceled them." The reduction in the value of my savings—10 Reichsmarks became 1 Deutsche Mark—did not cause me undue alarm. The small size of my account protected me from any substantial losses.

Agencies: Starting Point for Our Foreign Organization

In the early summer of 1948, the papers announced another political event of worldwide significance. The Marshall Plan agency, the Organization for European Economic Cooperation (OEEC), was set up in Paris. The Marshall Plan had been proposed in 1947 by U.S. Secretary of State George C. Marshall as an aid-and-reconstruction program for the peoples of Europe. It was one of those majestic undertakings, like the Hoover Plan after World War I, in which the Americans combined generosity with economic necessity. That some of those who profited from this aid did not always acknowledge their gratitude afterward is another matter.

Following the currency reform, the situation in Hoechst changed rapidly. More and more chemical traders, many of them German emigrants, came to see us from Holland, Belgium, France, Switzerland, and even Great Britain. We began to forge commercial links with them that later often developed into agency relationships. At the time, we had neither the money nor the legal right to invest abroad, so we were forced to rely upon agency arrangements for our foreign business. In many cases, these agencies became the embryos of Hoechst offices abroad.

No Master Plan

Journalists sometimes speculate that Hoechst had a kind of master plan specifying in detail its corporate policy and sales strategy in foreign markets. There was no such plan at the time. If there had been, it would have been a purely theoretical exercise. All our decisions then were entirely empirical and dependent on the approaches of people from abroad. We became great pragmatists, and this attitude changed only slightly in later years.

My chief at that time was Walther Ludwigs. Speaking several languages, he had made connections throughout the world. His particular interest was exports. Ludwigs was born in 1886 in the Rhineland town of Lennep. He began his career with Hoechst in 1912 in the dyestuffs export department. During the I.G. period, he was director in the chemical sales department. Since I was in dyestuffs sales, we did not meet during the I.G. period, when chemicals and dyestuffs were two different worlds. Usually only the top people had any contact with one another.

Walther Ludwigs had lost his wife early, and his two sons had died in Yugoslav prisons. A lonely man, he sought his life's fulfillment exclusively in work. A sister looked after the household.

Since I was unmarried at that time, it didn't matter to me that Ludwigs expected me to stay in the office until 8 o'clock at night, or to go through files and prepare meetings on a Saturday afternoon. Usually, we would drink a bottle of apple wine together on such afternoons and have long conversations about business affairs.

A Good Memory for People

Then as now, one of my strengths was a good memory. My colleagues were often taken aback when, having asked them to dig out a certain matter, I could tell them how long ago it had occurred and where the papers might be found. If one is supported by a secretary with an equally good memory, work can become a real pleasure.

My memory for events, and particularly for people, is not quite matched by a similar faculty for figures. However, even today I can still carry out fairly complicated calculations in my head. I am therefore determined not to resort to pocket calculators. In any case, the third digit to the right of the decimal point is of no interest to me.

Ludwigs was of the opinion that, unlike the former I.G., Hoechst should concentrate its activities in each country in a single company only. Any other policy would lead to fragmentation, greater expenditure, and, of course, jurisdictional problems. It should not, therefore, have come as a surprise that, later, I too consistently aimed at establishing only one organization in a country. With a few exceptions, I succeeded. The price was the odium of the dyed-in-the-wool centralists.

Ludwigs would be extremely eloquent and disarm his discussion partners with his charm. Much of his success stemmed from these abilities. He would frequently engage a visitor in a general, hour-long discussion. By the time the visitor left, he was often so dazzled by

Ludwigs's charm that he failed to realize that his business proposals had actually been rejected.

With all my great reverence for Ludwigs, I have always taken a different tack. I have never left either my colleagues or my business partners in any doubt about my opinions. On the other hand, it is not my habit to enter into business conversations with a preconceived plan or unshakable beliefs. Usually, I form my final judgment only after a discussion in which the other side has had the first say.

The Winnacker Era

After the founding of the Federal Republic of Germany in 1949, Chancellor Konrad Adenauer steered the little Bonn steamer unerringly on a Western course. Against the resistance of the opposition, he forced the acceptance of the Federal Republic into the Coal and Steel Community, proposed by French Foreign Minister Robert Schumann. Earlier, on March 8, 1950, the Federal Republic had become a member of the European Payments Union. Adenauer and Economics Minister Ludwig Erhard finally succeeded in persuading the Western Allies to desist from further dismantlement of Germany.

From 1950 on, fundamental changes took place in Hoechst as well. The release from Allied control and the change of the company's structure were prepared. The same control officers who not long before had been vigorously pursuing "decartelization" were now directed to help construct a new, viable organization for Hoechst. On December 7, 1951, Farbwerke Hoechst AG, Meister Lucius & Brüning AG was founded in Frankfurt.

Thus the era of Karl Winnacker began. He came to Hoechst with the clear intention of leading the company. The previous trustees were appointed as members of the board. Michael Erlenbach from Hoechst took over the pharmaceutical division, and Dr. Konrad Weil from Griesheim became the commercial chief. But neither the board nor the staff departments were left in any doubt of Winnacker's leadership.

A New Task: Creation of a Central Sales Management

The restructure of the company had no immediate consequences for me. My office had been located in a wing of the management headquarters. I came to know Winnacker better only in the course of time. Apart from Ludwigs, Konrad Weil was my immediate superior. As the top commercial man in Hoechst and also responsible for sales, Weil thought that a central sales management was needed in addition to the commercial directorate. He entrusted me with the job of creating one. Since I had been the head of the commercial directorate, I might well have regarded this further responsibility as a kind of demotion. Instead, I considered it a challenge. With the help of several reliable associates from the directorate, I finally succeeded. On this occasion, as before, I realized how much success depends on the selection of able colleagues. The ability to find them is one of the most important qualifications of a company manager.

One of my close collaborators during that time was Willi Hoerkens. Born in 1916, he was a vociferous, hardworking Rhinelander. He later became head of domestic sales within central sales management, then moved into the dyestuffs division, and finished up as sales director for fibers. Since 1969, he has been a member of the board of management with special responsibilities for fibers and for relations with the Eastern bloc nations.

For many years, I have not only taken an interest in recruitment for leading positions in the sales organization, but also concerned myself deeply with the development of a commercial training program at Hoechst. This program combines practical experience in the different departments with a strong grounding in theory. Before engaging an employee, a committee of senior Hoechst officials—of which I was a member for many years—talks with each applicant individually. The program has become increasingly sophisticated. The most capable young people in the company are assured that they will be educated in business management.

I am delighted to find again and again that many colleagues in responsible positions at home and abroad have undergone this training. There is a great deal of satisfaction as well in realizing that many of the leading men in our foreign sales organization have spent a few years under my tutelage.

No Standard Procedure for Choosing Leading People

I am sometimes asked about my criteria for selecting young people for senior positions. There is no simple answer.

Leave aside intelligence, professional competence, the ability to integrate, and persistence; all these attributes are expected as a matter of course. It seems to me, however, that the young people who want to advance to senior positions also must have a fundamentally optimistic, positive outlook. They cannot be too introverted, too meditative. They must be very flexible. In addition to physical stamina, they must have the emotional stability to cope with the inevitably greater physical and psychological stress imposed by high position. In spite of the most careful selection, however, there are still confirmations of the "Peter principle" at times.

I was fortunate enough at almost every stage of my career to find colleagues who gave me invaluable support. But I claim no infallible formula to ensure the right decision. However, failures must be few, and a degree of luck is indispensable.

A Further Task: Agriculture

In March 1953, I was appointed sales director for agricultural products. Actually, at that time it was not possible to become a director in Hoechst if one was in charge of a staff department. Winnacker, chairman of the board since May 1952, took the view that because of their involvement in all fields of commercial policy, staff departments already possessed a great deal of influence and power. As a young plant manager at Hoechst, he had learned from personal experience how staff departments in the I.G. central office tended to develop hypertrophic traits. After many years as head of various staff departments, I was very conscious of these facts.

Fertilizers and crop-protection sales were, therefore, added to my duties. This was a major challenge, particularly since it afforded me the opportunity of participating in the complicated initial negotiations for the formation of an export cartel for nitrogen-based fertilizers. At that time, fertilizers were a powerful pillar of European chemical exports. Although many overseas countries had inadequate production capacity, a ruinous competition soon ensued, especially since overseas customers frequently had a buying monopoly. Discussions dragged on for more than two years, but finally led to the formation of Nitrex, a coordinating office for exports outside Europe.

All the parties concerned were greatly relieved when this office opened in Zurich on October 12, 1962. Except for Italy, all important West European countries became members—with the blessing, incidentally, of the cartel authorities. In spite of all commitments to our market economy (I personally am a deep believer in free market principles), temporary cartels are sometimes essential on an emergency basis.

Nitrex and Complex, the sister organization for complete fertilizers that was founded later, are still in operation today. Their significance has been greatly reduced, however, as has that of nitrogen production in Western Europe.

You Cannot Get Along without Diplomacy

In the late 1950s I devoted most of my energies to developing our sales organization. At first, my task was not at all easy, because my colleagues in the managements of the various sales groups were generally older and more senior than I. Moreover, they held at least equal rank. The young man suddenly appointed to coordinate their activities was greeted rather coolly. Some made no secret of their skepticism. At first, therefore, I had to operate with borrowed authority until I was able to assert my own. It was good training in diplomacy and provided me with much valuable experience for my later career.

My trips abroad now became more frequent. No doubt my closer friends will find this difficult to believe, but it was just accidental that one of my first trips should be to France. This first rendezvous with *la douce France* was an unforgettable experience. I lived in a small hotel near the church of the Madeleine, using every spare minute to get to know Paris, practically "on the double." Nowadays my weekend schedule invariably includes a walk in the Taunus mountains near Frankfurt. But the distances I cover are a fraction of the many kilometers that I walked through Paris at that time. In order to learn that city really well, I later lived in hotels in many different quarters. I soon grew fond of France with its culture and *savoir vivre* philosophy. It became something like my second home. My wife had learned French as an *au pair* girl in Paris. And it was in Paris-Neuilly that we married in 1955. The ceremony was performed by a charming lady mayor wearing a tricolor sash; she quoted a chapter of the Napoleonic Code which states tersely: "With regard to domicile, the wife has to accept the professional life requirements of the husband."

In our married life, this stipulation caused no difficulties. We are both city people. Born in Mannheim, my wife spent much of her youth in Berlin, Dresden, and Ludwigsburg, a town near Stuttgart. We enjoyed living in Frankfurt, first on the Beethovenplatz and later on August Siebert Street. It was only when our first two children—Jutta and Peter—grew up that we moved into a house in the Taunus, where Holger was born. As our house was being constructed, I experienced the few pleasures and endless anguish of one who builds his dream house.

At first, we regarded living away from a large city almost as a sacrifice. But we soon came to love our quiet spot in the Taunus. Certainly, in the mountains it is easier to keep in shape. And when I am traveling, I try to choose hotels with a sauna and swimming pool.

The "professional requirements" to which the Napoleonic Code refers obliged me to travel, year after year, across Europe and around the world. The prime purpose of these journeys was to find people who could help us gradually build up our foreign organization. In the American zone, we were at first not allowed to fall back on colleagues from the old I.G. Initially, we regarded this restriction as a major obstacle. Later, we realized that making an absolutely new beginning presented greater opportunities. I believe we made good use of these new possibilities.

Truly a "globe-trotter in chemistry," Karl Winnacker once christened me jokingly. But as chairman of the Hoechst board and later of its supervisory board, Winnacker himself held this title. We made many trips together, especially after 1958 when, at the age of thirty-eight, I was appointed to the board to represent the sales organization.

Winnacker always demanded much from himself and his team—and not only on trips. Even those younger staff blessed with iron constitutions usually needed some kind of recuperation by the time they returned with Winnacker from a visit to the United States or Latin America. Every morning at 7:30, having already taken a long walk, Winnacker summoned us to breakfast to map out the coming day. Our bedtime the night before was of supreme irrelevance to him, and he was by no means the first to retire.

During our many years of close and trusted cooperation, I have learned much from Winnacker. I admire his astonishing tenacity of purpose and his unshakable optimism, so often vindicated. But unlike him, I never was an early bird. It is no secret among my associates that the early morning is not the best time to discuss touchy matters with me.

Although I am usually behind my desk by 8:30, I need some time to marshal my plans for the day and to put down on paper thoughts that may transcend the day-to-day business. Once the stream of visitors begins about 10 A.M., contemplation or informal discussion of the future is impossible until after the close of business late in the afternoon.

Change of Power

Winnacker used to call me his foreign minister at times. His "minister of the interior" was Dr. Rolf Sammet, born in 1920 and therefore a year younger than I. Dr. Sammet comes from Stuttgart where he studied organic chemistry; later, he worked as an assistant at the famed Max Planck Institute in Berlin. It was, in fact, the currency reform that caused Sammet to come to Hoechst. After the devaluation of the Reichsmark, the Institute could no longer afford his modest salary. He therefore decided to go into industry and applied to Hoechst, both to his and the company's lasting benefit. He first worked in the research laboratory. Later he became a member of the technical directorate, served as chief of Trevira production in Bobingen, and then returned to the technical directorate. Becoming its head in 1962 at the age of forty-two, he was appointed to the board of management. His specific responsibility was as deputy manager of the Hoechst plant and divisional chief for the films and foils division. He became plant manager in 1966.

After the retirement of Karl Winnacker, Dr. Sammet was named chairman of the board of Hoechst in 1969. I was appointed deputy chairman, retaining responsibility for sales and the coordination of Hoechst's foreign business. Nowadays, the main instrument in the latter field is the Central Regional Conference (CRC), a gathering of the commercial members of the board, the heads of the larger departments, and the directors of the staff departments. The domestic counterpart of the CRC is the Central Divisional Conference, headed by Dr. Sammet. This group is composed of the heads of the individual divisions of the company and the technically trained members of the board; of course, in both conferences, finance and administration are adequately represented.

Inter alia, there is also a Sales Conference, which I have chaired for many years. Meeting roughly at monthly intervals, it comprises the heads of the sales departments of the parent company and its major subsidiaries. The Conference affords its members a unique opportunity for discussing problems extending beyond the narrower issues of individual divisions. The Conference also ensures the maintenance of a well-coordinated sales policy in all the widespread departments of the company despite the great variety of our broadly based line of products.

Friends and acquaintances often ask me whether I still enjoy traveling. I must confess that I do, just as much and as deeply as before and even after 10 million kilometers in the air. I have not really changed in this respect since I was thirteen and began to tour the world on my bicycle. Today I do it by plane, and usually with some associates.

A trip by plane or by car often offers an opportunity for discussions for which there seems scarcely to be time at headquarters. I also usually have the leisure, especially on overseas flights, to try to keep up with literature, often in a foreign language. Lord Byron once wrote, "Life is tolerable only when you are travelling." His statement was, of course, valid only for himself and his age. Byron was married briefly and had no children. For me, a weekend without my family is something of a sacrifice in spite of all the attractions of traveling. No matter how many guests we entertain during the week, Saturday and Sunday are family days when I am at home.

In my travels, I am concerned not only with our own corporate problems. Together with the heads of our companies abroad, we maintain contact with other companies, subsidiaries, and affiliates, a practice that can be quite enlightening. In spite of all modern management techniques, it is hardly possible to get a complete overview of a company like Hoechst solely from its headquarters. Many strong and weak points of the central operation can be more clearly identified from a distance. On returning home, the "globe-trotter in chemistry" must then do his best to ensure that the strengths become stronger and the weaknesses weaker.

3

France: Chemistry with Esprit

There was to have been a celebration dinner at Maxim's. But the negotiations with Jean-Claude Roussel, the president of the second largest French pharmaceutical concern, had dragged on until late in the evening. Since lunchtime, a small Hoechst delegation had been sitting in Roussel's office, which looked more like a tastefully furnished study than a managerial command post.

In principle, the comprehensive partnership agreement between Hoechst and Roussel had long been ready for signature. Only some details still needed to be worked out. The French lawyers dealt with these no less patiently and thoroughly than did their German colleagues. Anyhow, the dinner at Maxim's had to be canceled around midnight. But at 3:30 the next morning we found ourselves a simple bistro, the Pied de Cochon, next to the legendary Halles, the "stomach" of Paris, with its mountains of fresh meat, seafood, and vegetables. In accordance with custom and along with the vast crowd of early morning revelers, we spooned down the famous onion soup. Since then, Les Halles has been replaced by the Centre

Culturel Georges Pompidou, where modern art has a chance to unfold fully.

Jean-Claude Roussel and his friends were as tired and content as the delegation from Hoechst. We knew that the approaching September 30, 1968, when the agreement was to be ceremoniously signed by Hoechst in the presence of the senior officers of both companies, would become one of the major milestones in the development of each firm. It might also introduce a remarkable chapter in French-German economic history.

I had made the acquaintance of Jean-Claude Roussel only a few months earlier. His name and his position in the French chemical industry were of course well known. His entry in the French *Who's Who* reads:

> Roussel, Jean-Claude, born 26th November 1922 in Paris, married since 27th May 1943, to Jaqueline Roussel, née Vachet, three children, Alain, Catherine and Oliver; president and director-general of Roussel-Uclaf since 1962.

Roussel's father, Dr. Gaston Roussel, was gifted in both business and science. In 1922, he founded his first small laboratory, the Institut de Sérothérapie Hémopoïétique. Its most important preparation was Hemostyl, an agent counteracting anemia and certain blood diseases. The preparation was so successful that Dr. Roussel was able to found more companies and even a foreign subsidiary in the following years.

In 1928, the Laboratoires Français de Chimiothérapie and the Usines Chimiques des Laboratoires Français (UCLAF) were formed. The first Uclaf factory was erected at Romainville near Paris. There, the stables of the former Serum Institute can still be seen.

More years of expansion followed, even in Latin America. In 1946, Dr. Roussel formed the Société Française de la Pénicilline (SOFRAPEN), the Roussel group's début in the domain of antibiotics. The founder of the company died in 1947, a year after Jean-Claude, then twenty-four years old and a graduate pharmacist like his father, had entered the group. The Roussel group, soon to diversify into the manufacture of products for agriculture and industry, continued to operate with great success.

For the founder's son, then barely twenty-five, the door to the executive suite did not immediately swing open. As so often happens in a company owned by a widely branched family, and in which majordomos have acquired powerful positions over the years, the heir is scrutinized for a long time before he is accepted as the chief executive officer as well as the owner.

How Roussel finally succeeded, through great effort, in forming Roussel-Uclaf in 1961 is a story all by itself. At the age of thirty-nine, Roussel became head of the company founded by his late father. The new group, Laboratoires Roussel, active in many countries throughout the world, represented the most important subsidiary in the pharmaceutical field. The years up to 1968 saw steadily increasing sales and the founding or acquisition of other companies.

In 1962, Roussel-Uclaf acquired Procida, the second largest French company in the field of agrochemicals. The most important result of the work during the last few years was the synthesis of the pyrethrenoids. The natural product had been used for many years in the production of household insecticides. Synthetic preparation of the very complicated chemical compound opened the way to the broad use of the product as an insecticide in agriculture. The products are nontoxic to warm-blooded animals.

In the meantime, Roussel-Uclaf had become the second largest French pharmaceutical manufacturer after Rhône-Poulenc. In the course of the last five years, total group sales have more than doubled.

Why, then, should Jean-Claude Roussel consider a partnership with a firm outside the frontiers of France, especially in view of his company's unusually rapid growth? He realized that pharmaceutical research and development in the future would require sums eventually beyond the resources of a family company. Moreover, it would become impossible for the company to pay equal attention to every pharmaceutical area. Successes in certain fields—hormones, for example—might necessarily mean reduced activities in others.

May 1968

But there was another essential factor: In May 1968, France came close to civil war. The student unrest in Paris, with which the other French universities, often including many of their faculty members, declared their solidarity, threw Paris into chaos and paralyzed industry and transport throughout the country for many weeks.

These developments may have been aggravated by the left-oriented individuals in the mass communications media. Before going on strike themselves, commentators on the government television and radio programs supported the student rebellion. So did many leftist newspapers as well as radical left and anarchist elements in industry, research institutes, and civil administration. De Gaulle procrastinated— whether from tactics or weakness will probably never be answered unequivocally. At any rate, the French citizens were deeply alarmed.

Eventually, however, De Gaulle, aided vigorously by Georges Pompidou and the support of the army, succeeded in reclaiming the reins of power. The powerful demonstration of the silent majority that marched down the Champs Elysées on January 30, 1969, with waving tricolors put an end to the immediate crisis. Had not the situation been stabilized at that time, the structure of Western Europe might have been deeply affected. Remarkably, and probably decisively in the failure of the insurrection, the communists and left-wing socialists, like the workers generally, did not sympathize with the anarchist students and intellectuals.

A Meeting in the Carlton in Cannes

For all these reasons Roussel instructed his bank to explore the possibility of selling a minority interest in Roussel-Uclaf to a foreign company. Our discussions with him began in Cannes, where he went each year to relax. From the start, the catalyst in our discussions was Henri Monod, the dynamic assistant of the *patron,* as Roussel was called. Since the reconstitution of the company, Monod has been one of the four members of its newly created executive board.

As for Hoechst, I had the support of François Donnay. He has been our "ambassador" in France from the beginning, focusing his great intelligence and diplomatic talent on what were at times very complex negotiations. However great the differences in our professional careers, Jean-Claude and I soon found our relationship in business matters ripening into genuine friendship. But unquestionably, the agreement succeeded because we, as Hoechst representatives, convinced Jean-Claude that our involvement in Roussel would be mutually fair and beneficial, without in any way compromising that company's independence. This was also the only approach to ensure the agreement of the French government to the French-German transaction. Normally, the French regard such mergers with great reserve.

From the beginning, Jean-Claude Roussel told us he was negotiating not only with Hoechst but also with another large German chemical company. After our agreement was completed, he made a point of personally visiting Professor Hansen of Bayer to explain why he had opted for Hoechst. Also, even before publishing the agreement, we informed Wilfrid Baumgartner, one of the great men of French industry, president of Rhône-Poulenc, the largest French chemical company, honorary governor of the Bank of France, and finance minister from 1960 to 1962. Baumgartner invited us to dinner in his elegant home on the Rue de Grenelle, and it was here that we explained our intention to him. He took our announcement very calmly, but could not resist telling us that in his view we had won only the first round *("Vous avez gagné la première manche")*.

Our agreement with Roussel had been preceded by many years of close scientific contact, many other negotiations, and the formation of personal links with the industry of France. Not only because of my function in business, but also because of my considerable affinity for the country, its culture, and its mode of life, abetted by my knowledge of the language and the mentality of the people, these activities afforded me many opportunities to gain a deeper insight into the culture, civilization, and—last but not least—the gastronomy of France.

Kehl on the Rhine, where I spent practically all my school years, had been a French bridgehead until 1932, under the Versailles Treaty. The occupation was a peaceful one, and the Kehl citizens did not take much notice of it. They could always board the tram for a fifteen-minute trip across the frontier to Strasbourg, where the inhabitants spoke mainly the Alsatian dialect, which is closely related to German.

For a young man like myself, Strasbourg and Alsace represented a rather happy blend of French and German culture and history. From the bell tower of Strasbourg cathedral, the Vosges and the Black Forest seem equidistant. Strasbourg, Alsace, Colmar with Matthias Grünewald's Isenheim altarpiece, and Sesenheim where Goethe fell in love with Frederika, are the intellectual heritage of my youth, spent on both banks of the Rhine.

A Chemical Industry Concentrates

France's chemical industry is no less traditional than that of Germany or England. Among its founders was Nicolas Leblanc, who developed processes for the manufacture of soda and who, as early as 1791, erected the first soda factory in the world. Soda was then used mainly by the rapidly growing cotton industry. Leblanc's discoveries ushered in the production of chemicals in the industrialized countries.

Louis Pasteur, France's most famous chemist, created the basis of enzyme chemistry, which is of great significance today, and developed protective immunization against rabies. The Institut Pasteur, founded in Paris in 1888, still enjoys a considerable reputation for its pharmaceutical research. Unfortunately, however, under state administration the institute has been unable to gain an economic importance to match its scientific achievements and resources.

For many decades, the French government has actively promoted both chemical research and the expansion of the chemical industry. After World War II, it spent billions on research and development in the field of oil and natural gas. Marseilles soon became the largest center for the transshipment of these fuels. Refineries sprouted near almost all the estuaries of France's large rivers, to be joined sooner or later by petrochemical plants. Upstream in the Rhône valley, for example, a pipeline supplies the refineries in St. Etienne and Grenoble and thus the industrial areas around Lyons and in Alsace.

Fine Chemicals and Culinary Arts

Lyons illustrates what good can come from decentralization in both culture and industry. This ancient Roman city at the confluence of the Rhône and the Saône has rediscovered its rich past. Buildings from the Roman-Gallic period and the Middle Ages have been restored. Magnificent works of early art are being shown in the beautifully laid-out new museum on the Colline de Fourvière. Under the direction of Jean Vilar, the cultural center of the industrial suburb of Villeurbanne has provided new inspiration for the modern theater.

Roussel-Uclaf's large factory for organic synthetic products is located in Neuville, another of Lyons' industrial suburbs. These complex facilities for fine chemical production were developed during the era of Professor Léon Velluz, Roussel's head of research. Visitors to Neuville will not fail to dine at the nearby restaurant of Paul Bocuse or at Mère Charles.

New Industrial Targets

Toward the end of the 1950s, French chemical industry turned its attention increasingly to oil and gas from the Sahara. However, great efforts to employ these resources in broadening the raw materials base of industry were sharply set back when Algeria gained its independence in 1962. Closer relationships with Algeria, as with most countries of French-speaking Africa, were reestablished only slowly. Fos, a new industrial port not far from Marseilles, became a new production center for the steel and chemical industries.

At about this time, many chemical companies clustered around the considerable natural gas deposits in Lacq in southwestern France. Since then, pipelines have radiated from these Pyrenean foothills to many important industrial centers throughout the country. Important chemical plants have also been erected around Lacq itself, as was SNPA (Société Nationale des Pétroles d'Aquitaine, S.A.).

Problems in the Pharmaceutical Industry

The importance of the French pharmaceutical industry also owes much to the support of the state. Finished pharmaceutical products cannot be imported into France. Therefore, anyone wishing to build up a pharmaceutical business in France has no choice but domestic manufacture. As a result, many partnerships have been formed between French "laboratories" and foreign concerns, very often to the benefit of both. The pharmaceutical price structure in France has always been unsatisfactory. The prices of finished products are rigidly controlled, as are the import prices of raw materials.

Ever since rising health-care costs began creating large deficits for social insurance, the pharmaceutical industry has been the target of ill-informed and politically motivated criticism in the mass media. Increased intervention regarding prices is planned, and arbitrary price-reduction directives have already been issued.

Unfortunately, no attention is paid to the basic truth that the products on the market must earn the money needed for research into new therapeutic agents. In the seventies, for example, Roussel-Uclaf spent 200 million Francs (US$40 million) per year on research.

Planning for Utopia

In looking at the chemical industry of France, one has to realize the exceptionally strong influence of the state, which in many respects regards itself as the directing authority. The planning network—the national five-year plans for the increase of productivity and the economic growth of the individual regions of the country—includes the chemical industry. The government's most effective control instrument is the channeling of bank credits to those branches of industry and to those regions which the state believes deserve special attention. As a growth operation, the chemical industry is in the forefront of these efforts. Planning has been the obsession of every recent French government—except for the liberal policy of Prime Minister Barre in the late 1970s. There have been five-year plans since 1947, including one for 1976–1980. Alas, 340 printed pages of "planned targets" are in great danger of being blurred by reality.

Private and State Enterprise in the Chemical Industry

Of the seven large chemical companies in France, no less than five are wholly or partially nationalized. Their influence on the affairs of this industry is hardly insignificant. Although only 20 percent of the output of the chemical industry is supplied by the nationalized concerns, this proportion will doubtless increase in the coming years, since investments by nationalized industry are less subject to cyclical forces. Oil and natural gas play a decisive role, and state intervention in this field is far greater than in its counterpart in the Federal Republic of Germany.

After World War II, the many coal-mining areas of the country were nationalized for economic and political reasons. The northern French and Lorraine coal-mining groups operated under the umbrella of the Charbonnages de France (CdF). They were soon joined by chemical plants, leading eventually to the formation of CdF-Chimie.

The process of state-promoted concentration in both private and public undertakings is characteristic of the postwar French chemical industry. It has been underway since the mid-1960s, and its end is not in sight. As in other industrialized countries, concentration is accelerated by the rising costs of research and technology. It has reinforced the position of French chemical industry in international competition.

An outstanding example of the concentration of production potential and capital is the development of Rhône-Poulenc S.A., the largest French chemical company. Founded by Marcel Poulenc in 1801 and combined with Société Chimiques des Usines du Rhône in 1928, it is the sixth largest industrial concern in France, with 300,000 shareholders.

In the boom year of 1974, Rhône-Poulenc achieved record sales of more than Fr20 billion, thus matching the size of the three leading German companies. Initially, Rhône-Poulenc produced basic organic and inorganic products, especially dyestuffs and pharmaceuticals. But at the beginning of the sixties, the company acquired the Celtex group (Gillet), controlling some 80 percent of French synthetic fiber production. It is, therefore, hardly surprising that the company has been hard hit by the textile crisis of the last few years.

In 1969, during Wilfrid Baumgartner's time, Rhône-Poulenc acquired majority holdings in Péchiney–Saint-Gobain and Progil. This acquisition boosted total sales to more than Fr14 billion, a threefold increase in ten years. Baumgartner's successor, Renaud Gillet, whose family still has a considerable shareholding in the concern, was faced with the task of reorganizing the complex structure. New divisions were created and other activities were ceded to subsidiary companies. For example, fiber production went to Rhône-Poulenc Textile, and organics went to Naphtachimie S.A. in Lavéra, in which the French arm of British Petroleum (BP) has a substantial minority holding. Together with Naphtachimie S.A. and Ruhrchemie AG, Hoechst operates an oxo-synthesis plant (Oxochimie S.A.) in Lavéra.

After Rhône-Poulenc, the second largest private concern in France is the PUK group, the end result of the 1971 merger of Ugine Kuhlmann with that part of Péchiney not absorbed in the Rhône-Poulenc group. In the Péchiney-Ugine-Kuhlmann group, the holding company Produits Chimiques Ugine-Kuhlmann accounted for two-thirds of total sales of Fr4.6 billion (US$1.02 billion) in 1975.

Nationalized companies such as Enterprise Minière et Chimique (EMC), the largest producer of fertilizer, are continually increasing their activities. With the acquisition of Ripolin-Georget-Freitag S.A., CdF-Chimie has become involved in the paint sector. The largely nationalized oil groups, Société Nationale des Pétroles d'Aquitaine S.A. (SNPA) and Compagnie Française des Pétroles S.A. (CFP), are expanding their capacities to gain a foothold in processing. In 1971, their interests in petrochemistry were consolidated in Ato-Chimie and their polymer interests in Ato-Emballage. The nationalized oil concern, Elf-Erap, moreover, was fused with SNPA to form Elf-Aquitaine.

Against this bustle of state activity, the agreement between Hoechst and the Roussel-Uclaf group assumes even greater significance. Even after Hoechst gained a majority holding in the company, Roussel-Uclaf remained a private French concern, still consistently increasing its international activities.

Giscard Finally Decides

On April 9, 1972, Roussel, not yet fifty years old, crashed to his death while piloting his private helicopter. It was a Sunday, and I was in Lisbon when news of the tragedy reached me. François Donnay was with me in the Portuguese capital. Next morning, Donnay took the first plane to Paris. I followed in the evening after I had dealt with the most pressing matters in Lisbon. As is customary in French Catholic families, the body of the dead man was lying in state in his bedroom at home. It was there that we said farewell to our dear friend.

Soon after this tragic event, the Roussel family offered Hoechst a majority shareholding in the company. Because of the close scientific and commercial links between the two concerns, it seemed logical to accept the offer. However, the French antipathy to foreign majority holdings complicated the transaction. During a personal talk with Valéry Giscard d'Estaing in his office overlooking the Tuileries, the then finance and economic minister complained, *"Vous voulez filialiser Roussel-Uclaf"* ("You want to make a subsidiary of Roussel-Uclaf"). But that was precisely what we did not want. We had agreed with our French partner that each company was to retain its identity. This is how matters remained after we had finally succeeded in convincing Giscard of the sincerity of our intentions.

We were compelled at the time to reduce to less than 20 percent our interest in Nobel-Bozel, a medium-sized chemical company belonging to the same family holding.

For many years before the Roussel merger, Hoechst had maintained close economic relations with the French chemical industry. These ties were possible only because our French neighbors had quickly and purposefully pushed aside the political and economic resentments that had accumulated before 1945. Only a few years after the war, French Foreign Minister Robert Schumann, together with Germany's Chancellor Konrad Adenauer and Italian Prime Minister Alcide de Gasperi, laid the foundation stone for a politically and economically unified Europe.

Candid and eventually friendly relations were quickly established between the large German and French chemical companies. Many events of the past had to be forgotten, for example, the forced partnership of Kuhlmann and I. G. Farben in Francolor. Similarly, Rhône-Poulenc had been forced to yield its pharmaceutical production to Theraplix, a joint venture with Bayer.

It Started at Rue Richelieu 60

Dyestuffs, almost always the avant-garde in the reestablishment of Hoechst's foreign activities, spearheaded the company's new beginning in France. Although French commercial policy was intent upon the protection of its domestic industry, clearly the German chemical industry had much to offer in dyestuffs and other areas in complementing the French range of products. With the help of an offspring of the old I.G., Solytec, whose headquarters were in Lyon, a new dyestuffs business was built up, based on the products of the three I.G. successors. This business became part of Peralta, later known as Hoechst France, only many years after its foundation.

A different situation applied in industrial chemicals. At first we worked together with a number of importers who often fought one another at the expense of our profits. Eventually, an elderly gentleman, Dr. Paul Neumann, visited us at Hoechst in 1949. Until 1933 he had been active in the management of Schering in Berlin. Anticipating future political events, he then went to France, where he managed the Schering branch until the beginning of World War II. He survived the war under the most difficult circumstances in Vichy France, at first not occupied by Germany. His wife died shortly after her liberation from the Theresienstadt concentration camp.

During his visit, Dr. Neumann proposed the founding of a small Hoechst company with capital supplied by French friends. Neumann would make available his industrial experience, to which François Donnay, his young French assistant, contributed his élan and expertise. Gradually, we consolidated all our activities in France in this company. In 1969, when the time seemed ripe, the company name was changed to the present Hoechst France.

But it was a long road from the three-room office at Rue Richelieu 60 to the present headquarters of the company, occupying twenty floors of Tour Roussel-Nobel in the modern commercial quarter of the Point de la Défense in Paris. From these offices, there is a most wonderful view across the Seine of one of the world's loveliest cities. The panorama ranges from the Arc de Triomphe across the Place de la Concorde to the mighty Louvre complex. But back to the Rue Richelieu. Apart from wise old Neumann and his young partner Donnay, the office included at that time only a secretary who came three times a week. Peralta, incidentally, was an attempt to Latinize the practically unpronounceable word "Hoechst," and also somehow to avoid disclosing at once the new company's German origin.

Hoechst contributed none of the modest starting capital of Fr50,000. Our capital was needed only in the succeeding years of rapid growth, but this financing never clouded the friendly relationship with our partners.

When Paul Neumann died in 1970 at the age of eighty-seven, Hoechst France had a staff of 800. François Donnay actually had taken over the management of the company some years earlier. The small Mowilith factory in Stains on the outskirts of Paris had become Nobel-Hoechst-Chimie in Lamotte, south of Orléans. An ancient factory had gradually changed into a modern, complex chemical plant that significantly complemented the Hoechst range, especially in the field of surfactants or tensides and auxiliaries. Hoechst France administered sales of the subsidiary companies as well as a laboratory unit in Stains in which some sixty people were engaged in a search for products tailored to the French market. Only pharmaceuticals remained separate as Laboratoires Hoechst. Today it is united with Roussel-Uclaf, but both the administration and the gleaming plant which produces Hoechst pharmaceuticals in L'Aigle in Normandy operate autonomously. From the windows of the research laboratory there are fine views of well-fed cows grazing in lush meadows.

Thus, France had moved to the top of the Hoechst export countries. As our industrial involvement in the country grew, our position was further consolidated. Foremost among our friends in the French chemical industry was Raoul de Vitry, a true *grand seigneur* of industry. At that time he was chief of Péchiney, a large aluminum and chemical company. De Vitry had been particularly receptive to approaches from Hoechst. Our lively exchange of commercial, sociopolitical, and technological information contributed greatly to the growing rapport between the two companies.

Hoechst also entered into a temporary partnership with the nationalized company CdF-Chimie. The aim was to boost its Société Normande de Matières Plasticques (SNMP) by supplying it with our polypropylene know-how. Since then, this partnership, in which Air Liquide also participated, has been dissolved by mutual agreement. Our French company now operates the factory, located on a beautiful site near the mouth of the Seine at Lillebonne. Plant capacity has been increased to 70,000 tons per annum. An aldehyde plant has since been added to this growing industrial complex.

Elitist Universities

The managements of almost all the large companies in France have remained centralized in Paris. One among many factors responsible for this concentration is the location in Paris of the great educational institutes of the country, such as the Ecole Nationale d'Administration, the Ecole Polytechnique (founded by Napoleon I), the Hautes Etudes Commerciales, and other institutions of learning. Admission to these *grandes écoles* is extraordinarily competitive. But, it is claimed, their élite graduates need not worry about their future careers in either government or industry. For the academics who have attended one of the many other universities in the provincial cities, this preponderance of the élitist schools presents a serious problem. We believe this is one reason why there is little autonomy in the provincial plants of the chemical industry and why all essential decisions are made in Paris. No doubt, the same is true of other industries.

Changes in Social Structure

The German businessperson who manufactures in France is always surprised by the social climates of the two countries. The French trade unions are all politically committed. The Confédération Générale des Travailleurs (CGT), representing some 60 percent of the trade unions in France, is largely an organ of the French Communist Party. The Confédération Française Démocratique des Travailleurs (CFDT), originally a right-wing trade union, has tried to bolster its essentially weaker position by attempting to surpass the CGT on the left.

The existence of other trade union organizations makes shop-floor cooperation no easier. The degree of trade union influence among academically trained employees is equally astonishing, especially in research and allied fields. During the troubles of May 1968, Jean-Claude Roussel was greatly depressed by the great chasm, never before so evident, between company management and the *cadres,* the senior staff. He remarked one day that although Pompidou had regained full control over the political events of the country, the days were surely numbered for large, family-run industrial companies. Undoubtedly, Roussel said, he had followed his political convictions too closely by investing the larger part of the family wealth in Roussel-Uclaf.

Nationalization has remained a topic of discussion in France to this day. Part of the reason is, no doubt, that in France social contrasts between the various strata of the population are far greater even today than in German society, in which class distinctions have been eroded by defeat and inflation.

I question whether codetermination can be realized in French industry in the near future. Like the parties of the left, the trade unions are seeking confrontation rather than cooperation with the wealth-owning classes. For the present, they seem to have no desire to share responsibility with owners and managements. And yet, here too I feel the situation will change. I well remember our first participation in a meeting of the supervisory board of Roussel-Uclaf. The representatives of the Comité d'Enterprise (the plant council) sat at the lower end of the table. They did not greet the other members of the supervisory board, who in turn made no attempt to welcome the employee representatives.

Since then, these attitudes have changed, at any rate on the surface. Furthermore, there is a growing progressive element in French industry that is attempting to bring about fundamental improvements in the relationships between management and staff. Organizations such as Enterprise et Progrés are playing a valuable role in these efforts.

God in France?

In much of industry and society we are still a long way from a unified Europe. I believe, however, that even if the enthusiasm of the first postwar years has largely evaporated, the younger French generation thinks in European terms, believing that without France, there will never be a unified Europe in the sense of the Treaty of Rome.

My image of France in early youth was no doubt shaped by the essay *God in France,* written by Friedrich Sieburg and published in 1929. Another twenty years were to pass before I had an opportunity of really getting to know my own country's neighbor. If asked why I, a German, am writing about France, I would answer in Sieburg's words: "Because every interpretation of France arouses in us the hope, or at any rate the wish, that this country will join us on our voyage into the future, for its benefit and ours."

4

A Chemical Empire

The venerable London *Times,* even in limbo more of a British institution than a newspaper, recently confirmed—on the occasion of the Queen's visit to Germany—that the British and the Germans fit together well. "Great Britain can take an example," wrote the *Times,* "from the German capacity for hard work and the German attitude to such issues as industrial relations which are based on cooperation and consensus rather than confrontation. On the other hand, German visitors admire British tolerance, the easy human contact and the less materialistic attitude of life. Indeed, we have much to offer to each other."

At Hoechst's Europa Conference in May 1978 in London, we experienced once more the meaning of liberal self-confidence and the tradition of "royal merchants." After a long day of speeches, discussions, and personal conversations in London's Intercontinental Hotel, we sat down in the Merchant Taylors Hall in the City of London and soaked in the medieval aura of this historic building. Otto von Habsburg, one of the truly dedicated Europeans, presented a searching analysis of the global political role of Europe. In the final analysis, the

British, too, feel committed to Europe, even if at times they assume a reticent stance on the issue. During the conference I recalled Churchill's famous postwar call to forge a united Europe as quickly as possible. British liberalism and cosmopolitan thinking in both politics and commerce impressed me even during my first visits to the United Kingdom in the early 1950s. At that time I of course did not envisage that we would one day have a real Hoechst business in that country. The present Hoechst U.K. Ltd., with sales of more than £310 million and some 6500 employees, plays quite a significant role in Britain's chemical industry.

Golden Anniversary

The undisputed leader, of course, continues to be ICI, the "first lady" of the British chemical industry. It celebrated its fiftieth birthday a few years ago; the name Imperial Chemical Industries was entered in the London Company Register on December 7, 1926. The starting capital of the new company was £65 million, and the number of employees 46,000. Production consisted largely of inorganic chemicals—alkalis, fertilizers, chlorine, sulfuric acid, hydrochloric acid, soda, and even ammonia. In addition to these chemicals, nonferrous metals were important. Finally, there was the explosives division, which ultimately evolved via nitrocellulose into production of paints and lacquers. Dyestuffs and other products of organic chemistry, however, were at first rather neglected by the new company.

Anyone who, like me, has had the privilege of getting to know ICI, in Balzac's words, as *'la femme de trente ans,''* will instantly confirm that the lady has retained her charms. ICI's business is growing apace all over the globe. Research and investments are at a high level and the yields are remarkable.

At a time when England's economy is beset with both inflation and a decline in the value of sterling, ICI knows not only "how to survive but even how to flourish," as the English historian William J. Reader put it. His two-volume history of ICI is distinguished from many other obligatory jubilee publications by its cool and sometimes critical approach. Yet he writes, "ICI is sending a ray of hope through the clouded autumn of the year 1976."

In Germany, I. G. Farbenindustrie was regarded as an emergency cartel created in response to the loss of foreign markets and patents and a flood of competitive chemical products swamping world markets in the difficult days after World War I. But in Great Britain, as in other industrialized countries, I.G. was bound to give the impression of a supertrust aiming at reestablishing the dominance of the German dyestuffs manufacturers on an even broader basis. In particular, I.G. packed enormous research clout. Its coal-liquefaction achievements especially impressed the British because the United Kingdom, like Germany, had much coal but no oil at that time.

However, it was not only I. G. Farben but also the chemical industry in America, grown so dynamically during the First World War, that gave rise to competitive worries in the British Isles. Was the mother country, pioneer of modern industry, going to be overtaken and out-maneuvered by her daughter?

British Industry Gets Steam Up

Indeed, Great Britain is the oldest, and was for a long time the leading, industrialized nation in the world. The country was the first to experience the industrial revolution, as the onset of the machine age has been described. What happened in the British Isles then was repeated on the European Continent more than half a century later.

The chemical industry of Great Britain, by contrast, first borrowed technology from the Continent. For example, in 1814 the English textile industry adopted the soda process developed by the French firm of Leblanc. Soda factories were constructed which yielded chlorine and hydrochloric acid as by-products. Soda, the "white gold," was needed primarily for scouring cotton textiles. However, it also softened the boiler feed water of steam engines and found application in paper factories and chemical plants.

James Watt (1736–1819), the Scottish inventor, based his discoveries on another French development. He had heard of the process, discovered by the French chemist Berthollet (1748–1822), in which vegetable fiber was bleached with chlorine. Watt produced chlorine from common salt. In 1799 he started to manufacture chloride of lime as a bleaching agent. Artificial fertilizer has been produced in Britain since 1842.

But the British chemical industry can boast of its own creative firsts. In 1856, for example, William Henry Perkin produced the first synthetic aniline dyestuff, reddish-violet mauvein. Thus, as early as the middle of the nineteenth century, the chemical industry of Britain had achieved a leading position in the world. The founding of the I.G. in

1926, however, gave industry leaders pause to ponder whether measures were not needed to guard against loss in world markets.

Proposed Alliances with I.G. and the Americans

At first, Britain was not inclined to form a similar cartel. An alliance with the Germans and Americans seemed a more profitable idea. The outstanding proponent of these plans was Sir Alfred Mond, whose father, Ludwig Mond, had emigrated to England after studying at Marburg and Heidelberg. In Britain Ludwig Mond had started a soda-processing company which became highly successful. Together with the Swiss John Brunner, he founded Brunner, Mond & Co. This firm became a leader in nitrate and soda manufacture.

During his first discussions with I.G., Sir Alfred Mond—later to become Lord Melchett, statesman, Zionist, and industrialist—found little sympathy for his alliance scheme. Indeed, Allied Chemicals in the United States turned him down flat. Mond had not yet left the United States following the failed talks, when another British industrialist, Sir Harry McGowan, the chief of Nobel Industries, approached him with a plan for a national solution.

Nobel was an important explosives company which, during the 1880s, had joined with German companies in a holding organization under predominantly British management. These German connections had been severed in World War I, and in 1918 a number of British explosives manufacturers had joined together to form Nobel Explosives. The catalyst in this fusion and the chairman of its board was Sir Harry Duncan McGowan. Born in Glasgow, he left school at the age of fifteen to take a job for the princely sum of 15 shillings a week.

With other leaders of the British chemical industry, Sir Alfred and Sir Harry boarded the *Aquitania* in New York on October 6, 1926, to return to Southampton. During the crossing, Sir Harry finally succeeded in winning over Sir Alfred to the formation of a British chemical giant. An agreement was drafted on four pages of Cunard Line stationery, outlining all the necessary measures for the fusion of Brunner, Mond & Co., United Alkali Co., Nobel Explosives, and the British Dyestuffs Corporation, Ltd., into the Imperial Chemicals Industries, Ltd. (ICI). British Dyestuffs itself had been formed in 1919 as the result of a fusion of two dyestuffs producers. However, the British government was the largest shareholder in the new company.

Thus, only a brief courtship preceded this British wedding. It was

The Atomium in Brussels, symbol of the World's Fair in 1958. Brussels is also the home of CEFIC (Conseil Européen des Fédérations de l'Industrie Chimique). Kurt Lanz is the President of the European Chemical Association.

Production of:
Fertilizers
Plastics
Solvents
Synthetic Resins _____ Energy

Bookeeping
Accounting

Pharmaceutical
Quality Control
Microbilogy

Engineering
Purchasing

"100 Year
Hall" A
Cultural
Center

Sales
Management
Marketing

on of:

rs

ceuticals

ic Chemicals

Research
and
Development

Training
Center

Aerial view of Hoechst AG in Frankfurt. The research center in front is one of the largest and most modern in Europe.

In 1924 an administration building was constructed within the headquarters of Hoechst; its characteristic feature is a tower and a bridge. Today the "tower and bridge" is the symbol of Hoechst all over the world.

The American "High Commissioner", John McCloy, attended the inauguration ceremony, held in 1950, to celebrate the first penicillin production, which represented an important turning point in Hoechst Pharmaceuticals. From left to right: Dr. Oeppinger, Ober-Buerger Kolb, Dr. Soltner, McCloy, Lanz, and Dr. Erlenbac.

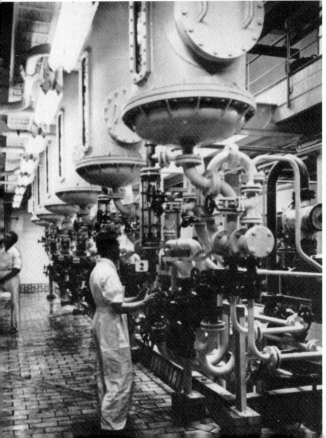

Penicillin manufacturing facility and fermentation vessel.

During the early years in England, the Hoechst dyes and pigments business was conducted by Industrial Dyestuffs located in Manchester.

The pharmaceutical activities of Hoechst U.K. have gradually increased. A research center has been built in Milton Keynes (below) approximately 100 km north of London. Radio diagnostic research is among the projects.

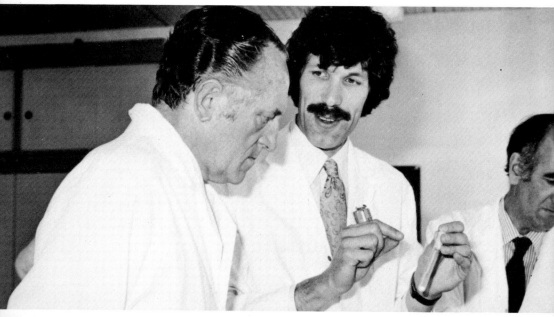

In Spain, Kurt Lanz at a state visit with former Crown Prince Juan Carlos and his wife Sofia.

A three-way handshake (among Kurt Lanz, Jean-Claude Roussel, and Wolfgang von Poelnitz) in September 1968 marks the start of a major partnership between Hoechst AG and Roussel Uclaf of France.

All four-color maps in this book reproduced by permission through the courtesy of Rand McNally & Co., R.L., whose written permission is required for further reproduction.
© Rand McNally & Co., R.L.
79-Y-99

APE PARRY
RAILL

JAMESON
LAND
oresbysund
Scoresby Sound
PE BREWSTER

JAN MAYEN
(NOR.)

Strait

Mufjordur Kópasker
Raufarhöfn
Akureyri Seydisfjordur
Neskaupstadur
ICELAND
Mt. Hekla
5,747 Ft.
Hornafjordur

Arctic Circle

906 Miles

840 Miles

1054 Miles

lgada

DEIRA IS.
(PORT.) Funchal

FAEROE IS.
F.(DEN.)
Tórshavn

SHETLAND IS.
(BR.)
Lerwick

ORKNEY
IS.
Kirkwall
C.WRATH Wick
Stornoway
HEBRIDES
UNITED
Aberdeen
SCOTLAND
Inverness
Dundee
Glasgow Perth Edinburgh
KINGDOM
Londonderry Newcastle
Sligeach (Sligo) ENGLAND
IRELAND N. IRE. Leeds
Galway Liverpool Hull
ÉIRE Manchester
Dublin Sheffield
Foynes Nottingham
Tralee WALES Birmingham
Cork Bristol
CAPE CLEAR Cobh Cardiff
LANDS END London
SCILLY IS. Southampton Dover
Plymouth Portsmouth Calais
Brest Boulogne
CHANNEL IS. Lille
St. Nazaire Caen Rouen
FRANCE Paris
Nantes Orléans Nancy
Loire Tours Strasbourg
La Rochelle Vichy Basel
Limoges Geneva Bern
Clermont-Ferrand Lyon SWITZERLAND

English Channel

Bay of
Biscay

Norwegian

Sea

North

Sea

Reykjank Arkhangelsk 2350 Miles

Hammerfest Kjelvik NOR
Tromsö Kautokeino
Inari Karesuando
Harstad
Narvik Kiruna
LOFOTEN IS. Bodö Jokkmokk Pajala
LAPL Kemijärvi
Mosjöen Lulea Kemi Qulu
Namsos Skelleftea
Steinkjer Umeå Raahe
Trondheim Levanger Östersund Kokkola
Kristiansund Harnösand Vaasa
Ålesund Sundsvall Kuopio
Lillehammer Söderhamn Kristiina
Gjövik Hamar Gävle Pori Tampere
Bergen Oslo Uppsala Turku Helsi
Haugesund Drammen Karlstad Hangö G. of Finland
Stavanger Fredrikstad Örebro Stockholm Tallinn Ra
Kristiansand Halden Norrköping HIIU ESTONIA Parnu
THE NAZE Vänersborg Linköping SAARE Tartu
Borås GOTTLAND Valmiera
Göteborg Jönköping Visby Riga R
Ålborg Halmstad ÖLAND Ventspils LATVIA
Randers Karlskrona Liepaja Jelgava Driss
DENMARK Helsingborg Klaipeda LITHUANIA
Århus Malmö (Memel) Sovetsk Kaunas
Odense Copenhagen Gdynia Kaliningrad Vilnyus
Flensburg Kiel RUGEN (Danzig) Crodno BYELO
HELGOLAND Gdansk Elblag Bialystok
Bremen GER. Szczecin Bydgoszcz Brest Pinsk
Hamburg DEM. Poznan R
Groningen Hannover Berlin U Warsaw
NETHERLANDS FED. Halle POLAND Brest
Haarlem Essen Leipzig Łódź Lublin
Amsterdam Cologne Dresden Kraków Zhiton
Rotterdam Bonn REP. Wisla Lvov Ternopo
The Hague Frankfurt GER. Prague CZECHOSLOVAKIA Kosice Ostrava
Antwerp BELGIUM Mannheim Nuremberg Danube Brno Bratislava
Brussels Luxembourg Stuttgart Vienna Debrecen
LIECH AUSTRIA Sopron Budapest Oradea
Strasbourg Munich Innsbruck HUNGARY Brasov
Basel Zürich Venice Ljubljana Szeged ROMANIA
Bern Milan Rijeka Zagreb Braila
SWITZERLAND ALPS Turin Genoa YUGOSLAVIA Zadar Danube Buc
Geneva Bologna SAN MARINO Belgrade Ruse
Bordeaux St. Etienne Nîmes Florence Ancona Dubrovnik Craiova BULGARIA
Montauban Toulouse Toulon Nice Livorno Adriatic Sea Durres Sofia Plovdiv
Marseille MONACO ITALY Rome Tirané Skopje
C. FINISTERRE Oviedo PYRENEES CORSICA Vaticàn ALBANIA GREECE
Gijón Burgos ANDORRA Ajaccio City Brindisi Ioannina Salonika
Vigo Valladolid Barcelona SARDINIA Naples Taranto Patras
Porto Zaragoza Tortosa (IT.) Cosenza KERKIRA Agean
Salamanca MINORCA Palma Cagliari Tyrrhenian Sea Messina Piraievs Sea
Coimbra Toledo MADRID MAJORCA Sea Ionian Athens
Lisbon SPAIN Valencia BALEARIC IS. Palermo Reggio Calabria Sea Roc
Évora Córdoba Murcia Med-i SICILY (IT.) Kalamai
C. ST. VINCENT Beja Alicante Catania Khania Iráklion
Faro Cadiz Sevilla Cartagena Constantine Bizerte C. PASSERO CRETE
Str. of Gibraltar Malaga Algiers Bejaia Tunis C. (GR.)
Tanger (Tangier) Gibraltar Oran Kairouan Valletta terranean
Port Lyautey Tétouan Sidi-Bel-Abbès Biskra TUNISIA MALTA
Rabat Tlemcen MTS. Sfax Tripoli Darnah Alex
Casablanca El Jadida Ain-Sefra Gafsa Gabès Bengasi Tubruq
Safi Meknès ATLAS Chott Djerid Al Khums Misrätah (Tobruk) CIRENAICA
Essaouira MOROCCO Ghardaia Zuwärah TRIPOLITANIA Ajdäbiyah Sallüm
Agadir Marrakech Béchar El Goléa ALGERIA Sirte

Bar

BRITISH
ISLES

EUR

EUR

M

hardly a true love match, but more a marriage of convenience. Indeed, Sir Alfred and the magnates as well had flirted passionately with the I. G. Farbenindustrie until just before the formation of ICI. Even afterward, Sir Alfred, the first chairman of this truly imperial company, toyed with the idea of an understanding with the I.G. Like the I.G., ICI too integrated the former individual companies, but with some difficulty. A far more serious threat were the storm clouds of a deepening world economic crisis that began to gather on the horizon. In fact, several supranational cartels were formed, especially in the dyestuffs sector. In 1932, ICI joined a cartel of German and Swiss dyestuffs manufacturers which had been formed in 1928 and 1929. The agreement ran for forty years and it covered the whole world market. Supplementary agreements were concluded with companies in Italy, Poland, and Czechoslovakia.

For I. G. Farben, the thirties brought decisive progress in pharmaceutical research. The discovery of the sulfonamides by Gerard Domagk at Bayer is one of the outstanding breakthroughs in medical progress. For the first time, these chemotherapeutic agents offered an opportunity for effectively fighting dangerous bacterial infections. Although ICI could not match I.G.'s historic discovery of the sulfa drugs in the 1930s, the company had great successes in other fields. In 1940 it achieved an important technological breakthrough with the high-pressure synthesis of polyethylene, one of the decisive milestones on the highway to the age of plastics.

Since the end of the 1920s, an ICI group in Winnington had been studying the effects of extreme physical conditions on chemical processes, for example, very high temperatures, pressures, or even vacuums. Abandoned for a time, this work was resumed in 1935 by Michael Perrin. By December of that year, he was able to produce 8 to 10 grams of polyethylene. Several more years were to pass, though, before the possibilities inherent in polyethylene were fully recognized. By the beginning of 1940, the new plastic had passed through the development stage and was introduced to the market, now shrunken by the war.

The Birth of Truly Man-Made Fibers

An equally important milestone was reached by ICI in the closely allied field of synthetic fibers. The gateway to this large development had been opened in 1934 by the American scientist Wallace Hume

Carothers. Du Pont, in the United States, had hired the young, genial chemist straight from Harvard University and provided him with a generously equipped research laboratory. Carothers proved to be an outstanding investment. He found a way of preparing a hitherto unknown polymer from a group of polyamides. This polymer could be drawn into filaments in the molten state which could then be stretched to three or four times their length, imbuing them with incredible strength.

Not until 1938 was Du Pont able to market this new fiber under the trade name of Nylon. A new era in textiles had begun. In 1937, the German I.G. chemist Paul Schlack, scientific director of the plant in Berlin-Lichterfelde, had read several papers in which Carothers dismissed a compound called caprolactam as a suitable starting material for the synthesis of polyamides. But it was precisely this substance with which Schlack was working in the preparation of synthetic fibers. And indeed, he was able to prove that Carothers was mistaken. In January 1938, Schlack produced Perlon fiber from caprolactam. When, a year later, the managers of Du Pont visited I.G., they were shown to their astonishment a fiber whose properties competed with those of nylon.

A Bible Expert Invents Polyester

Following the appearance of nylon and Perlon, Great Britain soon came along with a third major synthetic fiber, polyester. This fiber, however, was born, not in ICI, but in the laboratories of the chemist John Rex Whinfield and his assistant James T. Dickson at Calico Printers. This relatively small Manchester company was involved mainly with the development of dyeing processes.

A colleague of mine who visited Whinfield in 1965, a year before his death, told me that the invention of polyester fiber was really due to the biblical knowledge of its inventor. In the middle of the economic crisis of 1923, Whinfield was interviewed by the well-known chemist Charles Frederick Cross. Although Cross was one of the discoverers of viscose fiber in 1891, he closely questioned Whinfield, not about his chemical expertise, but about his biblical knowledge. Apparently Whinfield passed the examination, since he got the job with Cross, who later recommended him to Calico Printers. There he worked on bleaching agents and starch, but continued with his fiber research without the knowledge of the company. "They probably thought," recounted Whinfield, "that I was still working on starch. At that time we were not too closely supervised, and indeed we had a certain amount of freedom to occupy ourselves with projects outside the company's directives or interests."

There was only one large concern able to provide the means for translating the production of the new fibers from the laboratory to the manufacturing scale. Whinfield therefore turned to ICI, which quickly recognized the potential of the new material. The inventor moved to ICI, and the production of polyester began in 1946.

The History of Trevira

It was polyester that brought Hoechst and ICI together after World War II. The Bobingen textile fibers plant near Augsburg had been allocated to Hoechst from the I.G. assets. Paul Schlack, inventor of Perlon fiber, and some of his co-workers had turned up at Bobingen soon after the end of the war and started up Perlon production on a limited scale. Looking to broaden the product line, Hoechst chose polyester as the most suitable starting point, particularly as it soon realized that ICI was willing to grant a license.

ICI's head at that time was Sir Arthur Fleck, a Scotsman and chemist like his predecessors McGowan, Bain, and Rogers. I can well remember my first meeting with him. He received Professor Winnacker and our party in the ICI headquarters at 9 Millbank Street, London. During the visit, Fleck pointed to a map and showed us the production plants and sales offices of ICI throughout the world. It was an impressive demonstration which brought home to us how much work still had to be done before Hoechst could hope to gain similar status. I was struck by the fact that ICI was achieving no less than one-third of its sales abroad.

Fleck's successor in 1960 was, for the first time, not a person from the ranks, not one of the powerful dukes of the constituent companies of ICI, not a chemist, but a former government tax expert, Paul Chambers. His reign as chairman is linked not only with the company's commercial and financial reorganization but also with an operation that rather damaged the public image of ICI: the breaking off of ICI's proposed takeover of Courtaulds Ltd. Courtaulds had been founded in 1816 by Huguenot immigrants as a silk-weaving factory. Over the years, it became one of the largest manufacturers of rayon fibers and textiles in the world. This was one reason why ICI turned to synthetic fibers. An agreement, made in 1928 but long since lapsed, provided that Courtaulds would generally refrain from production of chemicals if ICI would keep out of artificial silk.

But then came 1962, when ICI ambitiously tried to incorporate Courtaulds within its empire. Efforts were made to gain a majority holding in the company through a very tempting share offer. This set off a tense public tug-of-war, with public sympathies lying largely with Courtaulds. ICI was able to buy only a bare 40 percent of the Courtaulds capital and had finally to abandon the siege of the Courtaulds fortress without result. One can scarcely imagine today how much these events affected both the general public and those directly involved. Contrary to the legendary English inclination to understatement, Sir Alexander Kearton, chairman of Courtaulds, even arranged a thanksgiving service in London's St. Paul's Cathedral. In his view, that was the only appropriate way of expressing gratitude for the danger that had been averted.

At any rate, Courtaulds remained independent and retained second place in the British chemical industry for many years. It produced all the man-made fibers: polyacrylonitril, polyamide, polyester filament, rayon, rayon high tenacity, spun rayon, and modular fibers. Courtaulds produces about one-third of all Britain's domestic textiles. In its home country, it employs some 110,000 people. At the last count, sales were about £1.2 billion.

Record Profits

From 1926 to 1975, ICI's sales increased about 80 percent, more rapidly than those of its German competitors, Bayer, BASF, and Hoechst. The output of these three German companies—on a Deutsche Mark basis —has edged beyond ICI's only because of the 50 percent drop in the value of the pound during those four years.

The structure of its production program has enabled ICI to steer a successful course in spite of strikes and the wage and price freezes. The company program is both deeply and broadly based. Backward integration extends to crackers, refineries, and the wellheads themselves. ICI has, for example, an 80 percent share in the Ninian oil field in the North Sea.

ICI, because of its size and success, sometimes overshadows the other important chemical companies of the United Kingdom. During a reception in its Queen Anne guest house, a British delegate once summarized the attitude of the public at large with the words: "If the British talk about the chemical industry, they talk about ICI, oil, and whisky." He was referring, of course, to British Petroleum, Shell, and National Distillers, all with interests in the industry. More recently, National Distillers have confined themselves once more to those activities to which they owe their name. Eli Lilly has taken over their pharmaceutical activities, while British Petroleum has absorbed the chemical facilities.

Not by Oil Alone

Like many other oil companies, partially state-owned British Petroleum (BP) has not stuck to the oil business. Through its subsidiary BP Chemicals, the company is deeply involved in the production of chemicals, such as ethylene and the plastics produced from it, and especially the manufacture of paraxylene, a starting material for polyester fiber.

I must confess that when BP and other oil companies increasingly penetrated these fields in the mid-fifties, Hoechst wondered about future competition. As it turned out, its concern appeared to be largely unfounded.

"If you look at the panorama of the chemical industry today," a top executive explained, "you will find that we have basically confined ourselves to petrochemicals." Thus, the importance of BP's partial acquisition of the Veba activities in the summer of 1978 should not be overestimated. I believe that success in the complex chemical area requires more backup by experienced scientists than oil companies can presently muster.

Shell International Chemical Co. Ltd., an offspring of the large oil company, has four factories in Britain alone. In addition, it has twenty-seven subsidiaries throughout the world, all of them chemical manufacturers. Two of them are in the Federal Republic of Germany: Deutsche Shell Chemie and Rheinische Olefinwerke (ROW), of which Shell is a joint owner with BASF.

Success of British Pharmaceutical Industry

The postwar rise of the British pharmaceutical industry began with penicillin. But it was not ICI, the largest English chemical company, that launched the production of this first major antibiotic. Its phar-

maceutical division was created only in 1944. Since then, however, ICI
has secured a firm place in this field and has made a significant contribution to chemotherapy. Two decades earlier, in 1924, the Glaxo
group had become active in the field of pharmaceuticals. It achieved
worldwide fame through its cephalosporine antibiotics.

The most important contribution to the development of semisyn-
thetic penicillins has been made by research workers of the Beecham
group. In respect to its product line, Beecham is probably the most
broadly based of British chemical companies. Founded in 1928, Beecham, together with Boots, is one of the largest producers of cosmetics
and pharmaceuticals in the United Kingdom. It also manufactures and
sells animal feedstuffs, household products, and even processed food
and beverages. In continental Europe alone, Beecham has thirty associate companies, and it has many more in the rest of the world. Beecham
typifies the pronounced international orientation of British chemical
companies in sales and, to a large extent, in production. The European
Continent has so far little noticed this fact. There, one often hears that
the English chemical industry is confined to the British Isles or, at best,
to the Commonwealth.

An important position in the British chemical industry is occupied
by Albright & Wilson, Ltd. Formed in 1892 as a family concern, it has
since been transformed into a holding company with extensive international holdings. Albright & Wilson has a broad production range,
including detergents, agrochemicals, flame-retardant products, substances for water and metal surfaces treatment, fragrances, and paints.
Above all, however, Albright & Wilson is one of the few manufacturers of thermal phosphorus and its derivatives.

A word about Fisons Limited. From a tiny beginning in 1843, Fisons
became the largest mixed fertilizer producer in England. In many
world markets, Hoechst encounters Fisons as an important competitor
in fertilizers, crop-protection agents, and pharmaceuticals.

Although these companies do not match the size of ICI or of any of
the three I.G. successors, their scientific potential and market strength
must not be underestimated. Centuries of international experience and
a world perspective are great strengths in their favor.

Hoechst in England

Not surprisingly, then, Hoechst needed to plan carefully before establishing itself in England. Our beginning in London after World War

II was even more modest than that in France. Again it was a refugee, Léopold Laufer of Prague, who was instrumental in our postwar start in Great Britain.

Laufer, a naturalized English citizen and founder of Laufer Chemicals, visited us in the early 1950s to discuss possible cooperation. However, we could not be the hosts for the subsequent luncheon in the former Hoechst dining room. Laufer had to invite us. The restaurant had been taken over as an American officers' club, and Germans were allowed to enter only as guests of an Allied representative. Later, sitting in the restaurant with foreign visitors, I often recalled the incident with amusement. In any event, in those days we were glad to get an excellent turkey lunch. Moreover, the talks were also successful. Laufer Chemicals became the embryo of the later Hoechst U.K.

In the years immediately following the war, the dyestuffs business in the United Kingdom was carried out by Industrial Dyestuffs in Manchester jointly with Cassella and Bayer. However, this *ménage à trois* proved unsatisfactory to all the partners. Bayer withdrew, and in 1963 we founded Hoechst Cassella Dyestuffs Ltd.

Our pharmaceuticals were sold in the United Kingdom through Hoechst Pharmaceuticals Ltd., which had been founded as a joint venture with Horlicks Ltd. in 1956. Horlicks produced a famous malted milk drink and a line of nonprescription pharmaceuticals. Its pharmaceutical plant, which might serve as a nucleus for our own production, was a deciding factor in this venture.

The first major upsurge in our pharmaceutical business in England came with the introduction of Lasix, a diuretic agent which rapidly eliminates abnormal accumulations of fluid from the body at a hitherto unknown rate. Lasix quickly became known as an internationally renowned preparation for these indications. In 1965, we took over the Horlicks interest in Hoechst Pharmaceuticals.

Hoechst U.K. Is Formed

In spite of all the psychological barriers arising from the war, I found the British friendly and fair business partners as we started the reconstruction of our activities in London at the beginning of the 1950s. We deliberately adopted a low profile, keeping in the background as far as possible. Only in recent years, when our industrial involvement in the United Kingdom increased considerably, did we embark upon advertising and public relations campaigns to inform our customers and the public at large about Hoechst, and to improve the climate for our products in the marketplace.

In 1965, Hoechst U.K. was founded as a holding company for our various subsidiaries. Our next step was to reorganize this new unit as an operating company. The directors of the former subsidiaries became active board members. In this way, we secured an active management for the new central company. Of course, to me, London has always been more than just a business address. As in Paris, even the most unimpressionable visitor cannot escape the impact of this metropolis.

London: Entirely Private

The many palaces, the Westminster quarter, Hyde Park, and the City still have the trappings of an imperial metropolis. During my many visits, I have had repeated opportunities to get to know new aspects of this great city and to penetrate its history. I took a particular delight in showing my wife and our children London and its surroundings. We lived in an old-fashioned English hotel at 11 Cadogan Street, our headquarters for many excursions.

In the Tower, we were able to learn something of England in the dark and bloody Middle Ages, portrayed so vividly in Shakespeare's royal dramas on both stage and television. To see John Gielgud as Macbeth in the Old Vic theater was a unique experience. In Westminster, the history of English parliamentarianism comes to life. The children were delighted by our outings to Blenheim Castle, Sir Winston Churchill's birthplace, where even today thousands upon thousands of people pay their respects to the great man.

I well remember my first visit to the British Museum, accompanied by the head of our London office at that time. I was deeply impressed by the wealth of cultural treasures, especially the antiquities. Even in Greece itself it would be difficult to find anything more beautiful than the Elgin Marbles, originally part of the Parthenon in Athens. I was astonished to find that my companion, like me, was visiting the museum for the first time, although he had lived in London for over twenty years.

On the Continent, we often have a narrow view of English painting. Many of us know only the works of Constable or Turner. I found my visits to the National Gallery in Trafalgar Square and the Tate Gallery on the Embankment therefore all the more stimulating. I also well remember the 1976 Turner Exhibition in the Royal Academy, including all his works, whose versatility revealed the artist to me in an entirely new light.

The amusement quarter in Soho, with its small restaurants offering specialties from Budapest to Peking, is also a great attraction to the foreign visitor. So, too, are the nightclubs awaiting guests in the little side streets.

Everything is represented in London—splendor and misery, gleaming highlights of art and culture and dull expanses of mediocrity. At one time Piccadilly Circus, where the heart of London beats, appeared to have become the meeting point of youthful dropouts from all over the world. Happily, this phase, too, has passed. But vagrants still spend cold nights lying, wrapped in newspapers, on the ventilation shafts of the Underground.

London's surroundings are especially attractive. We made many a weekend trip to absorb the scenery, culture, and history of Windsor, Eton, and many other spots. Glyndebourne with its rustic summer festivals is a unique British experience. We attended a performance of Mozart's *Magic Flute,* directed by Sir Benjamin Britten. The guests traveled in evening dress on a special train from London, and during the long intermission they ate sumptuous picnics on the lawns of the splendid country mansion there.

Whether it is history, theater, or exhibition, I am never fully satisfied if I return from a short trip to London without having found a little time outside the business agenda to view one of the numerous cultural attractions.

An Important Acquisition

At the end of 1969, we were able greatly to increase our business in Britain and throughout the Commonwealth at a stroke. We acquired the British paint company of Berger, Jenson & Nicholson (BJN).

Even for us, this acquisition was not a normal transaction. Hoechst had a policy of not competing with its customers in the various processing industries. For example, we were never interested in acquiring companies directly involved in consumer products. For many years we also followed this principle in paints and lacquers, although our dyestuffs, pigments, and synthetic resins form the starting products of the industry. However, in the 1960s other important chemical companies adopted an entirely different approach and began to acquire paint companies. The interlocking of suppliers of starting products and processors, and the acquisition of raw materials largely from the parent company, threatened to restrict our customers in this field. Economics journalists even asked us what we intended to do to counter the trend. They did not know that we had been following the situation closely, and had already established contacts that would lead before long to close cooperation with German paint companies.

Eventually we discovered that Berger, Jenson & Nicholson, the second largest paint company in England after ICI, was ripe for new ownership. It looked like a unique opportunity. Gustav Bunge, then head of Hoechst U.K., and I visited a lady by the name of Vera Hue Williams, whose family was the largest private shareholder in BJN. Mrs. Williams, who I believe was of Russian origin, told us that the family was indeed interested in selling its shares. Her husband, with perfect English reticence, said barely a word.

We therefore made contact with Celanese, the American company which owned 50 percent of the BJN shares. The Americans, too, appeared ready to sell. In accordance with the rules of the London Stock Exchange, we next made a public bid for the small number of BJN shares still in the hands of individual shareholders. Such bids have often been made without the agreement of the management of the company concerned. This we wanted to avoid at all costs. We were vitally determined to get the full agreement of the board members of BJN. I well remember the first discussion with the chairman of the BJN supervisory board, Lord Kings Norton, the managing director, J. A. Hughes, who was born in Australia, and some of his colleagues in the board room at Berger House in Berkeley Square, London. At first, our reception was not exactly enthusiastic. Our initial discussion was more like a cross-examination rife with skepticism. Our English partners were especially interested in learning of our future plans for BJN: how we saw the future development of the company, and whether it would retain its autonomy and identity. The atmosphere became increasingly relaxed as the board members eventually recognized that merger with Hoechst would ultimately offer great advantages to all concerned. In particular, the ownership situation would be stabilized and affiliation with a world organization of great importance would open up a multitude of previously nonexistent possibilities. Finally, the acquisition of BJN by Hoechst was realized, although not before an American company had put in a counterbid that forced us to increase our offer to the shareholders substantially.

In BJN, we acquired a company firmly established in the paint and lacquer market, with special strengths in decorative paints, car finishes, and marine paints. The product line also included wood preservatives, sealing agents, wallpapers, and household adhesives. The company was active in world markets, especially in the Commonwealth countries of Australia and New Zealand, where the company has large factories. BJN also operated an extensive wholesale and retail chain.

As in the case of Roussel, we took great care not to interfere with the management structure of BJN. Hoechst believes that the management of its foreign subsidiaries should be staffed with as many people as possible from the host country. The arrogant view held in Western Europe notwithstanding, our British friends work no less hard than we do on the Continent, even if they do not proclaim it quite so vociferously. In any case, a manager's efficiency cannot be measured by a bureaucratic stopwatch alone, but rather, by creative resources and success. A good idea conceived during the teabreak or lunch at the club is worth more than an 8-hour day completed without a single creative impulse.

Trevira Comes to Ireland

Our decision to construct a plant for polyester fibers near Limavady in Northern Ireland greatly expanded our presence in Great Britain. When the plant was inaugurated in 1970, ICI's patents on polyester, for which we had originally taken out a license, had expired. We could now offer our polyester fiber, Trevira, throughout the world. The selection of Limavady as a plant site was governed partly by the fact that it provided us with additional production facilities in the Greater European Free Trade Area, into which exports would otherwise have been hindered by tariff barriers. It weighed equally in our considerations that the British textile industry enjoyed—and enjoys—an outstanding position in the world market. Moreover, the Northern Ireland government compensated us for location drawbacks with generous subsidies.

Hoechst Pharmaceutical Research in the United Kingdom

The pharmaceutical activities of Hoechst in England have expanded into important therapeutic areas. In Milton Keynes, about 112 kilometers northwest of London, a plant has been erected for the production of vaccines and serums for veterinary use. The Milton Keynes complex includes a Hoechst research center engaged in the development of radio diagnostics. These are radioactively labeled substances emitting harmless radiation that can be traced for diagnostic purposes as they pass through the body. Roussel also maintained a successful regional presence in England with its pharmaceutical production in Swindon, 120 kilometers west of London. There we were able to produce the Hoechst line of human pharmaceuticals. Since the pharmaceutical research group of Roussel cannot be expanded in Swindon because of lack of space, it will be transferred to Milton Keynes.

In 1967, Harlow Chemical Co. was founded in equal partnership with Revertex. Harco, as the company is known, has secured us an important place among the suppliers of the paint and adhesives industries as the manufacturer of plastic dispersions (Mowilith) and polyvinyl alcohol.

In 1977, the recession in the man-made fiber field forced us to close down the texturizing facilities acquired at Stainland near Halifax in Yorkshire. However, we have not withdrawn from this region around Manchester, so vital to our industrial business. Two plants for textile auxiliaries and pigment preparations were constructed on the Stainland site in 1978.

To round out the ethical pharmaceutical business with a series of diversifications, we decided in 1974, together with Roussel, to produce over-the-counter medicines for self-medication. We acquired Optrex, well known in England and other English-speaking countries for eye-care products and cough mixtures. We were also interested in the hospital supplies field. Although a start had been made in Germany, a proper springboard was lacking. This came with the acquisition of Euromedical, a small but highly specialized catheter producer with plants in southern England and Malaysia.

Given the importance of our pharmaceutical business for the company as a whole, these activities will gain increasing weight in the future.

Britain Joins the Common Market

Entry into the European Economic Community (EEC) has opened up a greater market for Great Britain. Unlike many of its politicians, Britain's industrialists have generally favored going into Europe, even when labor and the trade unions were less than eager about the prospect and De Gaulle's brusque double veto was more than welcomed by some in the British Isles. Many of the leading people in British chemical industry assured me at the time that "even if it does not look like it at the moment, we shall go into Europe." It was a day of satisfaction not only for those industrialists but also for us "little Europeans" when Great Britain's long march into Europe, which began with Prime Minister Harold Macmillan's application, was happily completed with the approving referendum of 1973.

British Chemical Industry in Continental Europe

The importance of the European market to the British chemical industry was proven by the many British subsidiaries on the Continent long before Britain joined the EEC. Many large British chemical companies have been established in the Federal Republic of Germany. First and foremost among them is ICI. From its European central office in Brussels, it controls some fifty associates in Europe, including several in West Germany.

Apart from the polyester plant in Ostringen near Heidelberg, the most important ICI installation is in Rozenburg near the Europort in Rotterdam, Holland, where fibers, polymers, and plastics are produced. At the mouth of the Rhône in Fos, France, ICI has an extensive plant for high-pressure polyethylene. Except for the United States, Europe is, today, ICI's fastest-growing market. Of course, many other major British chemical companies have both exported to Western Europe for a long time and established their own production and selling organizations in the most important markets there. In the future, they will no doubt exploit all the opportunities the European community offers them.

Future Perspectives

A year or two ago, it seemed that some positive developments would soon occur in the British economy as a whole and the chemical industry in particular. There was some hope—now questionable again, of course—that the trade unions might pursue a less harsh policy of confrontation. This possibility might well have led to a significant reduction in the number of labor disputes that have so harmed British industry in the last decade. The causes for these disputes are not only economic. The traditional class structure of British society is a major factor. In British society, the demarcation between upper, middle, and working class is far sharper than in many other industrialized nations. This may be because the English social structure has not been dissolved by inflation and political breakdown, as has happened in most European countries. Also, the deep British love of tradition opposes excessively rapid change. Adherence to existing tradition manifests itself in many small ways in the chemical industry; for example, in the permanence of satisfactory business relationships. Amusingly, since its founding more than fifty years ago, ICI's annual report has been produced by the same printer.

Nevertheless, a change, albeit a slow one, is clearly taking place in social distinctions. George Orwell, famous for his book *1984* (the terrible vision of a totalitarian state), observed in his 1947 book *The English People* that class differences, still very perceptible, had narrowed during the preceding thirty years, and that the war had accelerated this trend. Yet trade union leaders continue to resist the truth that the external labor disputes, often begun before the possibilities of negotiation have been exhausted, are not likely to serve the British working people in the long run.

Nationalization: A Sword of Damocles?

While trade unions continue to ignore economic realities, the left wing of the Labour Party insists on further nationalization, irrespective of the hardly encouraging experience with industries already nationalized. These account for some 60 percent of the British gross national product, be they coal, electricity, gas, railways, airlines, or the steel industry. But ideologists often happen to have this curiously distant view of reality. Edward Heath, the former Conservative prime minister, joins much of industry in decrying nationalization thus:

> Nationalized undertakings have all become monopolies; neither consumers nor workers have a choice any longer. Market forces can operate only in a minority sector of our industry; market forces in the economy as a whole and even in the private sector (because of the monopoly position of the public industries) are far less effective . . . nationalization robs the managers of these industries and organizations of the need for financial and commercial discipline. . . . Consumers no longer have any say.

Nationalization, the sword of Damocles that has been suspended for so long over many British companies, has not promoted greater investment. Whether further nationalization plans will actually come to realization in the long term is certainly doubtful, in view of the Conservative Party's victory and the great differences of opinion within the Labour Party itself. Many people with whom I have talked hold this view very firmly, including journalists as well as industry leaders.

Apart from the nationalization problem, price controls acutely inhibit urgently needed investment. The price code has prevented increases in a recovery phase when the market could and would stand their effects. As a result, profits needed for future investments cannot be earned by many companies.

The unsatisfactory domestic price structure is one of the reasons why the British chemical industry has made such determined efforts to increase its exports, especially in the last few years. Given the weakness of the British pound, satisfactory price levels can be achieved only in the export area.

Gas and Oil from the North Sea

In the final analysis, however, British industry, and above all the British chemical industry, is about to score a tremendous plus point whose full significance will be realized only in a few years' time. I am referring to the natural gas and oil finds in the North Sea.

This development began in the fifties when the Dutch discovered huge natural gas fields off the coast at Groningen. The suspicion that the fields might possibly extend far into the southern North Sea was soon confirmed. The British, primarily British Petroleum, started to drill in 1963, and the first success was recorded two years later. In the West Sole Field east of Hull, a large natural gas deposit was found at a depth of about 3 kilometers. A pipeline to the coast was constructed and gas began to be pumped in mid-1967. More successful drillings followed. The largest deposit found so far—on the Lemon Bank not far southeast of West Sole—will yield about 300 billion cubic meters of natural gas, according to rough estimates. In 1971, more than 90 percent of Britain's gas requirement could be met with North Sea gas. Since 1977, British gas consumption has been provided entirely from North Sea gas, and a surplus is being exported.

But not only natural gas was found. In the British off-shore areas a far more valuable treasure was discovered: oil. In the northern part of the North Sea, beginning with the Edinburgh latitude and extending far to the north of the Shetland Isles, BP discovered the Forties Field in 1970. A second find was soon recorded through joint efforts by Shell and Esso. Then followed a sequence of successful drillings. Meanwhile, deposits of some 2 billion tons of oil have been confirmed in the British North Sea. The first oil from these fields was pumped to Scotland in 1975. Most of the fields should be operational by 1978 or 1979. After 1980, production of 100 million tons of oil a year is expected, enough to fully meet the predicted British oil requirements.

Production costs of North Sea oil were at the time much higher than the prices set by the Organization of Petroleum Exporting Countries—before the recent cutoff of Iranian oil. However, they may be lowered in the course of time. Even now, the guaranteed oil supply on their doorstep is enormously attractive to both the British chemical industry and foreign companies in Britain. It will prove to be a powerful trump card in the future. Despite the shadows darkening the British scene for some time past—its political and social institutions and, above all, its economy—one must not lose sight of the strengths of the country. Among them is the worldwide political importance of Great Britain. Although somewhat reduced by American leadership, it still manifests itself internationally. In the United Nations, for example, Great Britain has veto power as one of the five permanent members of the Security Council. Britain's foreign policy is thus equipped with levers whose usefulness is beyond question.

The pragmatically gifted British have been able to wield—through the loose association of countries described as the Commonwealth—a considerable influence in international affairs. The results, as in the case of Rhodesia, have not always been entirely positive. Of course, over the years the Commonwealth members have been reduced in number. But in many Asian and African nations, the institutions of law, administration, and education established by Britain have survived. And these countries are linked by the common bond of the English language, which has powers beyond a purely communicative function.

A network of special commercial relationships has also survived. Although some have long since lost their profitability to the mother country, they still ensure many useful connections in the services field. The center of this nerve system is the City of London. This square mile of densely built-up land in the heart of the metropolis has few full-time inhabitants. But there are commodity exchanges, insurance companies, banks, and financial institutions. Without their complicated and efficient services, the world economy would be crippled. Anyone acting as a banker in the Eurodollar market or in the futures business will confirm this observation.

And, last but not least, there are the people, in the end always the decisive resource of any country. Molded by insular existence and therefore reserved, especially when dealing with strangers, the British are nevertheless international in outlook, sociable, and ready to make friends. The Anglo-Saxon, to the regret of many Hanseatics, is a species without parallel in the world. Perhaps these contradictory characteristics are what we like so much in the British: the mixture of self-consciousness and tolerance, the penchant for the eccentric and yet the pronounced sense of reality. Great Britain, therefore, is a country that can be trusted with mastering difficult future tasks. The British chemical industry, in particular, will also prove—perhaps more than we may like—to be a strong, innovative, and fair competitor in the world marketplaces of the future.

5

Scandinavia: Between Prosperity and Socialism

My first encounter with Scandinavia—actually, with Copenhagen—was accidental. In the summer of 1936, I applied to the Reichsbank for foreign currency for a vacation trip to Italy. I had managed to scrape together the princely sum of RM150. However, the Reichsbank more than took its time, and I didn't want to waste mine. The liras for my Italian trip were at my home when I returned from my holiday in Scandinavia.

While awaiting the Reichsbank's response, I set off with a school-friend, cycling north through Lübeck, Kiel, and Flensburg up to the Danish border, as in many previous summers. By then the temptation to cross into Denmark had become irresistible. It was not exactly easy to convince the German border guards that we seriously intended to finance our stay in Denmark with the standard foreign currency allowance of RM30 per person. But somehow we managed, and spent a beautiful week in this lovely country.

At the time, Denmark was still an enchanting idyll. We admired the bridge across the small sound, at that time the largest bridge in Europe, and journeyed across to the island of Funen where we visited Hans

Christian Andersen's house in the picturesque little town of Odense. We then took the ferry to Zeeland and eventually finished up in Copenhagen.

To the two of us, having grown up in the provincial town of Baden, Copenhagen seemed a metropolis, which immediately captivated us with all its charm. The marvelous world of the Tivoli, the little nymph from Andersen's fairytale world, the carefree middle-class atmosphere of a small country living in peace in the midst of not exactly the happiest period in the world all to be eclipsed so soon.

On our return, we attended the 1936 Olympic Games in Berlin. The excitement and splendor were well organized to push back the threatening shadows for a time.

These recollections were in my mind when I visited Hjalmar Lautrup-Larsen half a generation later in 1952. Recommended by our friend Olaf Thrane, he had founded a small company to reintroduce Hoechst pharmaceuticals to Danish doctors. Fun-loving, experienced, sixty-year-old Lautrup was successful in this endeavor, with the help of his young colleague, Allan Güldner.

Dansk Aniline, a former I.G. affiliate, joined Hoechst Danmark in 1969. Two years later, Hoechst Danmark moved out of the center of Copenhagen into a modern administration building and warehouse on the periphery. Naturally, the typically Danish style of interior architecture graces its offices.

The importance of Denmark's industry is often underestimated and the country is frequently considered purely agrarian. Of course, exports of meat and dairy products, especially to England and Germany, are an important source of income. They were also decisive in Denmark's entry into the European Economic Community. In the meantime, however, the processing industry has assumed greater significance in the economy. The chemical industry, too, occupies a remarkable position. Novo, for example, is Europe's second largest insulin manufacturer, with sales of about 300 million Krones (US$60 million). It is also the leading producer of enzymes from animal pancreatic glands.

Because of its pharmaceutical research, the Lundbeck group has achieved a more than local reputation. Through the granting of licenses, it has become eminent throughout the free world. Cheminova operates a successful business in agricultural chemicals on the wide western coast of Jutland. This company, founded in 1938, is one of the largest producers of phosphorus-based insecticides. Its export quota of 99 percent is the highest in the Danish economy. The location of this company, incidentally, has become a model for the development of industrial land while avoiding environmental problems. Its example was copied elsewhere only many years afterward.

For many decades, Denmark has pursued a liberal, free-trade economic policy. Therefore Hoechst had no serious export obstacles or other commercial difficulties to overcome. Nevertheless, we set up our own manufacturing facilities in the country, and these provided a valuable support for our import business. Dansk Ozalid continues the production of reprographic paper begun in 1937. Taking over Chemische Werke Albert in Wiesbaden in 1964, we also acquired Polyplex, its Danish subsidiary, supplying synthetic resins to our Scandinavian customers. Some years ago, Hoechst Danmark purchased the texturizing plant of Kaj Neckelmann A/B to expand polyester filament production in the Scandinavian countries.

Germany's Danish neighbors have tried the Scandinavian route to the welfare state, although in a less doctrinaire socialist form than found in Sweden and Norway. However, welfare benefits soon outpaced productivity, leading, of course, to excessive wage increases and social payments and, finally, to a loss of purchasing power through inflation. Competition in foreign markets has thus become all the more difficult. But which Western government has so far succeeded in convincing its voters that belts have to be tightened from time to time? Whatever the economic problems, Copenhagen has lost none of its charm or cultural vibrancy even if business talks have become a bit edgy recently. Our Danish friends, however, face up to problems more cheerfully than do our Nordic neighbors on occasion.

Sweden: Balanced Equality of Opportunity?

When first crossing the border, one thinks the Swedish landscape is very similar to that of Denmark. Fundamental differences are encountered as one progresses into the interior, however.

Infinity and solitude are the rewards for the traveler journeying through the Swedish countryside. Even traveling by car, one may not encounter a single human being for hours—sometimes not even another car. Travelers are alone with the forests, the lakes, and the infinite horizon. They can refresh themselves with the crystal-clear water of Sweden's lakes.

The warm gulf stream along the Norwegian coast both forms and favors the climate of Sweden. It is therefore a particularly popular holiday country for people who, instead of merely lying in the sun, seek the real relaxation and quietude that have become so rare elsewhere.

I got to know our Swedish customers as early as 1940. Alas, it was only at a distance, namely, from the desk I occupied, at the end of my training years, as correspondent in the dyestuffs sales department for the Scandinavian countries. At that time, Hans Th. Winckler conducted the business in the style of an elegant, reserved Hamburg merchant. For some years, Consul Helge Svenson had been the manager of the Anilinkompaniet in Göteborg, another diligent entrepreneur who was always friendly but distant. After the war, Svenson resumed the Hoechst dyestuffs business in Sweden together with his old colleague Friedrich Arnold. Since 1953, the Hoechst business has been conducted by a managing director, Sixten Arnold, the second generation of this name.

In the early fifties I often visited Sweden, accompanied by Hans Winckler, who had in the meantime become dyestuffs export manager in Hoechst. We gradually concentrated our Swedish interests first in Hoechst Anilin, then, in 1966, in Svenska Hoechst.

After more than forty years of Social-Democratic government, the 1976 elections produced a coalition government of the nonsocialist parties. Remarkably, five women were appointed to the new cabinet. Those forty years of socialist government changed not only the structure of industry but also the way of life of the population. Through steadily rising and steeply progressive taxes, the state carried out increasing redistribution of the national income. Salaries ascended, as did the rate of inflation. Education, housing, and codetermination by the trade unions in the factories are the laboratories of middle-class socialism. Although respecting the rules of democracy, Swedish socialism strove not only for equality of opportunity but also for a Scandinavian variation of *egalité*. It certainly did not promote its subjects' enjoyment of work.

From Kirsten to Ralph

Egalité takes on a quite different social nuance in Sweden. Swedish employees normally address one another by their Christian names. If, for example, secretary X has to write to her boss, she will address the note quite formally to, say, "Director Ralph D.," but the letter itself will start with "Dear Ralph" and will be signed "Yours, Kirsten." In this context, I recall many discussions with the management in Göteborg concerning the salaries of the leading staff members. None of them was interested in routine salary increases because, as a result of the steeply progressive taxes, these would in fact have reduced their incomes.

For example, the Swedish tax rate for an income of approximately Kr50,000 is 64 percent. For an income of Kr150,000, the rate is 87 percent, leaving only 13 percent to the tax payer. On the other hand, since old-age pensions and national health contributions depend on the gross income, in theory no one ought to object to an increase in income. In fact, if income does not keep pace with inflation, old-age pension and sickness contributions are reduced.

Family income consists of salary or wage income, often supplemented by large social subsidies. For example, 60 percent of all the families with children receive a housing subsidy.

The Swedish tax system works in strange ways. For example, if the parents of a family with four children both work for only half a day at a wage of Kr25,000 each, they receive the same net income as they would if the husband earned Kr100,000. The husband in a family with five children earning Kr110,000 a year has a net income of Kr62,300, but he would receive the same net income from social security benefits alone without having to earn a penny. If salaries are increased, many of the social contributions may be forfeited or so greatly reduced as to leave employees with less than they had before. Salary increases must be fairly high, therefore, to ensure that the net income is indeed improved. These examples explain why nobody is really interested in working overtime. Under certain circumstances, such diligence can mean the loss of social benefits.

A Welcome Need for Rationalization

The Swedish industry was faced by a basically healthy need for extensive rationalization. And where even reorganization no longer availed, many textile and apparel businesses had to close down. Between 1960 and 1976, the number of workers in this branch of industry was reduced from 95,000 to about 45,000, and people were able to meet only 20 to 34 percent of their need.

The government intends to solve the problem through founding a government-owned textile and apparel concern initially embracing fifteen companies. The remaining firms will be eligible for support through favorable loans and subsidies. Furthermore, by concentrating the cotton spinners into three companies, the government expects greater efficiencies.

Whether these measures will be successful remains to be seen over the next few years. Several textile concerns have moved their plants into countries with lower wage rates. While relief work and retraining have kept down the number of unemployed, inflation has grown steadily.

The conservative coalition now in power must contend with a grave inheritance. Its main preoccupation must be to steer developments into calmer waters. Sweden's stabilizing influence on Western European foreign policy must be maintained.

Forests and ores make up the large natural wealth of the country. It has been estimated that the reserves will last another 150 years. These raw materials were also the starting point for Sweden's chemical industry. Indeed, the largest chemical concern in Sweden is Kema Nord A/B, whose main shareholder for many years was Stora Kopparberg A/B, the oldest mining concern in the country.

Since before World War II, I. G. Farben had cooperated with Mooch Domsjö in northern Sweden. This company produced acetic acid, acetaldehyde, and various solvents as well as polyvinylacetate. Perstorp A/B utilized the wood carbonization process which Degussa employed in southern Germany until well after the war. Before the war, I.G. too had closely cooperated with Perstorp in the solvent production. Thanks to the barbecue rage with which the United States has blessed us, charcoal production once more has become a thriving business in many parts of the world.

Some years ago, a state-owned holding company, Statsföretag, was established to promote industry and ensure that the state could exert its influence. Berol-Kemi, Kabi, an important pharmaceutical producer, Syntes, and Katalys have in the meantime become members of this group.

State-Produced Pharmaceuticals Are Not Cheaper

The state acquired a majority holding in Kabi, the large domestic pharmaceutical manufacturer, for reasons of health service policy. The measure was primarily intended to complement the earlier nationalization of pharmacies. However, pharmaceuticals did not become cheaper as a result. On the contrary, they became more expensive.

Sweden's largely nationalized health service, however impressive its statistics, could provide German critics with much food for thought.

The national health service of Sweden succeeded neither in limiting the consumption of pharmaceuticals nor in suppressing rising costs. Indeed, Swedish patients have had to pay a high price for the progressive control of medical services. In their model welfare state, there is practically no free choice of doctor; physicians' home visits are becoming a rarity; and the time spent waiting for both in-patient and out-patient treatment in hospitals is lengthening. The extent to which the expectations engendered by a national health service have remained unfulfilled is strikingly illustrated by the nationalization of the pharmacies. Sweden's politicians expected a reduction of 10 to 20 percent in the cost of pharmaceuticals once the corner pharmacy had been taken over by the state.

However, when the pharmacies were finally nationalized on January 1, 1971, the selling price of pharmaceuticals not only increased but actually rose more rapidly than the prices of manufacturers and wholesalers. As a result, therefore, the retail markup on the final price of Swedish pharmaceuticals has risen instead of declining. The nationalized pharmacies of Sweden show that costs may rise in inverse proportion to the restrictions placed upon personal initiative. The loser is usually the patient who has to pay more for less service.

Overall, Swedish industry does not lack dynamism. Many able managers are not very impressed by the ultimately illusory appeal of equality, because they love their country and its way of life more than anything else. This became evident when, some years ago, we founded a joint company with Perstorp A/B, in the vicinity of Malmö, to supply the markets in the north with plastic dispersions. This joint venture functioned well from the first day and a significant extension of the product line is now underway. I paid many visits to Stockholm, the beautiful metropolis of the north, although the headquarters of our company has remained in Göteborg, since the processing industries and agriculture are located mainly in the south, where 80 percent of the population live.

Previously, one traveled from Göteborg to Stockholm by rail, crossing the most beautiful part of central Sweden with its boundless dark forests and lakes. Our pharmaceutical business and an essential part of the reprographics business are conducted from Stockholm. The "leisure value" of the two cities is probably the same. Both are located between sea, rock, and forest. A summer weekend among the skerries outside Stockholm is an unforgettable experience. The Swedes, so calm and relaxed, know how to enjoy life whether it be outside in nature or together with friends around a well-stocked table. Smorgasbord and gravlax are among the many delicacies to which guests readily return.

El Dorado of Recreation

Stockholm, the "Venice of the North," has a wealth of prospects for utilizing spare time between business meetings. For example, a visit to the small town of Mariefred in Södermanland becomes a unique encounter with the history of Sweden. One of the little steamers that tie up on the quayside in front of the town hall takes you across Lake Mälar to, say, Gripsholm Castle. Built in the late fourteenth century, it acquired its present appearance only in the seventeenth and eighteenth centuries at the time of the Vasa kings. Gustav the Third later built a charming small theater in one of the two round towers of the castle, where "political theater" was produced long before the term was known elsewhere.

Of course, like most localities between 60° and 62° of latitude, Stockholm also has its rainy days. Perhaps its rainfall is why there are so many museums in Stockholm—literally no fewer than fifty-two, including a post museum, a historical museum, and an army museum. A lot of rain will have to fall in Stockholm before one can do justice to all the institutions that document the country's way of life and history.

My first encounter with Swedish literature was via the charming story *Nils Holgersson's Wonderful Journey with the Wild Geese.* My children also enjoyed the book. In any case, Scandinavia's literature has a considerable following in Germany and other parts of the world— from Andersen to Ibsen, from Kierkegaard and Strindberg to Knut Hamsun. The same is true of music: the Norwegian Edvard Grieg's *Peer Gynt,* and especially his piano concerto, once were the highlights of the programs in concert halls. And the Swedish films, including the major, and sometimes controversial, works of Ingmar Bergman, continue to be shown in cinema and on television.

Norway—Waterpower and Oil

Viewed from the European Continent, Norway excites interest today mainly because of the fundamental structural changes which may be engendered by the vast oil and gas finds in Norwegian territorial waters.

These finds are hardly ten years old. It was a real Christmas surprise when the Phillips group, which had drilled for some years for off-shore oil, announced tersely, one December day, that the Ekofisk field had an oil potential that might well make it one of the fifty largest in the world. Since then, the North Sea as a whole has developed more quickly than any of the world's other oil fields. More than twenty finds have been made in the territorial waters of Norway alone. The Ekofisk field is by far the most profitable. Its production potential is 1 million barrels of crude oil per day, in addition to some 15 billion cubic meters of natural gas each year. Thus, this field alone is capable of an output equivalent to six times the present annual Norwegian oil production. Clearly, there will be a large and growing net export of oil, enabling Norway to settle all its state debts by 1980.

There can be no doubt that these new riches will provide Norway with enormous wealth and transform the economic life of the country. Of course, a boom of these dimensions harbors dangers, as well. Full employment in the country and increased investments may lead to rapid inflation. The national oil company, Statoil, therefore also has the task of applying the brakes at the proper time. In this connection, some people talk of "hurrying slowly."

World War II provided painful evidence of Norway's strategic importance. Narvik, the most important ice-free Norwegian ore port in the far north, was one of the major targets of the German invasion and occupation. In spite of the bitterness evoked by the German actions, once the conflict was over, the old economic connections proved more durable than the memories of unhappy days.

At the beginning of the 1950s, Hoechst had renewed contact with Norsk Hydro, the large energy and chemicals company in which the Norwegian government has a majority holding. Although the head-quarters of the company is in Oslo, it owns energy sources north of the Arctic Circle. Norsk Hydro employs much of its vast hydraulic potential for the manufacture of aluminum and magnesium.

After the war, some people who had produced magnesium in the Bitterfeld works of the I.G., then occupied by the Russians, gathered at Knapsack, the Hoechst subsidiary near Cologne. They formulated a plan to manufacture magnesium for West Germany in combination with a Norwegian partner and possibly with an office in Norway. Despite good intentions on both sides, the project was never realized. In essence, the Norwegians probably did not want too close connections with foreign companies. Concern over excessive foreign influence is always present in a country of only 4 million inhabitants.

A similar situation arose after the discovery of the rich oil and gas fields. Petrochemical projects were soon discussed, to be realized with both foreign know-how and outside capital. In the end, however, the Norwegians decided to run the operation themselves.

Norsk Hydro, Statoil, Borregard (a private company originally engaged in woodworking), and Saga Petrokjemi are playing decisive roles in this endeavor. A natural gas pipeline to Emden has been constructed. From Emden, the gas is distributed through the European network and the petrochemical companies. Norway should encounter few problems in selling its natural gas. On the other hand, it seems to me rather more problematical how this nation is going to dispose of its petrochemical products. The small domestic market is in no way large enough to absorb all that is going to be produced.

Oil Bonanza, Pro and Con

Norway, blessed with natural wealth in ores, wood, pulp, hydroelectric power, fishing, shipping, and now oil, nevertheless has gone through difficult economic straits in the recent past. Although the country did not adopt socialism in the Swedish manner, the ideal of social equality and a free economy, rendered largely illusory because of state intervention, has had its dark side. Wage policy did not differ from that in neighboring countries. In Norway, too, wage policy is quite openly regarded by the unions as a way to redistribute wealth.

The new oil and gas finds are not solving these economic problems, although they understandably caused a kind of gold rush for a time. But it is probably not a very tragic matter if Norway is currently living in a slight euphoria and somewhat above its means. It seems certain that future income will restore the balance and that the difficulties caused by the economic utilization of the new oil riches will be resolved.

A Fresh Start in Norway

After the war, Hoechst also had to start afresh in Norway. As was so often the case, our interests were much fragmented in spite of the relatively small market. Finally, however, the focus of our efforts became our contact with Consul Erling Lind, who was primarily interested in Gersthofen waxes. Lind was a merchant of world experience whose offices reflected spartan simplicity. When he entertained me at lunch for the first time, each guest received three sandwiches and a glass of light beer. I put that menu down to his thriftiness, but I did him an injustice. As I was to learn later, the Norwegians have a second breakfast for lunch and a hot meal only in the evening.

Lind was not ready to invest the amount of money in Hoechst that reconstruction would have required. Accident would have it that soon afterward, returning from India via Pakistan, I met Magne Hjelde in Karachi, where he managed a branch of the Calcutta company of his Norwegian compatriot and old member of the I.G., Thorolf Böge.

With the agreement of his friend and chief, Hjelde agreed to return to Oslo to work for us. In 1957, therefore, we founded Norsk Hoechst with Hjelde as managing director. Our interests in Norway were consolidated in this company so as to be able to profit from the opportunities of the Norwegian market, especially promising in the textile field.

Oslo is one of the most attractive Scandinavian cities. Its bay provides a natural harbor that could hardly be more ideal, and it is a major reason why the city has become the most important center of the country for commerce, industry, and shipping. With half a million inhabitants, Oslo is the social center of the thinly populated nation.

Finland: With the Giant at Its Back

During the first decades of our century, Finland repeatedly had to fight to win and preserve its national independence. During World War II, it suffered the fate of a buffer state which became a theater of war. Since the war, however, Finland has managed to secure a measure of autonomy and has skillfully defended itself on all sides. From the twelfth century until 1809, it was part of Sweden. For the next 108 years, the country was a grand duchy of Russia. The Finns gained their independence in 1917, and since then their nation has formed part of Scandinavia.

As a result of its common cause with Germany during the last war, the country has suffered bitterly. It lost a large slice of its territory to Soviet Russia and its economy was strained to pay an indemnity to the Soviet Union of $444 million. But these misfortunes have not eroded the people's absolute reverence for the memory of Field Marshal Mannerheim, Finland's greatest military and political leader.

The reparations paid to Russia resulted in the development of new industries, especially in the metal and woodworking fields. As reparations payments gradually decreased, new markets were opened up in both the East and the West, minimizing injury to the nation's economy.

In the postwar years, the internal politics of Finland have never been calm. The Communist Party commands a fifth of the country's votes, but is not strong enough to form a government of its own. Dr. Urho Kekkonen, the wise old president who is without doubt friendly to Russia, has always been able to form coalition governments acceptable to the giant at the back door.

This political situation has had economic consequences, including extremely high inflation. During the last twenty years they have not prevented the development—frequently with considerable state involvement—of new industries in machinery, steel, paper, textiles, and chemicals.

Enso-Gutzeit, for example, is an internationally known company in the pulp and paper industry; Finland actually is the largest pulp supplier in the world. State-owned Kemira Oy is the largest chemicals manufacturer. It includes the fiber manufacturer Säteri and, among others, the two partially state-owned plastic manufacturers, Stymer Oy and Pekema Oy.

Suomi—Not Only for Its Sauna

Trying again to gain a foothold in Finland early in the fifties, we found a greatly changed industrial landscape. The trading companies that had developed out of the old I.G. agency were no longer viable. However, Holger Bergenheim, who conducted an extensive business in chemicals and cosmetics, proved to be a very receptive partner. Together, we founded Oy Holger Bergenheim & Co. A/B in 1951. In 1957 the name was changed to Hoechst Fennica, and within a few years we concentrated our activities in Finland in this company. Since 1969, we have been operating an administrative and distribution center on the outskirts of Helsinki. This complex also includes production facilities for pharmaceuticals and reprographics.

Despite the harsh Scandinavian climate the Finns are fun-loving as well as tough. It is still noticeable today that even at the lowest level, they have never been serfs. The soil belongs mainly to small, independent owners, who have become an especially vital element of the nation.

Conversations about Finland inevitably turn sooner or later to the country's widespread and versatile folk art, regarded abroad as highly individualistic. This folk art can be studied in the National Museum in Helsinki or in the eighty or so folk museums, all lovingly maintained. The most original are the open-air museums; for example, in Seurasaari on an island near Helsinki, entire farmyards have been reconstructed in their original design. On the monastery hill in Turku, one can view a complete city quarter dating from the seventeenth century.

Throughout the world, Finnish industrial art has contributed lavishly to contemporary living. Textiles, costume jewelry, wood carvings, lamps, and designs in ceramics and glass provide interior décor with an exquisite accent, especially against a background of Finnish furniture. Knotted work, including Rya carpets, are found wherever good modern taste prevails.

It is characteristic of the Finnish theater and literature that both the Finnish and the Swedish languages are employed without conflict. A language quarrel that lasted until recent times has been decided in favor of the Finns. Today only 6 percent of the Finnish people—those living along the coast of western and southern Finland—use Swedish as their mother tongue. They operate their own schools. The traditional business world speaks Swedish, but outside the capital either Finnish or, to some extent, English must be used.

The country's natural landscape, its forests and lakes, and above all, its appearance of infinite space, have always held a particular fascination for me. I enjoyed my first sauna in 1951 on Lake Vesijarvi, near the town of Lahti, and I have spent many delightful evenings with Holger Bergenheim in the Fiskatorpet near Helsinki.

Scandinavia Conference on a Holiday Isle

Some years ago, we held a small Scandinavian conference in Helsinki at the Hotel Haikko on the Gulf of Finland. The weather was so beautiful that we never entered the conference rooms. Instead, we took a boat to the small holiday island of one of our colleagues and held our discussions there. The talks were most fruitful, even though sauna and lake were part of the program. I very much hope that this small country with its brave people will succeed in its balancing act between its desire for independence and the Russian bear at its back door.

6
Switzerland, A Model Country for the Chemical Industry

My home in Kehl on the Rhine in southern Baden is about two hours by rail from the Swiss frontier. But, during my schooldays in the thirties, Switzerland seemed very far away. At that time, distances were measured in terms of politics rather than of geography.

I think it was Alfons Paquet who once described Switzerland as "the land of chocolate and freedom." At the time, this bon mot was frequently cited in Germany, where both items had become rare. Indeed, Switzerland, for many young Germans, was then something like a *fata morgana* of an outward-looking society. Even now, decades later, the country still conjures up something of that image of a fairy enchantress, although its enviable economic development is no longer quite so far ahead of some neighboring states.

In one respect, however, these neighbors are unlikely ever to match the Swiss. And that is their natural aversion to pompousness, their healthy mistrust of all extremes—especially of ideology—and that unshakable tolerance which has made Switzerland so often the refuge of the oppressed and persecuted. It is this cosmopolitan life-style that makes my trips to that country so enjoyable and rewarding. It is precisely because of this human potential that Switzerland continues to play a world role often far exceeding its economic power.

A Welcome Guest at an Early Hour

Among the prominent Swiss visitors who came to see us soon after the reconstitution of Farbwerke Hoechst was Dr. Robert Käppeli, president of Ciba AG in Basel. I well remember both the guest and the visit in those early postwar years. I had come to know and appreciate Dr. Käppeli during European conferences in Switzerland. He was known far beyond the boundaries of his native country as a collector of ancient art. The excellent museum of antiquities in Basel contains many fine examples of Greco-Roman art from the Käppeli collection.

In the immediate postwar years, Dr. Käppeli had seldom represented his company, even abroad. This function was performed by his deputy, Dr. Arthur Wilhelm, who established and nurtured the first contacts with German chemical industry. Dr. Wilhelm died prematurely in 1962. The large gathering at his funeral in the old Münster Abbey church of Basel included people from all over the world, and bore witness to both the human and commercial bonds that this extraordinarily sympathetic man had been able to forge for his company.

Dr. Robert Käppeli and the Swiss chemical industry: one must devote a few words to this relationship and to the personality of the man. Without the signature, as careful as it was independent, of this grand old man, the history of the Swiss chemical industry in the last half-century would scarcely have been the same, and certainly not so successful.

But to know Käppeli's work for Switzerland's chemical industry is to know only half the man. During a celebration of Käppeli's seventieth birthday, no less a figure than Carl Jakob Burckhardt praised his artistic talents. Käppeli's literary output was of poetic rank, and he was as much a deep thinker as an open-minded music lover. Yet he managed to blend all these activities in harmony with his professional life.

Käppeli was born in 1900 and grew up in Lucerne. His love of solitude may be explained by his passion for mountaineering—or perhaps the other way round. Doubtless, the solitude of the mountains replenished the immense strength needed to cope with a work load that earned him a reputation of indestructibility among his colleagues.

Käppeli took his degree in national economics. In 1934, he joined

Ciba as secretary of the administrative council. In 1944, he became chairman of the directorate committee, and was elected to the administrative council in 1946. Ten years later he succeeded Professor Max Staehelin as president of the administrative council.

During his visit with us in 1954, Käppeli proposed that we revive the excellent relationships that formerly existed between German and Swiss chemical industry. In our discussion, he made a remark, almost incidentally, which I have since often pondered. On going through our annual report, Käppeli said he had found that the DM250 million (US$59.5 million) distributed to our shareholders in 1953 was approximately equal to the share capital of his company at that time. Nothing illustrates more clearly the vast differences between the capital structure of Swiss and German companies resulting from two lost wars and two inflations in Germany. American companies also meet most of their money needs through internal financing. But with fiscal conditions as they are in the Federal Republic of Germany, the situation is not likely to change in the foreseeable future.

CH—Not Only a National Identity

As already indicated, it is not just material well-being that characterizes the Swiss people. Their position in the world has grown out of a culture molded by a unique history. One hundred and thirty unbroken years of peace have allowed the organic growth of an economy that is the envy of far larger countries that have again and again wasted their resources in military follies.

Clocks and watches strike many of us as the typical Swiss products. But these articles account for less than 10 percent of Swiss exports. The chemical industry, on the other hand, has an export share of 20 percent, and in very good years, 25 percent. We might even regard the "CH" symbol (for "chemistry") as standing for both Swiss national identity and the importance of the Swiss chemical industry. This high export share reflects a highly developed industry based on only a small domestic market. This, in turn, explains the astonishing export quota, which has been over 80 percent since 1975. There is nothing comparable in the world. At the same time, the figure is characteristic of a successful attempt to carry out an unusual amount of research although no adequate domestic market exists.

Another essential factor is the support of the Swiss chemical industry by its massive production facilities abroad. Even before the turn of the century, the industry had constructed chemical plants in Germany, France, and Poland. Today, the overseas production potential of the Swiss chemical industry is overshadowing its domestic capabilities.

It Started 160 Years Ago

Although the sixteenth-century scientist Paracelsus is celebrated in Basel as the pioneer of modern chemistry, the industry really began in 1818 with the establishment of the first chemical plant in Uetikon near Zurich. As almost everywhere else in the world, sulfuric acid ushered in the vast range of chemical raw materials. By 1860 or so, when today's major European chemical concerns were starting off as dyestuffs manufacturers, the industry in Switzerland was already important enough to warrant the formation of the Swiss Society of the Chemical Industry.

This solid background furthered the establishment in Switzerland of the first supranational European Chemical Association in 1959. The Swiss society was godfather to CEFIC (Conseil Européen des Fédérations de l'Industrie Chimique), and Zurich was its headquarters until the recent move to Brussels. I am sure that Zurich's hospitality and devotion will never be forgotten.

In 1972 and 1973, I personally experienced how the liberal attitude of the Swiss representatives helped to gather the European associations under one roof. No simple task this, because the organizational structure in the different European countries varied so widely. But eventually the representatives of industry and the boards of the associations concluded that the European Commission in Brussels must be granted a discussion partner who could speak on behalf of the entire chemical industry, and who could thus ensure soundly based opinion making.

Basel and the "Big Three"

According to the residents of Basel, Switzerland would be unthinkable without their city. Those who know it speak of a "city state," reflecting the self-assurance of a patrician tradition that grew up in Basel over the centuries, always at a measured distance from the rest of Switzerland. Among the many typical Basel families, the generations of Burckhardts, for example, which have become famous in so many different fields, are rather more than just a flattering advertisement of the city.

For almost 200 years, Basel has also been the well-guarded money box of a country whose business leaders have always been diplomats as well. An epigram discloses the flair permeating the Basel patriarchy: It is regarded as indelicate to live off the interest of one's savings; one should live off the compound interest.

And if this quip reflects a substantial reality, it is not merely because, in a move as timely as it was clever, the Basel families changed from the somewhat slender silk-ribbon industry of the nineteenth century to the chemical industry, so that the money box filled up year by year. There were happy times even when it overflowed with money.

Toward Germany, just north of their city, Basel exhibits an often described and quite terrifying indifference. There is a story told by citizens of Basel: After the collapse of 1945, Frau Sarasin of Basel was visited for the first time in years by German relatives. One mother tells that her son had fallen in Russia; a cousin reports that her husband died in a concentration camp; and an uncle laments that his wife was burned during an air raid. After the relatives leave, Frau Sarasin remarks to her maid: "Germans always seem to have the misfortunes."

For a time, the coal-tar dyestuff industries of Switzerland and Germany dominated the world export markets. The Swiss pharmaceutical industry, especially, has played godfather to the establishment of many pharmaceutical concerns throughout the world. Even today, a market without a Swiss presence can hardly be found even in economically highly developed countries.

If ever there were a Swiss I.G., it was terminated in 1951 by a court judgment. For all practical purposes, however, it had ceased to exist long before. Over the years, the common interest had become far less important than the individual concerns of the member companies. The decisive factor may have been contemporary history itself. Close economic links with the "large canton" (Germany) were no longer meaningful—and doubtless rightly so.

Of the "big three" companies in Basel, Geigy is the oldest. The name of Johann Rudolf Geigy is mentioned as early as 1758 as the founder of a drug-trading company in Basel, then a town of 15,000 inhabitants. Dyestuffs were imported from France and Germany. Marriages between the Geigy sons and the daughters of silk manufacturers created familial links with the prospering silk industry. In 1833 Carl Geigy built a dyewood mill on the grounds of the cloth manufacturer J. H.

Linder-Stehlin. Twenty-five years later, his company began the production of the coal-tar dyestuff, fuchsin. In 1861, production of synthetic dyestuffs began at the same time that similar plants were being established in Germany.

It was probably Carl Koechlin-Iselin, a nephew of the Geigy-Merian family, who advised against too close a connection with the German competition. He staked his money on the expansion of the company's own research, a decision that promoted independence and encouraged the construction of production facilities in Russia, France, and Germany. At the turn of the century, the company gained considerable recognition through the synthetic wool dyestuffs developed by Traugott Sandmeyer. In 1901, the private company became J. R. Geigy AG, with a share capital of 4 million Swiss francs.

For far longer than in comparable companies, dyestuffs were Geigy's stock in trade. Eventually, however, the production of insecticides began, sparked by the discovery of DDT by Geigy scientist Dr. Paul Müller. This achievement won Dr. Müller the Nobel prize. A pharmaceutical division was established later and produced the first specialities at the end of the thirties. Since then, the pharmaceuticals and insecticides have become the foundation stones of the Basel chemical industry.

Ciba's Start with Dyeing

The founder of Ciba, Alexander Chavel, was a silk weaver from Lyons, France, who established his own silk dyeworks in Basel in 1859. The first dye used was aniline red, one of the most famous coal-tar dyestuffs discovered by A. W. Hofmann in 1858, and popularized under the trade name Fuchsin. Even then, the city of Basel seems to have had a highly developed environmental conscience. At any rate, a magistrate banished Alexander Chavel from the city to prevent any further nuisance to its inhabitants. In 1864, therefore, Chavel acquired property on the Klybeckstrasse, then on the outskirts of Basel, which has remained the company location to this day. The new company, Bindschedler & Busch, founded in 1873, had a dyestuffs output which soon outstripped the company's requirements. In 1884, silk dyeing was abandoned altogether and the much more lucrative dyestuffs production expanded. In the same year, the company went public with a capital of SwF2.3 million. The company name of Ciba, an abbreviation for Society of the Chemical Industry in Basel, was adopted and registered as a trademark.

Within five years, the company began its production of pharmaceuticals, soon after the discovery that trypoflavin and certain other dyestuffs had a disinfectant action. Diversification into cosmetics, textile auxiliaries, plastics, and color photography followed at roughly ten-year intervals. Corporate expansion proceeded steadily. Fusion with Anilinwerk vorm A. Gerber & Co. in 1898, acquisition of Basler Chemische Fabrik in 1908, takeover of Tellko AG in Fribourg, and the acquisition of Lumière in France and Ilford in England are some milestones in Ciba's development. In 1949, Ciba established in London the Ciba Foundation for the Promotion of International Cooperation in Medical and Chemical Research, an institution of worldwide renown.

A Marriage of Giants

When, in 1970, J. R. Geigy AG and Ciba merged to form Ciba-Geigy AG, by far the largest chemical company in Switzerland came into being. The new company, in fact, was the largest in the country after Nestlé. The father of the merger was Dr. Käppeli, who overcame many internal obstacles in both companies.

The joining of two such giants calls for a brief discussion. With world sales in 1969 of DM2.9 billion (US$743 million) for Geigy and DM2.8 billion (US$718 million) for Ciba, each company was a leader in world chemistry. And yet, rising costs of research, investment, and administration were outstripping potential growth. Experts had become increasingly difficult to recruit, not least because of the restrictions on the admission of foreign workers. The merger promised improved efficiency in all areas, increased profitability, and a comparatively complementary product range. To minimize the fiscal problems of founding a totally new company, it was agreed that Ciba should take over Geigy.

For a time, there was justified concern about an initially top-heavy management and the rivalry between managements of equal rank and within administrative and staff departments. A book of humorous anecdotes, *Letting the Cat out of the Bag*, records what happened internally. Unfortunately, circulation of this unique book was confined to the two companies. Under the direction of the president of the administration council, Dr. Louis von Planta, who has headed the concern since the merger, the problems that accompany elephantine weddings have been mastered within a few years.

The production programs of the two companies have now been coordinated. This phase, lasting seven years, temporarily retarded corporate growth. However, the company now is among the top producers of agrochemicals, dyestuffs, and research-based pharmaceuticals.

Same Industry, Same Problems

But soon Ciba-Geigy too had other worries. In the seventies, the large Swiss and German chemical companies encountered worsening export problems. The export business of these companies far exceeds their domestic sales. Since sales to weak-currency countries are considerable, the Swiss, like the Germans, suffer from the continually increasing value of their currency. This problem is assuming ominous proportions. From 1972 to 1976, the Swiss franc increased in value by 52 percent and the Deutsche mark by 31 percent, in comparison with other world currencies. The trend was magnified in 1977 because the inflation rates of the other partners could not be reduced. In the midst of hard worldwide competition this has two bitter consequences: the balance sheets of the parent companies, in Swiss francs or Deutsche marks, show only modest growth rates in comparison with competitors in the soft-currency countries. At the same time, the higher value of the Swiss franc and Deutsche mark raised prices in foreign markets. For example, if Ciba-Geigy had kept its Swiss franc prices unchanged since 1970, American customers would have had to pay some 75 percent more in 1976, simply because, between 1970 and 1977, the exchange rate dropped from 4.37 to 2.39 Swiss francs per dollar. Since Ciba-Geigy, like large-scale German chemical industry, could not fully pass on this increased cost, profit margins had to be drastically reduced.

For the industries of both countries, the only possible way out is to transfer research, development, and production to foreign markets to reduce expenditures in hard currency. In part, therefore, the members of the Basel chemical industry are forced to assume the role of holding companies for international undertakings.

Sandoz Goes to Seeds

In 1886, the merchant Edouard Sandoz, from Les Ponts-de-Martel in the Neuchâtel Jura, and the chemist Dr. Alfred Kern, from Bülach, canton of Zurich, built a dyestuffs factory. Following Kern's early death, Sandoz continued the rapidly growing business on his own until 1928. Ten years later, the business was converted into a public company, with the smallest share, 20 percent, of the Swiss I.G., while Ciba held over 52 percent and Geigy over 26.5 percent. With the acquisition of Dr. Wander AG Bern in 1967, Sandoz expanded its major share of the pharmaceutical market and diversified its product mix.

Although Wander had sizable pharmaceutical sales, its main business was in dietetics. Ovaltine, its best-known product, became the starting point for a whole line of foods marketed in Europe as energy supplements for athletes. In the United States, in contrast, Ovaltine was promoted as a children's drink and a relaxing beverage at bedtime. Twenty-five years earlier, the other partners of the Swiss I.G. had attempted to prevent Sandoz from branching into the lucrative pharmaceutical market. Today, its pharmaceutical division accounts for more than half the company's sales. About one-third originates from dyestuffs. Through the 1976 acquisition of the American company, Northrup, Kind & Co. of Minnesota, Sandoz became one of the largest seed producers in the world.

The current head of the company is Yves Dunant, who has close cultural ties with neighboring France. One of his forebears was Henri Dunant, who founded the International Red Cross after the terrible carnage of the battle of Solferino in 1863.

Agrochemicals and dietetic specialities make up the remaining sales. Like Ciba-Geigy and Hoffman-La Roche, Sandoz achieves more than 90 percent of its sales abroad and, with these other companies, will suffer the same consequences of an appreciating currency.

From Near-Liquidation to World Success

Today the largest pharmaceutical manufacturer in the world, Roche avoided liquidation at the beginning of the century by the skin of its teeth. In 1897, a Basel bank called in a loan that it had granted to the one-year-old company. If T. Hoffman, the father of the company's founder, had had his way, son Fritz would have ended his pharmaceutical venture and taken a job in a cement factory. But Fritz Hoffman obtained new financing and reestablished the company in 1898. With one of his first employees, chemist Emil Christoph Barell, Hoffmann laid the technological and scientific foundations of the company's future. By taking up pharmaceutical production, Hoffman-La Roche streaked ahead of the other Basel chemical companies.

World War I brought with it the occupation of the plant in Gren-
zach, Germany, and the loss of its Russian market, which had accounted for a fifth of its total sales. In 1919, the company went public
but remained under family control. A year later, Hoffman died and the
management of the concern was taken over by Barell. Healthy profits
enabled the share capital to be repaid in full—and in cash—by 1943.
A Canadian holding company, SAPAC Corporation Ltd., was founded
in 1927 to manage Roche holdings outside Europe. SAPAC and Hoffman-La Roche & Co. are sister corporations, both owned by the same
shareholders, the majority of whom continue to be the Hoffman,
Sacher, Sarasin, and Oeri families.

In 1976, Roche pharmaceuticals accounted for 88 percent of sales,
equivalent to some DM4.5 billion (US$1.8 billion). However, the lead-
ers in the United States and Europe are catching up fast. Hoffmann-La
Roche is noted not only for its psychopharmaceuticals, vitamins, and
food dyestuffs. It also occupies a leading position in the field of es-
sences and aromas as the result of its acquisition of Givaudan in
Geneva.

Hear the name Hoffmann-La Roche and two drugs immediately
come to mind: Librium and Valium, the world's leading psychopharmaceuticals. Librium almost instantaneously conquered the international market in the early 1950s. A few years later, Hoffman-La Roche
marketed Valium, a compound related to Librium. Valium is perhaps
the top-selling pharmaceutical in the world. In the United States, it is
the most frequently prescribed drug. Many psychologists and sociolo-
gists, especially those taking a pessimistic view of civilization, believe
that psychopharmaceuticals are symptomatic of our age. Indeed, it
often seems that present-day humans set out deliberately to develop
a need for sedatives, tranquilizers, and antidepressants. In reality,
however, these phenomena are not all that new. The lives of earlier
generations were hardly less, though perhaps somewhat differently,
harassed than ours. Those generations would probably have been just
as grateful for drugs against emotional distress as thousands of us are
today. This to say nothing about the use of psychopharmaceuticals in
psychiatric hospitals as replacement for dreaded electric and insulin
shock therapy, straitjackets, and locked rooms.

This new class of medicines was born not in Switzerland but in
France, in the laboratories of Rhône-Poulenc. The first modern psychopharmaceutical was the classic chlorpromazine. This was followed
by reserpine, an alkaloid isolated from the rauwolfia root of India by
two Ciba chemists, Emil Schlittler and Johannes Muller.

It took a very long time, incidentally, before Hoechst joined the psychopharmaceutical producers. More than ten years of research and development, and an investment of almost DM100 million were needed before our antidepressant Alival was introduced into medicine. This unique drug is intended for a specific indication in the treatment of depression.

As worldwide leader in psychopharmaceuticals, Roche has often been attacked. In any case, since the period of high growth and profits in drugs is waning while research expenditure and financial risks are growing steadily, Roche, like many others, is diversifying its business. Bioelectronics, diagnostics, and cosmetics are new areas of activity.

The Chemistry of Swiss Culture

Hoffman-La Roche is linked to a musical renaissance in Basel that has contributed much to this city's renown. Following the accidental death at age thirty-six of Emanual Hoffman, one of the sons of the founder Fritz Hoffman-La Roche, his widow, Maja, married the young musician Paul Sacher. This marriage soon helped Basel to become one of the best-known music centers in Europe. Backed by the considerable wealth of his wife, Paul Sacher founded the Schola Cantorum Basiliensis, the music academy of Basel. The Schola also houses a unique collection of old instruments. In his home on the Schönenberg near Pratteln, Sacher maintains one of the most important collections of musical scores in the world. Himself an extraordinarily gifted and highly regarded conductor, he commissioned musical works by Richard Strauss, Paul Hindemith, Béla Bartók, and Bohuslav Martinü.

For her part, Madame Maja collects modern art, as a true devotee. Moreover, the family generously uses much of its yearly dividend income for cultural purposes. Without the Hoffman wealth, the city of Basel would be poorer today in cultural resources.

Art Is an Elixir in Switzerland

The story of the acquisition of a famous Picasso painting typifies the attitude of wealthy Swiss citizens to art and culture. Years ago, the Basel Museum of Art was eager to acquire the world-famous *Harlequin.* A public collection yielded a fraction of the astronomical sum that was demanded for the picture. Basel's chemical industry came to the rescue, and since then, the *Harlequin* has been exhibited in Basel.

If one recalls the many great European intellectual figures who settled in Swiss towns and villages or were given shelter in this hospitable country, one begins to understand what Conrad Ferdinand Meyer meant when he described Switzerland as "little Europe," with the "supra-nationality of the heart" which is a more important part of human beings than nationality is.

Perhaps only in this way can we understand the extent to which art and artists, literature and writers, intellect and intellectuals have been sheltered and nourished by the Swiss. From Erasmus of Rotterdam to Ulrich von Hutten, Richard Wagner, and Friedrich Nietzsche, from revolutionaries like Lenin to poets like Rainer Maria Rilke as well as to many strange species of less definable quality, Switzerland has taken under its wing a human gathering from abroad whose variety could hardly have been greater.

By the Fireplace in a Chalet

Lonza AG, the fourth largest chemical company in Switzerland, was established in 1897, and was long a family business. The Basel chemical industry bought into the company in the 1960s with the intention of having it become the joint producer of basic chemicals. Competing with the large capacities of the foreign manufacturers was found to be uneconomic, however. The only way out was to convert to production of more highly finished intermediates. In 1973, the Lonza shareowners sold their holdings to Alusuisse, which became the parent company. Lonza has remained autonomous, however, and now also carries out the chemical operations of Alusuisse.

The changeover was achieved under the careful guidance of Dr. Jurgen Engi, formerly general counsel at Ciba. He has looked after the contacts with European electrochemistry for many decades. I remember a visit to Visp in the lower Rhône valley at his invitation. It was in Visp that Lonza, as the only chemical company in the country, took the first steps in petrochemistry and, moreover, erected an important nicotinic acid factory.

In a little guest chalet located within a company-owned vineyard, we were served a hearty raclette prepared over an open fire. The melting slices of cheese came with potatoes, boiled in their skins, silver onions, and pickled cucumbers. We washed the raclette down with the delicious white wine of the Rhône valley.

After lunch, we drove to the nearby little town of Raron. On a rocky hill, dominating the valley, stands a castle church dating back to the early Middle Ages. Once in ruins, the church has been expertly restored by Lonza to commemorate the company's centenary. In the churchyard is the grave of Rainer Maria Rilke.

Lonza is the only Swiss producer of fertilizers and polyvinyl chloride plastics. With 1976 sales of about SwF1 billion, Lonza has achieved dimensions requiring production facilities outside Europe. A complex built in Bayport, Texas, in 1977 produces organic intermediates for the United States, the biggest chemical market in the world. Lonza has had installations in Waldshut on the Baden banks of the Rhine for many decades.

A Quick Dash to Reykjavik

Our association with Lonza has been strengthened by reciprocal visits to factories and laboratories which provoked many technological discussions. Since both companies have been pioneers in electrochemistry, we had much in common.

One of our recent visits, however, was somewhat out of the ordinary. We flew to Reykjavik in two small jets to inspect the aluminum plant erected by Alusuisse in cooperation with the Icelandic government. The Isal plant has a capacity of 75,000 tons a year. Iceland is barren island. Its economy is based primarily on fishing and the export of fish products. There is some breeding of wool sheep, which also contribute high-quality meat to the Icelanders' diet.

The other resources of the country are hot springs and waterpower. Providing wells are drilled deep enough, hot water can be obtained in any amount because the country is geologically relatively young. Volcanic eruptions, fortunately of comparatively small dimensions in recent decades, no longer cause the serious damage suffered in earlier times.

Iceland has gained a new source of wealth—its almost unlimited potential for producing electric power. The mountains with their bizarre volcanic shapes are covered by glaciers, so that the construction of hydroelectric plants is rendered easy. We inspected two such plants, built to meet the high current requirements in aluminum production. Their capacities can be greatly increased without much effort, providing reserves beyond those needed for expanded aluminum production. It would be surprising if, sooner or later, an electrochemical industry were not established in Iceland. Electric power cannot be transported over large distances, but energy-intensive raw materials present no such problems. Shortages and high costs of energy will in the coming years become overriding factors in the economic development of Iceland—and the rest of our planet as well.

Plüss-Staufer and Hoechst in Switzerland

The large Swiss family company Plüss-Staufer, closely linked with Hoechst after the war, was founded by Gottfried Plüss-Staufer in 1884 as a putty factory. The chief ingredients of putty are chalk and linseed oil. The company produced both, selling its surplus in Switzerland and abroad. The next step was to convert linseed oil into varnish and to produce lacquers. Plüss-Staufer erected its own manufacturing facilities abroad at an early date, usually where there were large deposits of chalk. Thus factories were established in France, Austria, the Federal Republic of Germany, Spain, and England.

During and after World War II, Plüss-Staufer also began marketing foreign chemical products. Its interest in Hoechst pharmaceuticals was awakened through the connection of the company owner with Dr. Peter Peiser, known to me from his Bayer days. Until the middle of the thirties, Peiser held a leading position in the medical and scientific areas at Leverkusen. He then went to Shanghai for the I.G. There he met his future wife, a native of Basel, with whom he returned to Switzerland. Soon after the end of the war, he renewed his old links with Germany. He was more than sixty-five when we met in Hoechst. A gentleman through and through and full of entrepreneurial spirit, he had the great gift of transmitting his optimism to others.

In 1947 this was not easy. The trustee of Hoechst was still traveling daily to report to the U.S. Control Office in Frankfurt. The agency agreement which Hoechst and Plüss-Staufer concluded in 1948 had to be ceremoniously approved by the Americans. It was, incidentally, the first agreement of this kind, and I had the honor of typing it myself. Willy Kaiser, for many years manager of the Plüss-Staufer business, ensured that dyestuffs and chemicals, that is, the entire Hoechst range, were also covered in the agreement. Plüss-Staufer soon succeeded in obtaining the release of the Hoechst pharmaceutical trademark, which had been confiscated in Switzerland also, and contributing it as a dowry to the new liaison.

Before uniform representation by Plüss-Staufer could be assured, certain problems relating especially to dyestuffs had to be resolved. For example, postwar links with trading companies had to be canceled. Thanks in part to an accident, we cleared up matters with dispatch. My old teacher Walther Ludwigs, at that time commercial manager in Hoechst, had traveled to Basel with me to meet our new partner. Ludwigs, who could boast of a respectable corpulence, collided with a pram on the narrow pavement of the old city, fell, and broke his arm. What was to have been a one-day visit turned into a stay of almost three weeks in the local hospital. We had plenty of time not only to discuss how to expand the business, but also to cement human relationships.

I am often asked why there is no Hoechst Helvetia. Why have we so firmly maintained our links with Plüss-Staufer, family-owned with no Hoechst delegates in the management? The answer is simple: our friends in Oftringen are doing such a good job for Hoechst that I believe we could do no better. The only exception is pharmaceuticals where, because of international goodwill and the critical importance of domestic research, we founded Hoechst Pharma AG with headquarters in Zurich. But even this concern is owned fifty/fifty by the two companies.

Our friendship of many years with Max Schachenmann, the head of the company, and Gustav Baumann, his present director-general, has withstood all the inevitable strains. Schachenmann, once glider champion of Switzerland, still pilots his own plane throughout the world. He built up the family concern he now heads to its present dimensions. Factories for Omya fillers (a brand name for calcium carbonate), by far the largest sector of the business, were erected in many countries of Europe. Two important acquisitions in the United States in 1976 have brought the company to the New World. A large marble pit, together with the necessary processing plant, was acquired from a family concern in Vermont. Plüss-Staufer also bought a similar concern with installations in California and Alabama. The company, therefore, now has broadened its previous import business in Omya fillers.

Foreign Presence

How does the Swiss chemical industry differ from its West Europe counterparts? The distinction is found more in the product mix than in commercial practices. Early on, the Basel industrialists recognized that the main chance lay in high technology products. Their share of pharmaceutical specialties, therefore, is far higher than in the neighboring European countries. Together with dyestuffs, these products account for almost two-thirds of the entire production value. Their concentration on sophisticated products is what makes them such good customers for the rest of the European chemical industry. Our Basel friends were never interested in the major synthetic fibers or in mass-produced plastics. On the other hand, Swiss chemical industry meets 10 percent of the world requirements for pharmaceuticals, and an even greater share of the market for dyestuffs. We have good reason to be impressed by the international progress of the Swiss chemical industry. As late arrivals after the war, we had to struggle to recover our market shares in Europe and North America. But competition is conducted fairly on the basis of mutual research.

At the end of his book, illustrated with his own watercolors, Dr. Käppeli described Switzerland thus:

> My fatherland—often praised and abused, risen through its own strength on the principle of freedom, coming into full bloom in ominous calm of the last century, finally matured and thus forced, under an unalterable law, into renewal, with changed signs of the time and without the safe conduct of a strong, new idea.

7
Bella Italia: On the Road to Eurocommunism?

Anywhere else, it would have been just a formal reception—a few courteous but inconsequential words of greeting, and nothing more. But not so in 1976 in the Palazzo Chigi in Rome. Premier Guilio Andreotti had just returned from a parliamentary debate. Once more the fate of his government coalition had been in doubt. Nevertheless, the head of the government now gave his entire attention in the most charming personal manner to his guests: Karl Winnacker, Rolf Sammet, our wives, and myself.

Turning to Winnacker, Andreotti said he regretted being unable to attend the Italian celebration introducing Winnacker's memoirs. However, he had read portions of the book. Above all, he had been impressed by its German title (which, in English translation, is *Never Lose Your Courage*), a good motto not only for the president of a company but also for the premier of Italy.

The conversation shifted to the economic problems of the country and projects to further its industrialization. In acknowledgment of Hoechst's contribution to Italian economic development, Andreotti awarded Rolf Sammet and myself the Gran Ufficiale de Ordine al

Merito, a high Italian order. Professor Winnacker had received the same honor a year before.

After the stimulating encounter with the *homo politicus* Andreotti, this high award delighted me even more. An hour later, walking through the streets of the Eternal City in a slight November rain, I recalled my first visit to Italy, five years after the end of the war. Although we had been able to revive the traditional dyestuffs business with the aid of the trading company, Colea, founded by former I.G. friends, we were unable to enter the pharmaceutical market. The German trademarks had been seized as enemy property.

The branded Hoechst products from the joint product line, sold successfully under the Bayer cross, had quickly found a buyer in Rome —a Mr. Pine. He had first come to Germany with the American occupation troops, perhaps to trade in Italy by this roundabout route. In partnership with Dr. Otto Weber, the former Bayer chief in Italy, Pine had introduced the trademark into a new company with the name of Emelfa. The name was derived from *M* for Meister, *L* for Lucius, and *Fa* for Farmaceutici. An ingenious scheme, but, of course, one requiring Hoechst to cooperate as supplier. However, this we were not prepared to do quickly. After all, Italy had always been our most important export country—a position it has retained to this day.

At the very least, we wanted a partnership with Emelfa. Dr. Weber was quite sympathetic to this proposal. However, he died suddenly, at a relatively young age, during the final phase of our talks. The subsequent direct negotiations with Mr. Pine were unpleasant, almost intolerable. That we concluded the matter satisfactorily with the acquisition of Emelfa was owed to the fair attitude of Enzo Avanzini, president of Emelfa until 1969. We encountered the same fairness from his delegate, Georg Wörn. Active for the I.G. in Italy before the war, Worn found his way to Emelfa after being released as a prisoner of war.

In the succeeding years, we concentrated all our other activities in this company in conformity with the policy of one country, one company.

Economic Miracle, Italian Style

As we moved to reestablish Hoechst in Italy, the country was taking its first steps toward democracy. Premier Alcide de Gasperi, together with Robert Schuman and Konrad Adenauer, was a vigorous protagonist of the concept of a European community. He cleared the way for the rapid elmination of all obstacles to European trade, greatly facilitating the Italian economic miracle of the 1960s.

Thanks to the remarkable economic developments south of the Alps, we were able to expand our business there rapidly. Those years also saw the founding of our pharmaceutical division in Italy, which helped us to reestablish the positive image of our company in the south.

I have often wondered why Hoechst, despite the great volume of its business, established its own production facilities in Italy only at a late date, and even then to only a moderate degree. There are two reasons for this. First, Italy has for centuries been a liberal, largely export-oriented country, with few import restrictions. Second, an efficient industry had developed in Italy, especially between the two World Wars when Mussolini pursued a strict policy of self-sufficiency.

The chemical industry experienced a considerable upsurge after the war through the exploitation of the natural gas deposits in the Po plain. In addition, as the result of the subsidies for Italy's under-developed south, new chemical plants also flourished south of Rome, in Sicily, and in Sardinia. They were linked with the new state-owned refineries and operated to only a limited extent by the normal rules of a free market.

The Autobiography of an Elephant

The largest Italian chemical company is Montedison, formed in 1966 through the fusion of Montecatini, originally a mining company founded in 1888, and the private electric company of Edison, established in 1884 and nationalized in 1966. Edison invested its compensation in the chemical industry. Its compensation payments were originally earmarked for the rescue of Montecatini, then already suffering from a chronic lack of funds. However, not much came of this plan. The management of former Montecatini President Dr. Giorgio Valerio was unable to cope with conditions. The badly needed reorganization brought few improvements, and certainly there was no adequate modernization of the antiquated Montecatini installations.

One strategy, however, the company did pursue with great energy. This was diversification, which, as a journalist once wrote, soon ranged from bras to nuclear power stations. In the end, the company had 150 subsidiaries. Chemistry itself provided about 60 percent of sales.

No wonder that success eluded the company and that for many years dividends could be paid only by dipping into the reserves. The Montecatini elephant was floundering.

At the beginning of 1971, Eugenio Cefis arrived on the scene. Popularly described in the industry as a "Prussian amongst managers," his mopping-up operation was remarkably thorough. His motto was, "Back to chemistry." With a block of Montedison shares, the company acquired Rhône-Poulenc's majority holding in Snia Viscosa, the largest man-made fiber producer, as well as complete ownership of the subsidiary Rhodiatoce. Montedison also took over Farmitalia in which Rhône-Poulenc had a 50 percent share. Finally, Montedison acquired Carlo Erba, one of the largest Italian pharmaceutical manufacturers. Under the new chief, Cefis, there was no longer any talk of diversification outside the chemical industry.

But before long the company was confronted by difficulties that the state was creating for Italian industry in general. There is an intraministerial committee for economic planning—CIPE—which is quick to hold out the hand of state intervention at the hint of the slightest crisis. Even today, the aim of the 1966 merger—to stabilize a private company—has not been achieved. Caught between the need on the one hand to create new jobs through huge investments, and staunch, almost chronic losses on the other, the company's only solution was to seek state financial aid. This, of course, meant state control. Although Montedison belongs to more than 200,000 shareholders, its influence is slight compared with a control syndicate in which the state, through the ENI and IRL companies, and private industry are equally represented.

Since Montedison's foreign organization was inadequately developed, the company pursued an export policy that frequently undermined the price structure in many markets.

Government on the Brink

In the autumn of 1977, almost 700 of the 800 state-owned concerns in Italy, including world-renowned companies, were bankrupt by conventional economic standards. Company capital had been largely exhausted; some companies had succeeded in using up their capital twice in three years. Many concerns were so far in debt that they had difficulty in keeping their heads above water. The major creditors of these companies were banks and other financial institutes that, in turn, were now also partially owned by state holding companies.

To rescue the public companies, enormous funds would be needed. A good example is Alfa-Romeo. It is claimed that each Alfa Sud car leaving the factory means a loss of DM2700 (US$1400). Nobody knows how to cope with such a situation. But it is difficult to imagine an industrialization of the south that would conform to the rules of a free market economy.

As for Montedison, fresh attempts are, of course, always being made to reorganize this company, now operating with only 10 percent of company capital. The ultimate answer may be a state-promoted merger including ANIC, SIR and Rumianca, possibly also Liquigas and Liquichimia, and, of course, Montedison.

ANIC, affiliated with Montedison in the chemical field since 1976, is 70 percent state-owned. It employs 20,000 people in petrochemistry and synthetic fibers. Some operations are joint ventures with domestic and foreign partners. Our friends from Wacker-Chemie, for example, have operated Quimica Ravenna jointly with Anic since 1958. This company produces vinylacetate monomers and polyvinyl chloride. The venture has been fruitful for many years.

SIR and Rumianca are outsiders in the Italian chemical industry. Exploiting the low-cost state-financing possibilities, as well as subsidies from the Cassa per il Mezzogiorno (Southern Development Agency), new plants have been erected in the south, especially in Sardinia. The oldest of the twenty manufacturing plants are in Lombardy. Producing mainly petrochemicals and man-made fibers, they share the worldwide problems of this branch of the chemical industry. Most critical of all is the situation in man-made fibers. Using subsidies, partially from the regional funds of the EEC (through the European Development Bank), the companies built new capacities in the man-made fiber field at a time when, quite obviously, only curtailment of production and the elimination of obsolete plants could again lead to economically viable operations.

No doubt the Italian industrialists realize this too. But they live in an economic and sociopolitical climate in which it is simply impossible to close down even the most uneconomic plant, to dismiss workers even temporarily, or to shorten the paid workweek to overcome the crisis. The continual strikes and their many variations have discouraged both small and large Italian companies from making new investments.

The Shy Foreign Investors

I do not propose to try to analyze the present social, economic, and political situation in Italy. But I would mention one aspect: the Italians' astonishing gift of improvisation, the individual adaptations to apparently hopeless situations, which often helps them to cope with their present painfully low ebb of fortune. Indeed, away from the large centers such as Milan or Rome, one is at times barely conscious of the deep crisis either in daily life or in industry.

In the years 1956 to 1973, foreign investments amounted to DM3.3 billion. In comparison, the Federal Republic of Germany invested ten times this amount, DM35.3 billion during the same period. This circumstance is one of the causes of the chronic deficit of the Italian trade balance. It is rendered even worse by the negative balance sheet of the agricultural industry, whose age-old structure has up to now rather successfully resisted feeble attempts at reform.

The out-of-date structures of agriculture and industry and the unbridged gap between the industrially highly developed north and the still underprivileged south are probably the real reasons for the long-festering crisis in Italy.

Italy's problem, which to a certain degree also applies to West Germany, derived in some measure from an unmastered past. Tomasi di Lampedusa's novel *The Leopard,* which Visconti made into a 1962 film with the Italian title *Il Gattopardo,* has helped me to understand these issues far better than any historical work or social criticism.

Trucking and Tourism

In 1976, Hoechst transported some 160,000 tons of material across the Alps. This is equivalent to one freight-train load a day, or to about 10,000 truck loads a year. It staggers the imagination to think how this can be managed in the summer months when hundreds of thousands of tourists are jamming the same roads to the south.

With this in mind, the discouraging political developments of the last few years have not restrained Hoechst from consolidating its position in Italy. We have begun to explore ways of making some products in the country itself.

The production of Ozalid paper, begun before the war, has been augmented by the acquisition of IMG (Industria Materiali Grafici) near Verona. When we acquired the majority holding in Cassella, we also obtained its Italian dyestuffs factory, Lombarda Colori Anilina, outside the gates of the beautiful town of Bergamo.

Novacrome, a handsome factory in Lomagna near Como, supplemented our domestic German production of raw materials for the plastics processing industry, whose exports we have been able to help substantially. Finally, through our liaison with the Turnauer group in Austria, we gained the latter's Italian subsidiary, Vernini Lalac, which provided us with an operational base for our new paint division in Italy.

Italian from Head to Toe

The Italian fashion sense has always been highly developed. There is hardly another country whose national costumes, for example, are more varied and colorful than those in Italy. This charming playfulness in matters of clothing developed into a flourishing industry in the decades after the war. Rome, Florence, and Milan have become fashion centers successfully competing with Paris. When window shopping in the cities of the world, whether large or small, I always come across fashion goods whose design immediately identifies their Italian origin.

Where fashion is being created, colors and dyestuffs are never far distant. These products, along with processing chemicals for textiles and leather, become an important activity in such an environment. As a result, Hoechst had joint ventures with AIC (Approvigionamenti Industriali Chemici). Finally, we acquired this company in the seventies. AIC operated from Turin for the fashion industries concentrated in this region. In the meantime, we are making every effort to extend the range with those products that are in special demand in industry.

In the field of plastic dispersions for binders, paints, and adhesives, Hoechst was competing successfully against domestic producers, largely through a well-organized technical advisory service for customers. However, falling market prices and the high transportation costs for the long haul from Germany made importing uneconomic in the long run.

We therefore decided, a bit late, to secure the remaining market through local production. We entered into a partnership with Sara S.I.A., a family business in Romano d'Ezzelino, near Vicenza. Since then, the synthetic resin plant there has been expanded and a costly effluent problem resolved.

In the field of carbon electrodes, we built upon the close cooperation existing between Siemens Plania and the present Elettrocarbonium before the war. With the friendly help of our partner, Guiseppe Azzaretto, this old association was soon reestablished. Through Sigri Elektrographit, owned fifty/fifty by Siemens and Hoechst, we acquired a share of Elettrocarbonium, one of the most important companies in Europe in this field. Azzaretto, whom we grew to appreciate greatly over years of close cooperation, became not only partner and president of Elettrocarbonium, but also president of Hoechst Italia, in 1973.

Back Home in Milan

Our manufacturing plants, accounting for a quarter of our business in Italy, are scattered throughout the country because we acquired them only gradually. While this dispersion has many disadvantages so far as centralized administration of our Italian organization is concerned, it has enabled us to solve social and political problems on a local basis and in regions of a manageable size.

Emelfa, our old agency in Milan, developed quickly after the initial difficult years. In 1957, we moved to our own office building in Milan. But this modern building soon proved too small for our requirements. We were growing out of the import business. For example, the import volume of pharmaceuticals was valued at DM68 million (US$26.2 million) in 1976. We could not see how we could continue to conduct this booming business through imports alone. We simply had to think in terms of a pharmaceutical production plant in Italy.

As a start, we founded Albert Farma, with a small product line, and an Istituto Behring in Scoppito near L'Aquila. These were the first steps on a road that led to the opening, in 1979, of the Hoechst pharmaceutical plant in Aquila. We now no longer have to rely on complex import planning.

Our location in Milan is in the center of Italy's major industrial landscape and a region rich in history and culture. Milan is always worth a trip, quite apart from any business duties. It is fascinating to compare cities like Milan and Vienna in regard to their art treasures. I am always impressed by the breadth of view presented as the Italian Renaissance developed from the Middle Ages while the artists on the Danube celebrated the "happy Austria" of the aristocracy.

Leonardo da Vinci's *Last Supper,* in the refectory of the former monastery of S. Maria della Grazie permits a unique encounter with the great Western faith of the sixteenth century, when the Christian view of life had not yet been contested. And, naturally, one has not really seen Milan until one has paid at least one visit to the world-famous La Scala opera house. This great lyric theater has more than 3500 seats and 146 boxes in six tiers; the soaring grandeur sets your emotions tingling even before the curtain has risen. Milan is also justly described as Italy's financial and commercial metropolis—half the industry of the country is centered there. And Milan may have more cars per person than London, more luxury restaurants than Hamburg, and more private airplanes than Düsseldorf.

The delightful little town of Monza, renowned for more than its motor racing, is almost a part of greater Milan. In the Middle Ages, it was the seat of the Lombardic princes whose coats of arms are exhibited in the town museum. Not far away is Bergamo, where we have taken root with our subsidiary Cassella. Architecturally, Bergamo is one of the most uniform and complete towns of the Italian Middle Ages. The inner city is almost untouched, and industry has had to establish itself on the outskirts of this proud town.

One Road to Rome

My travels are likely to take me to Rome more often now that our newly erected pharmaceutical production in Aquila is fully operational. In the past, our main activities centered in the Milan area, in spite of great efforts to promote the industrialization of the south. Nevertheless, Rome remains an important commercial city. Because all the provincial and government offices are located here, Rome is more crucial to the country's industry than other European capitals, except perhaps Paris. Moreover, Rome has always been a popular center for meetings and congresses.

I realized this when the Italian Chemical Association arranged a meeting of the Conseil Européen des Fédérations de l'Industrie Chimique (CEFIC) in Rome two years ago. The magnificent rooms of the Castel Sant'Angelo provided a setting of great dignity. A later meeting of CEFIC took place in Venice. During this event, I thought of Hemingway's novel *Across the River and into the Trees,* set in such a landscape. Perhaps this novel is not one of Hemingway's greatest, but its description of postwar history greatly impressed me by its individuality and atmosphere.

Our modest branch office in Rome is concerned primarily with pharmaceuticals. The health service of Italy is in a difficult transition phase. A price ceiling on pharmaceuticals, in effect now for ten years, has kept prices at an unacceptably low level. When depreciation of the lira is subtracted, the import of finished medicaments is a losing proposition to the producer.

In May 1973, Italy hosted the first of the Hoechst Europa Conferences. These gatherings, held at intervals of about eighteen months, bring together the divisional heads of the parent company and the leaders of Hoechst's major organizations abroad. Although there is ample opportunity for discussing current events, the main purpose is to consider the future targets of the company.

Personal relationships are also renewed. At the end of the 1973 gathering at Santa Margherita near Rapallo, we sat together in a beautiful seafood restaurant above the port of Portofino, toasting in good Italian wine a farewell to our old friend Georg Wörn. With the company during all the years of reconstruction, he was now approaching his well-deserved retirement. Carl G. von Asboth, who had joined Hoechst in Vienna, took over the reins of the company. He had pleaded for years for the establishment of production facilities in Italy.

Enchanting Sounds

Portofino's cuisine is another of Italy's many gastronomic attractions. I must confess that my wife and I, as well as our children, have a pronounced preference for Italian pasta. Canelloni is a favorite dish of the whole family.

No doubt, Italian dishes appeal to the ear as well the palate: tortellini, tagliatelle, rigatoni, fettuccine, ravioli—what a melodic world of sound as well as taste. Perhaps the Italian language was especially designed for gourmets.

I have followed with great pleasure the global spread of the Italian pizza during the last decade. This empire now extends right up to our front door and into the smallest towns in the Taunus. I am not sure whether, for me, the Italian cuisine has edged out the French, my former favorite. But I am still enticed by Paul Bocuse's *nouvelle cuisine.* From his restaurant near Lyons, he has introduced a light touch into the traditionally rich French cooking.

New Pharmaceutical Pathways

Our plant in Scoppito near Aquila may bring to actuality the plans of our long-time pharmaceutical head, Dr. Agostino Caradente. Assuming that some degree of normality returns to Italy, Hoechst Italia will challenge its pharmaceutical team with new tasks. We intend to introduce new formulations conforming more closely to the prescribing habits of Italian physicians. We will also develop in Aquila new products specific to the country. We have been encouraged in these plans by the single-minded industrialization policy in the south.

During my travels in Italy, I have often observed with pleasure how satisfactorily commercial cooperation between Italians and Germans can be conducted once the language barriers have been overcome. In spite of repeated encounters with foreign soldiery, including the Germans, our southern neighbors have retained their friendly attitude to foreigners.

Italian Hospitality

Hospitality is a fundamental characteristic of the Italians, but one found most often outside the tourist centers. In the company of Italians, one can still observe traces of the former feudal state in the present social structure. But the picture is changing. In any case, the courtesy, especially of the simple Italian and frequently in welcome contrast to the visitor's own demeanor, usually bridges any gap. Unfortunately, and I regret this most when I meet the unsophisticated people of the country, I have only an incomplete command of the language. But I have not given up hope of one day reading Dante's *Divine Comedy* in the original.

A Concerned Look across the Alps

In the last few years, I have had fervent discussions with my Italian friends about their nation's desperate internal political situation. With some justice, they explain the crisis, and it is more than just creeping anarchy in the large cities, as a consequence of recent history. The individualism of the Italians, suddenly released after the collapse of Fascism, resulted in a large array of parties in the first democratic election. The surprisingly high Communist vote may have reflected Socialist weakness. Among the conservative parties, both monarchists and indirect successors to the Fascists were admitted. But originally these were only marginal phenomena. The only large conservative group that appeared to be viable was the Christian Democratic Party. As a broadly based people's party, it was, however, characterized from the beginning by widely diverging factions. The weakness of the Socialists and the growth of the Communist Party, despite the splitting off of left-wing groups, and finally the loss of Christian Democrat votes to other conservative parties, work against the formation of clear majority governments. Attempts to form a coalition have failed repeatedly. Government decisions have become, therefore, increasingly dependent upon interparty agreements.

As to industry, the large banks, widely subject to state control, long ago created industrial holding companies. Consequently, essential parts of Italian industry in nearly every field have come under the indirect influence of the state. In many cases, there is virtually a state industry.

Italy has been suffering for centuries from chronic unemployment. Industrialization has really been achieved only in the north, where almost all the large industries are concentrated. Although the feudal structure of agriculture may have disintegrated in the south, few viable small or medium-sized companies exist there.

In the central and southern regions, therefore, the state is the most important employer. Candidates are, of course, selected according to political considerations, so that the region is governed by a motley of officials reflecting the prevailing power situation. Since state employees cannot be dismissed, almost any attempt at cost cutting through reducing the number of jobs has failed.

All this, of course, has long since affected private industry. The capital resources of the companies were insufficient from the start. And since rationalization had often become impossible in private industry, its meager capital resources were soon exhausted. The result is that only the banks—especially the large state banks—have any liquidity. But attempts to use these funds even to rescue industry usually are crushed by Italy's elephantine bureaucracy. It is remarkable, incidentally, that in Italy only bearer shares are issued by joint stock companies. The trend to hide wealth, born entirely out of mistrust toward the state, means that private savings are hardly ever invested in industry. Under such circumstances, it is easy to understand that a prerequisite for a new start must be the normalization of the relationship between citizen and state, the restoration of mutual trust. One must hope that this will be achieved after an unavoidable transitional period.

Relaxing in Rome

A free day in Rome makes it easy to forget political and economic difficulties. Even if one is frequently confronted with reality in the large Italian cities, Rome has retained its old attraction. It offers innumerable opportunities of escaping from the most urgent problems, at least for a few hours.

Whenever I can manage it, I take a stroll along Via Condotti. And I usually have a quick look at Gucci, the world-famous fashion temple of conservative elegance for men and women. And, of course, there's also Pucci, court of first resort, so to speak, for ladies' fashions.

I once had a charming experience in Rome while looking for a knowledgeable *cicerone.* I was directed to the Biblioteca Hertziana, one of the most important institutes supported by the Federal Republic of Germany. There I found a young art historian who conducted us through Rome for a day and a half.

We were especially impressed by a visit to the German cemetery in Vatican City. Our guide showed us, among other things, the grave of Ludwig Curtius, the German archaeologist. Before the war, he was the first director of the German Archaeological Institute in Rome. He died there in 1954.

Our young guide remembered hearing Curtius lecture at the University of Rome. From time to time, she said, he took small groups of his pupils on art expeditions. On one of these, they visited the Thermae Museum, which is full of antique works of art. Curtius had led them to a sculpture from the fourth century. It shows an exceedingly beautiful female figure emerging from the water, a scene which has been interpreted as the "Birth of Venus." According to our guide, Curtius said, in his soft voice, "If old Rockefeller came today from heaven and said to me, 'Curtius, I only have fifteen minutes to spare, show me the most beautiful sight in Rome,' I would lead him to this Birth of Venus."

This, too, explains why Italy, —why Rome—has been a dream for many people at least since the Middle Ages, and why it will continue to be this dream.

8

Espana Es Diferente

It was an unforgettable moment of my life in 1955 when I stood in Alcalá de Henares, a small village near Madrid, before the birthplace of the most famous Spanish poet. When I had started to learn Spanish at seventeen, my eventual aim was no less than to be able to read, in the original, *Don Quixote,* by Miguel de Cervantes, the master of the Spanish language.

In 1936, no one could foresee whether I would achieve this. The immediate reason for studying Spanish was a planned vacation trip with a friend to the Iberian Peninsula. In the middle of our preparations, the Civil War in Spain broke out. All I could do, therefore, was to pursue my studies of the language with the aid of Ullstein's *A Thousand Words of Spanish* and, later, the Toussaint-Langenscheidt Dictionary. I believe I have learned Spanish fairly well. As already mentioned, during World War II I passed an interpreter's examination in Spanish. When, many years afterward, I stood finally before the birthplace of Cervantes and was privileged to feel something of his personality, my sentiments were mixed with not a little pride.

My first excursion into Spanish literature, however, was not with Cervantes but with the popular novel *Sangre y Arena (Blood and Sand)*, which Blasco Ibañez had written in the thirties. It describes so beautifully the art of bullfighting and the mentality and life of the toreros outside the arena that later on, when I attended major bullfighting events, I did not sense the cruelty which repels so many others.

From Outcast to Modern Industrial State

What I learned with the aid of Ullstein and Toussaint-Langenscheidt was of great use to me in my business life after the war. Human contacts are so much easier to cultivate if one has at least some knowledge of the language of the country.

This proved true in 1952 when I first visited Spain. Two impressions from this trip have remained deeply engraved in my memory. First of all, I had acquired an entirely wrong idea of the conditions in the country from behind my desk at Hoechst. I became aware of how severely the Civil War from 1936 to 1939 had devastated almost every province. There were more than a million dead, and widespread destruction in cities and factories. The paralysis of the country must have been very extensive. But that Civil War had taken place thirteen years before my visit. Germany's defeat in 1945 had occurred only seven years earlier, yet reconstruction, especially of its economy, was well underway.

With regard to Spain, I had overlooked certain important factors. The Civil War had ended, but World War II started just five months later. And even though Spain stayed neutral, this war did not provide it with an exactly fertile climate for rapid reconstruction. In addition, the country had always been primarily an agrarian state. Ever since the resignation of Primo de Rivera in 1930, Spain, in its industry, had lagged far behind the rest of the Western world.

After 1945, Spain found itself in disfavor among the Allies, in part because it had remained officially neutral. The United Nations organization ostracized the country, or perhaps more precisely, the Franco dictatorship. This ban lasted until 1950. During these barren years, the Spaniards were only slowly able to lead their country economically and industrially out of the doldrums.

Glimpses of the Past

Spain's economic underdevelopment was also responsible for my second impression of the cities that I visited at the time: vehicular traffic consisted of old-timers of every type and make. The infrequent traffic jams looked very much like photographs from the twenties. The cars testified more poignantly to the long stagnation of the country than almost anything else. The taxi driver patiently trying to start up his vehicle with an old-fashioned crank handle symbolized the urgent need to catch up in almost every field.

There was something ingratiating about the horse-drawn cabs that could occasionally be seen in front of the hotels, many of which did not want to subject their guests to the often uncertain fate of a taxi journey. These idyllic carriages, incidentally, fitted in well with the appearance of the hotels at that time. There was little purpose in going window-shopping in the evenings. What one saw in the windows was the kind of display encountered in Germany before the currency reform.

And yet, conditions for the development of a chemical industry were not altogether unfavorable in the 1950s. The Spanish authorities gave some priority to the construction of chemical plants. The country had most of the important raw materials. Oil could be imported through a free trade policy. So-called export refineries were erected along the coast or on the islands of the Grand Canaries.

Gasoline supplies for Spanish motorists were in short supply for a surprisingly long time. Only when the tourist flood increased and proved to be an important source of foreign currency did gas become more freely available. Its quality left no doubt that distribution was controlled by a state monopoly (Campsa). Obviously, the Spanish sunshine was expected to compensate for this failing, as well.

Between Civil War and Costa del Sol

Spain and tourism: an almost inexhaustible subject. The sun has become the biggest capital asset of the country. The tourist flood poured across Spain's border. It became the dominant force behind the Spanish economic miracle.

Spain accepted this unexpected boom with almost hectic abandon. Its enchanting beaches became catchment areas for the sun-seeking hordes from the less benevolent climates of the European Continent, especially Germany. Spain's gastronomes and architects, as well as investors from various countries, conjured up one hotel after the other along the Mediterranean coast. Holiday camps and apartment blocks sprouted like mushrooms. And the Costa del Sol, the huge transit center of the most modern tourist area, became the symbol of a development that has begun to divide the world into areas of light and shade: here they work, there they take their vacations. The tourist boom brought prosperity to Spain. Today, the country occupies eighth place amongst Western industrial nations, a respectable position indeed.

The unexpected boom produced unexpected conflicts. In Spain, as elsewhere in an increasingly environment-conscious world, economic necessity and the ecological balance must be carefully weighed against each other whenever industrial expansion is contemplated. The importance of the tourist trade in Spain almost precludes industrial processes that damage the environment. As in the highly developed countries, authorizations for the construction of chemical factories often take months.

We experienced this very acutely in 1975 and 1976 while planning a new project in Vilaseca near Tarragona for our subsidiary, Hoechst Ibérica S.A. Since this area includes the seaside resort of Salou, the local council consists essentially of hotel owners. Predictably, the mayor at first roundly rejected the construction of a chemical plant in his area. He was not even willing to discuss whether or not the plant would emit any troublesome or harmful waste gases or effluent at all.

The position of a Spanish mayor is almost invulnerable. He is appointed by the civil governor of a province or, in the case of larger localities, by the minister of the interior. He thus has almost unlimited administrative power within his municipality. From a purely legal point of view, even the minister of industry is powerless against such a strong position.

Our Spanish Interest in Transition

Until the end of World War II, I. G. Farben played an important role in the Iberian Peninsula. Two highly capable sales agencies represented the company. One handled pharmaceutical sales and operated a Behring Institute and a laboratory for the formulation and packaging of its specialties. The other looked after the sales of dyestuffs, chemicals, and other products.

I. G. Farben had also acquired holdings in other profitable manufac-

turing undertakings in the country. The Ludwigshafen works of the
I.G. had licensed and assembled Spain's first two large nitrogen fertilizer plants, operated by the high-pressure process. After the war, German property in Spain was confiscated. Most of it was released a few
years later, but most of I.G.'s assets were auctioned off to interested
parties in the country.

Before the war, Unicolor was the leading company in Spanish industry, specializing mainly in sophisticated textile and leather technology. It collaborated closely with the dyestuffs factory of FNCE, in
which I.G. had held a parity interest since 1926. Fernando Birk, who
hails from Zeilsheim near Hoechst, was chief of Unicolor for many
years. It is thanks to him that Unicolor was spared the disasters of
requisition and auctioning. He also maintained its close connections
with FNCE. After the dismemberment of I. G. Farben, its share could
thus be recovered from the founder families of FNCE who had a great
deal of influence in Madrid.

Birk kept this part of the I.G. inheritance intact. It seemed natural,
therefore, that after the war Unicolor should continue the business of
the I.G.'s successor companies in Spain. For many years, it successfully
defended their position against Spanish competitors that, of course,
had a much easier life both during the war and in the period immediately afterward.

It must have been a heavy blow for the Unicolor management,
therefore, when Hoechst, and later BASF in Ludwigshafen, decided,
for commercial reasons, to withdraw from the old Unicolor. But in the
end they did so in a most loyal way that did justice to the personal
relations and also to the friendships that were newly concluded after
the war.

Recovery of Our Trademarks

Both before and during World War II, pharmaceuticals were a major
activity in Spain. However, as so often after the war, Hoechst faced
the trademark problem again. During the era of the I. G. Farben, the
company's pharmaceuticals were sold under the Bayer cross through
a joint central sales group in Leverkusen. Its agency in Spain, La
Química Comercial y Farmacéutica, had been requisitioned in 1945 as
a wholly German company. After liquidation of the I.G., it was auctioned off to the Banco-Urquijo group. The new owners took the view
that all the trademarks registered on behalf of Química Comercial

were among the assets that they had acquired. Included, of course, were the trademarks of all the specialties produced by Hoechst. Years of negotiation were needed before Química Comercial was prepared to sell our Hoechst trademarks back to us.

In the long intervening period, we had attempted to sell a number of the classic Hoechst products under new trademarks—alas, with only modest success. This was not only because new product names had to be used, but also because the Hoechst symbol of the "tower and bridge" was unknown in Spain. Under these conditions, the sales organization needed a disproportionately long time before achieving any success.

In the negotiations with the Urquijo group, we were very greatly assisted by Don Francisco Ripoll, lawyer and president of the Cros and Flix companies, the latter an old I.G. participant. Until the end of the war, Ripoll was closely befriended by a number of the leading I.G. personalities. Now almost eighty, he was an imposing figure, with whom conversation was always an intellectual pleasure. Don Francisco realized that Flix would not remain viable in the long term without the help of a partner strongly based in technology. Cros, a company active mainly in the fertilizer field, had acquired the I.G. share in Flix, so that it now held 90 percent of the capital of the company. At this point, Don Francisco sent the managing director of Flix, Dr. Wolfgang Just, on a trip to Germany.

Human sympathies and inclinations determine the fate of a company. Since Don Francisco had had close connections only with the I.G. head office in Frankfurt, he thought it wise to find out through Dr. Just whether one of the I.G. successor companies would be interested in acquiring a share of Electroquímica Flix. For reasons of convenience as well as personal relationships, the choice fell upon Hoechst. Don Francisco eventually retired from Cros and Flix, effecting in the course of a generation the friendly separation between Hoechst and the Cros group. Thus Hoechst was able to establish its own agency with the support of a strong Spanish partner, and to begin operations in all the traditional fields, first in sales and later in manufacture.

Barcelona a Generation Later

Walking along the Traversa de Gracia in Barcelona today and looking at the impressive administrative building of Hoechst Ibérica, one finds it hard to imagine the trials of those first years. During that time the foundation was laid for a future that gained us the importance in Spain which we have achieved in the rest of the world.

The main thrust of our activities today is in cooperation with Unión Explosivos Rio Tinto (ERT), the great complex formed through the merger of the former Unión Explosivos and the old mining company of Rio Tinto. With this largest chemical concern in Spain, we have erected petrochemical plants in the new chemical center in Tarragona, one of the zones for the chemical industry promoted by the state. These plants have been expanded during the last few years.

Pharmaceutical Industry under State Pressure

Both before and after the war, there was no large-scale organic chemical industry in Spain to support modern pharmaceutical research. However hard the government tried to establish pharmaceutical research and production in Spain, most companies, apart from the antibiotics manufacturers, confined themselves to formulating and packaging.

The pharmaceutical sector in Spain is characterized by the construction of modern factories, mainly with the aid of licenses from foreign pharmaceutical producers. The subsidiaries of the German chemical companies, too, have long since overcome the handicap of the years of isolation.

Of course, the problems faced by the pharmaceutical industry worldwide are felt particularly acutely in Spain. There, as elsewhere, the greatest volume of pharmaceuticals is distributed through the national health insurance system. As in other countries, the system cannot meet the increasing costs.

Even if the share of pharmaceuticals in total medical costs is significantly less than is usually claimed, measures of the centralized state bureaucracy are nevertheless directed primarily against the pharmaceutical industry. The authorities obviously believe that this is where they can intervene most readily. Because the industry in Spain is largely owned by foreign companies, they are unjustly accused of earning too much from the raw materials supplied to their Spanish manufacturing plants. So far as Hoechst is concerned, the company in reality is facing nothing but losses from its Spanish pharmaceutical business at the present time.

Indeed, the problems are becoming even more serious. The authorities are now granting approval of new pharmaceuticals only on condition that domestic manufacturers conduct their own research and produce an important part of the base materials in the country itself. At the same time, manufacturers are supposed to export a proportion of their products, fixed by the authorities, in order to earn foreign currency for the import of needed raw materials.

We have already begun Spanish production of some important base materials. It is hard, however, to imagine that such a concept, which seems to have been borrowed from dusty corners of outdated economic theories, can be successful on the whole. It can only be hoped that the charming Latin ability of giving an individual interpretation to cumbersome bureaucratic directions, thus rendering them tolerable, will manifest itself again in this case.

The Brakes Are Released

Administrative perfectionism, of course, ensured that the state had complete control for many decades over the development of the chemical industry, as well as of many others. During the era of self-sufficiency, the initiative of private industry was inhibited more than anything else. Its capital was often tied up in plants that could no longer operate economically.

The state became an entrepreneur. One step in this direction was the foundation of the Instituto Nacional de Industria (INI). Its task was to finance large capital-intensive projects with state money. The first chemical complex constructed by this organization is in Puertollano. It was designed for the production of organic base chemicals and later enlarged by a cracker and, in a partial partnership with foreign companies, a plant for the production of plastics based on petrochemicals. At present, a new chemical production center is being developed in Huelva, Tarragona. The participation of Spanish and international companies is being canvassed. Dow, which already has an extensive involvement in Spain, is mentioned as a participant and our partners from ERT might be involved. The complex is to be completed in the 1980s.

The various experiments to provide the country with a state-controlled industrial backbone are largely past history. Tourism and the establishment of a modern, increasingly competitive industry will continue to advance Spain economically. The income of large sections of the population has increased manyfold over the years, and the

Spanish standard of living is approaching that of other West European nations.

Spain's problems, however, must be seen not only in the light of its industrial and economic development. Forty years of the Franco regime have had other consequences for its position today, tomorrow, and the day after.

November 1975: Zero Hour

When Franco died on the night of October 25, 1975, "Francoism," the ideology of the generalissimo and his faithful, had already begun to dissipate. Old age and serious illness had forced him to let go the reins some time before. The transition from the zero hour of his death to the post-Franco era was accomplished remarkably smoothly.

Franco's political stance after Spain's Civil War was sometimes described as an ideology of law and order. Admittedly, his government succeeded in bringing his country peace and its first economic prosperity in modern times, even if this was distributed rather inequitably. Francoism also meant that power was exercised by a self-perpetuating minority. The majority was told that it was not ready for liberty and needed a period of preparation. This basic attitude dominated the political life in Spain for forty years. Just how ready for freedom the Spanish people were, in reality, was revealed soon after Franco's death. The Spaniards have handsomely acknowledged both the skill of young King Juan Carlos I and the reform plans of their minister-president, Adolfo Suárez. The elections of June 1977, the first free parliamentary elections in forty years, were a vow by the population to work for a modern, outward-looking Spain. The democratization program has gotten underway and needs now to be consolidated economically. The difficulties are, however, enormous.

Franco and his system have left their imprint on the economy of the country. Take the labor market, for example. There were no free trade unions under Franco. Now there are ten, all in intense competition with one another. Their combative attitude and unrealistic demands do not exactly help stabilize the economy.

Another characteristic of the Franco era was the dominance of the banks. Whether state-owned or private, they were all under state control. Industry was not financed through the capital market but exclusively through bank credits, so that the state was able to exert considerable influence. In addition, there are the state-supported industries which are partly merged in the Instituto Nacional de Industria, the state holding company, and partly operated directly by the state. This close-knit system is not easily or quickly liberalized. Suárez's government, however, has made the attempt, through an economic action program, to loosen up the old system by promoting greater competition and reducing controls. Unfortunately, this program has largely failed in its purpose because the government fought shy of the radical measures that were needed.

When, after the parliamentary election, the peseta was devalued by 25 percent, the population was greatly surprised. For many it was clear proof of the general ignorance about the worsening of the economic situation.

Spain faced three immediate problems: the trade balance was in deficit, inflation was accelerating, and unemployment increasing. The deteriorating situation eventually brought all parties around the table. Many agreements were concluded: anti-inflationary fiscal and monetary policies, stringent price controls, ceilings on wage increases, fiscal and agrarian reforms, and so on. Although how long the government can succeed in implementing these unpopular measures remains a troubling question, they have had some beneficial effects. Increasing exports strengthened Spain's external position, bringing the trade balance into surplus. Inflation was cut from 25 percent to the present 16 percent, but is creeping upward. Unemployment has risen by one-quarter to over 7 percent, prompting government programs to boost the economy.

EEC: The Great Hope

Spain recognizes its precarious economic situation. The country is conscious of the limitations to its own initiative. The desire to join the European Economic Community (EEC) has, therefore, become all the more urgent. I find that the present members of EEC are not making things easy for Spain. Now that the political objections no longer apply, national egoism—especially in the agricultural sector—is proving to be the inhibiting factor. In addition, Spain's auto and raw materials industries can readily compete with companies in EEC countries, as can the chemical, textile, and consumer goods industries.

On the other hand, certain EEC countries already compete against the most important agricultural products of Spain. Olive oil, citrus fruits, sugar beets, wine, legumes, grain, and potatoes are available in ample quantities from other EEC countries. The Spanish recognize that they must expect resistance from France, Italy, and possibly also Holland in regard to these products. They place their hope, therefore, in the Federal Republic of Germany and the noncompeting EEC countries.

Industrial Nation with Weak Points

Although Spain has reached tenth place among the industrialized nations, it still faces large problems. Unemployment stands at about a million, or 6 percent of the country's labor force. The inflation rate is over 15 percent, and salaries and wages are increasing correspondingly.

Spain has few energy resources of its own. Repeated droughts make waterpower fairly unreliable. Almost every year, dependence on oil imports has greatly strained the balance of payments, although the income from the tourist trade and remittances from the Spanish labor force abroad have to a good extent redressed the situation. The most important Spanish question, however, is: What will the urgently needed financial and fiscal reforms achieve? Although the government already has taken certain steps, these will not be sufficient. Far-reaching and unpopular measures will be unavoidable.

Spain's Chemical Industry Is Moving Ahead

A 1976 cross section reveals how actively the chemical industry of Spain is expanding its positions. Some 50 million tons of crude oil, or 7.5 million tons more than in 1975, were processed in the refineries of the country. Some 285,000 tons of ethylene have been produced in the petrochemical plants, about half the domestic requirement. An expansion phase, characterized by investment-intensive projects, started in 1976. Spanish and foreign chemical companies have earmarked a total of 150 billion Pesetas (US$2 billion) to be spent by 1980. A further Pts70 billion (US$1 billion) have been designated for the construction of a petrochemical plant in Huelva by Unión Explosivos Rio Tinto, ICI, Dow Chemical, and the Spanish company, Cepsa. However, all these plans have some way to go before realization.

The last years have been marked by great plans for heightened investment in research and development. Research expenditure of Spanish industry has been about Pts14.5 billion, or only 0.34 percent of the gross national product. That amount is to be spent in 1980 on chemical industry research alone. By 1980, chemical output in Spain will reach an estimated Pts900 billion (US$13 billion approx.). A concomitant need for trained chemists will rise from about 330,000 in the early seventies to 470,000 in 1980. In May 1977, production by Joint Ventures Taqsa in Tarragona, founded jointly by ERT and Hoechst, gradually went on stream. The plants are producing vinylacetate, polypropylene, and low-pressure polyethylene with technical perfection, but are encountering price problems in export markets.

Away from Industry and the Costa del Sol

Many phenomena in Spain can be reconciled only with difficulty, at any rate at first sight. The recent past and very recent events have in a sense juxtaposed eras that do not immediately appear to be compatible. A sensitive observer once remarked that immediately beyond the apartment towers of the Costa del Sol are the strongholds of Moorish culture. Indeed, travelers in Spain would know the land only incompletely if they confined their interests to industry or the Costa del Sol. Spain will be understood properly only by those who not only enjoy the sunny coast and the blue ocean, but who are also prepared to scour the interior of the country and to allow themselves to be captivated by the lavish beauty of old Castile, Estremadura, or the refined remainders of Moorish culture in the south.

Spain's wealth of artistic and cultural creations is almost inexhaustible. The splendor of those great historical epochs is reflected in collections like the Museo de Arte de Cataluña in Barcelona. This museum is a major center for the art of the Middle Ages, full of relics of the Romanesque and Gothic periods, as well as famous works of the nineteenth and twentieth centuries. No visitor can fail to be deeply impressed by a collection of priceless wall decorations from the Romanesque churches of Catalonia.

A visit to the Picasso Museum in Barcelona is no less rewarding. Picasso hailed from Málaga, and the Spanish now are proud of him. They will admit that they were not always so. It took a long time before Picasso's art surmounted years of antagonism and argument about his importance and became an accepted part of the Spanish heritage. Anyone admiring his *Harlequin, Nana,* or *Margot* today can hardly understand the aversion, and indeed indignation, displayed by Spanish citizens against Pablo Picasso for so many years.

To listen in the refectory of an old monastery or the courtyard of a historic mansion to a candlelight concert by Andrés Segóvia, the famous Spanish guitarist and master of the lute, is to catch something of that Spanish spirit which, behind these walls, has helped mold the Western world throughout the centuries. This spirit does not imbue the Costa del Sol. And, no doubt, it will play no part in the sober, factual discussions to decide Spain's industrial, economic, and social future. But this Spanish spirit remains a positive aspect of the country and must not be underestimated.

Spain will doubtless find its way into Europe, both politically and economically. The millions of West Europeans who each summer discover the country afresh will help integrate its austere and reserved people into the community of nations.

9
Portugal Needs Integration into EEC

My finest memory of Portugal is linked to the year 1955. After my marriage at Neuilly, and beautiful days at the George V Hotel, we headed for a honeymoon trip on the Iberian Peninsula. Our first stop was Barcelona, which I knew well from previous visits. Estoril was to be the base for showing my wife some of Portugal's art treasures and buildings that bear witness to the great past of this country.

At a time when cars were a rarity in Spain, we set off for Portugal in a rented Peugeot. Our route took us along the Mediterranean through Granada and Seville to Jerez de la Frontera, the locality near the Spanish-Portuguese frontier where Spain's famous sherry originates. From Jerez to Estoril, the refuge of exiled kings, was not very far. We stayed in Estoril for fourteen days, making daily excursions into the surrounding countryside.

At that time, Portugal's territory was still intact. Besides the parent country, it included the Azores and Madeira in the Atlantic, the rich overseas provinces of Angola in West Africa and Mozambique in East Africa, a few islands off the West African coast, the Cape Verde Islands, Macao in China, and Timor in the Indonesian archipelago. The

three provinces of Goa, Diu, and Daman also were Portuguese possessions, later to be annexed by India.

The country had been ruled since 1932 by Antonio de Oliveira
Salazar, whose role in Portuguese history has yet to be defined. While
he saved the country from economic and political collapse, he did so
at a very high price. Under his aegis, Portugal remained an undemo-
cratic, rigid corporate state whose economy made slow progress in the
hands of a few rich families. While the United States and West European nations had to include this strategically important country in
NATO, the Portuguese have frequently felt that from a democratic
point of view, they have not been regarded as entirely presentable.

The City of Henry the Navigator

Our first Hoechst agency was in Porto, an old commercial city in the
north where Prince Henry the Navigator was born in 1394. There has
always been a certain degree of rivalry between Porto and Lisbon.
Porto once regarded itself as the secret capital, particularly because it
gave its name to the country. The Lisboans take revenge for the occasional presumptuousness of Porto by calling its inhabitants *tripeiros*
("tripe eaters"), although *tripas à la mode* is one of the tasty dishes of
Portugal.

Porto by no means lives on its famous port wine alone. In addition
to shipping, the city boasts a long-established textile industry. Our
first partner, Señor Lopez, who had already cooperated with the I.G.,
came from an old Porto family. His trading business was conducted
in a beautiful old building. In the center was a large store, surrounded
by five floors of offices. Customers came directly into the store, exam-
ined the goods, drank a glass of wine, and completed their orders.

A Lopez daughter married a young man named Cezanne, who came
from Sindlingen near Hoechst, and whom we had sent from the dyestuffs sales department to Porto. Soon after the wedding, his father-in-
law entrusted him with many business transactions. Eventually,
young Cezanne became the head of the Hoechst organization. By the
time of my 1955 visit, we were able to establish a Hoechst sales
organization in Porto with eleven employees. In those days, I usually
commuted between the two cities. Much to our distress, our young
friend Cezanne died in an airplane crash.

As the industrial business began to flourish in Lisbon, we decided
to move our headquarters there. With a basic capital of DM10 million

(US$2.5 million), we founded Hoechst Portuguesa in 1965. Wolfgang Kemper, a young man full of ideas whom I had met in Brazil, took over the management of the new company. In the north of the country, where the textile industry is largely located, we expanded the production of textile auxiliaries. We erected the only plant yet existent for formaldehyde and glue resins. It remains one of the most modern in Portugal. In the industrial area of Mem Martins near Lisbon, we built a production complex for pharmaceutical specialties, cosmetics, and perlon filament.

Bloodless Revolution

During the night of April 24, 1974, revolution broke out in Portugal, a country where many people believed time had stood still. After the death of Salazar in 1970, an out-of-date government did not manage to initiate the kind of sensible evolution which later succeeded in neighboring Spain. How the revolution proceeded in a practically bloodless manner cannot be fully explained. Like others in our Western world, we followed the spectacle presented by the different revolutionary groups with concern, especially in regard to the welfare of our staff and the future of our factories. Kemper advised us to stay in the country, believing that Hoechst Portuguesa would have a future as a private company. Yet, it often appeared that the situation was hopeless. The revolutionaries began to dissolve even weak industrial groups and to expropriate heavy industry and landed propertyowners without compensation. The unconditional retreat from the overseas provinces, including Mozambique with its rich oil reserves, affected industry equally seriously.

While many qualified entrepreneurs and technicians left Portugal, the *retornados* returned from Angola and Mozambique to the mother country. Some 400,000 people must have been involved, increasing Portugal's population to more than 8.5 million. Many of the *retornados* were young people, not all of whom were able to find work. In 1977, a tenth of the working population was unemployed. In addition, there was virulent inflation and an oppressive mountain of debts. Yet, credit negotiations with the International Monetary Fund (IMF) were regarded with great distrust by the left-wingers. They considered such activities as intolerable interference in Portugal's internal affairs.

What will happen to Portugal's chemical industry in the future? At the center of this development is the Sines Chemical Complex. Its establishment goes back to an action of the government in 1972. A European port was developed 150 kilometers south of Lisbon and, together with Companhia União Febril (CUF), the largest industrial group in the country, the foundations were laid for a comprehensive petrochemical complex. At that time, Portugal still had access to the great oil reserves in Angola. Moreover, there was every indication that the demand for mass-produced plastics in Europe would continue to grow rapidly.

Future Overcapacities

When the revolution broke out in 1974, the boom in mass-produced plastics was abating throughout the world. Perhaps distracted by the turbulent internal political events, perhaps guided by the need for economic success, the new rulers of Portugal did not appear to have taken note of this development. Although Sines was nationalized together with CUF, the construction of the refinery in Sines was continued in its original large dimensions. The refinery is designed for an annual output of 10 million tons. Sines will supply three times as much liquid energy as the country needs. Therefore, large amounts will have to be exported. Perhaps in recognition of this one-way street, a cracker was built for the production of 350,000 tons of ethylene annually. Only a small part of the derivatives from this ethylene is required in the country itself. They too will place a great burden on state coffers and will present an additional worry for the Western European chemical industry. Further, a large aromatic chemicals complex currently is being erected near Porto, and further plants for ammonia and fertilizers are to be established in spite of the existing overcapacities.

The state is supplying more than half the proposed funds for the expansion of the chemical industry, although there is neither an adequate domestic market nor cheap raw materials or energy. It is an undertaking with considerable risks for the national economy. Although Hoechst was frequently approached to participate in these large projects, it has not yielded to the temptation. Against the background of a not very labor-intensive, large-scale chemical industry, Portugal did not appear to be a particularly attractive location, especially not for exports. We felt our investment plans had better be guided by the requirements of the country.

Little Contact between Neighbors

Most visitors to Portugal, unfamiliar with the history of the Iberian Peninsula, believe there is a close cultural and economic link between this country and neighboring Spain. This assumption is founded on the similarity of language, the identity of religion, a likeness in world policies for centuries, and in the last decades, a related political system of advanced dictatorship: Salazar in Portugal and Franco in Spain. In reality, however, the two countries have often lived in open animosity. The Portuguese have felt that too intimate a contact with their larger neighbor might result in an embrace from which there would be no escape.

Further, Portugal has sometimes seemed to turn its back on all Europe. It has looked across the Atlantic and elsewhere overseas. Its lack of significant domestic resources, a poor soil that cannot feed a population with a traditionally high birthrate, and the missed chance to participate in the nineteenth-century industrial revolution of central Europe led to continual emigration and to stubborn retention of the relatively rich African colonies. At the end of a revolution that was not thought out properly in economic terms, Portugal is in the middle of a great struggle for the social advances expected by its brave population. It will require a sensible, and perhaps at times unpopular, economic policy to reactivate the earlier, traditional industriousness of the workers—to provide agriculture with new impulses and prepare for future membership in the Common Market. The mortgages of the revolutionary era must be discharged; doing so will be a slow and painful process.

The indebtedness of the country can be reduced only if increases in agricultural and industrial output are achieved. To do this, the confidence of private investors has to be regained. The government appears to have recognized this need, for it is prepared to grant flexibility to private initiative.

In the meantime, the nine Common Market countries have on principle approved Portugal's future membership. This sets the signals for a favorable economic development. Still, it will probably take years for the Portuguese to attain a standard of living equal to that of the central Europeans. It calls for a rapprochement between the two Iberian peoples, for which conditions have never been so good nor the need so obvious. The economic and friendship treaty signed at the end of 1977 between the Portuguese and the Spanish heads of state appears to provide a good foundation for such a development.

There are also signs that political conditions will stabilize through further democratization. It should not be forgotten that, during the years of the most violent political unrest, the Portuguese military did all in their power to avoid bloodshed. When real civil war threatened, they kept away. Tolerance, one of the most valuable characteristics of the Portuguese, prevailed. After all, it is one of the peculiarities of Portuguese bullfighting that the bull is forced to its knees but it is never killed.

10

Greece—on the Shores of Light

Athens still looked rather provincial and many things reminded me of the privations of World War II and the ensuing civil war when I paid my first visit in 1951. My first stop was at the bank, in order to change my traveler's checks into drachmas. I had been granted the usual moderate amount for a stay of about four days, but the equivalent in Greek currency filled a whole briefcase. Inflation was in full swing and the currency had been devalued, a situation reminiscent of my childhood days in Germany after World War I.

What brought me to Athens was not a search for "the land of the Greeks with the soul," but a visit to a candidate for a Hoechst agency and an exploration of the possibilities of restarting the Hoechst pharmaceutical business in Greece. A few months earlier, two members of the former Greek pharmaceutical organization of Bayer had visited us in Hoechst. Messrs. Peripinias and Papaspirou had taken over the stores of the former Bayer agency and continued the business as best they could in the first postwar years. During their first visit to postwar Germany, both men had to realize that a joint Bayer/Hoechst agency in Greece would no longer be feasible. They had to choose, as indeed had we.

Finally, we remained with Stamatios Papaspirou, who had studied chemistry in Germany early in the 1920s and knew German. He was a nice, straightforward man, and we were sure that we had been lucky to work with him in rebuilding a market from which we had become completely estranged.

I found Papaspirou occupying modest offices and storage facilities at 14 Socrates Street, in one of the Athens suburbs not very far from the airport. Of course, we wished to establish an overall Hoechst company also in Greece. So Hoechst Pharmacrom was founded, which became Hoechst Hellas in 1967. At last, all our products were again under one roof.

Of course, this caused quite some trouble with the former I.G. representative, Demetrios A. Delis. He had an excellent organization for the industrial part of the business, and he had been well supported by the parent companies. In the end, we came to an agreement with Mr. Delis and maintained a peaceful coexistence with him for many years thereafter.

In those early postwar years, there was hardly any tourist traffic. For a visitor on business, such as myself, this provided a chance to visit many of the famous ancient ruins without being too much bothered by the crowds of tourists, whom you can barely avoid today.

Although I was not educated in the humanities, I am quite familiar with the world of Greek gods and the myths of Delphi, Olympia, and other superb sites. And atop the Acropolis I met many "acquaintances" from my school days.

The National Archaeological Museum in Athens, though not outstanding either architecturally or in terms of presentation, will remain an inexhaustible source of new discoveries for any historically and artistically interested person. Since that first visit, my wife and I have spent many holidays in Greece and its islands, especially Rhodes and Crete. Visitors to these islands are fascinated not only by the landscape and the sea but, above all, by the unique quality of Greek light.

The Gods—and the Police—Are Angry

The establishment of our Hoechst business progressed well. We moved from suburban Socrates Street to a modern, centrally located office on Constitution Square. Soon this building too proved inadequate, and we sought land to accommodate not only a new administrative and storage complex, but also a pharmaceutical plant. In 1968, we found a suitable location in the little wine village of Kifissia above Athens. There we erected the required buildings that gave us a great deal of satisfaction from an architectural as well as a practical point of view.

About 150 kilometers from Athens, on the right-hand side of what is today the superhighway to Salonika, we built a factory devoted initially to the manufacture of Mowilith, and joined later by a synthetic resin production plant. As so often happens when starting up a plant in a foreign country, we encountered stiff local resistance.

The initial phase was almost farcical. Construction work had hardly progressed beyond digging the foundations when a policeman arrived on a bicycle and decreed that work must stop immediately. Our protests were in vain. The policeman had a telegram from Athens, issued by the Office for the Protection of the Antique Sites of Greece, a particularly powerful authority, as we were to discover. It had intervened because minute remains of a wall of ancient appearance had been discovered in the adjoining field. Reference to authorization of the project as published in the Greek *Legal Gazette* and a few discussions with other appropriate authorities in Athens soon enabled construction to resume.

The second attack by the opponents of the project was not so easily beaten off. A press campaign, largely inspired by the left, mobilized public opinion against the Germans. The fact that the building site was located on the Bay of Aulis, from which Agamemnon had once set out to conquer Troy and where Iphigenia had barely escaped sacrificial death, was enough to generate acute feelings of indignation over this desecration of Greek soil.

We learned the true reasons for the opposition to this project only gradually. An adjacent poultry farm was afraid that it might no longer be able to retain its extremely cheap labor force. Also, a well-known politician had acquired land on the same bay for the erection of summer-houses. Furthermore, the local member of parliament had accidentally not been invited to the ceremonious laying of the foundation stone. In the end, the project overcame these hurdles. Since its completion, the complex already has been enlarged.

The anger of the gods over the profane activities on this history-drenched site may well have exhausted itself on a previous *casus belli*. On the same bay, a large cement factory, in existence long before we invested there, presents a far greater threat to the environment than our installations are ever likely to.

A Shift in Focus

The Greeks, especially the younger generation, are friendly toward the Germans. This attitude is no doubt due as much to the close economic links as to Germany's large share in Greek tourism. At any rate, we had no difficulties in securing for Hoechst Hellas an adequate number of academically trained staff who were fluent in German, and who adopted the Hoechst cause as though it were their own. During my regular visits to Greece, it is a main concern to strengthen this commitment to the company and jointly to forge plans for the future.

The Colonels Take Over

Life in Greece changed suddenly when its democratic government was replaced by a military dictatorship in 1967. This event deeply affected commercial life in the country for seven long years. If Papadopoulos saved Greece from Communism, he did not bring freedom to the people. Petty measures angered and provoked not only many Greeks but also the Western democracies, which would not accept a military dictatorship in Greece, especially since that country had been the cradle of Western democracy.

Historical truth is best served by admitting that King Constantine II had no real support in the population, and that the royal house, especially Queen Mother Frederica, intervened excessively in internal politics. Indeed, few Greeks regretted the abdication of King Constantine and his family. A referendum has since confirmed that the people do not desire his return, which probably is one of the reasons why Constantine Caramanlis achieved such an outstanding victory. He had contributed much to the development of Greek economy from 1955 to 1963, and was instrumental in bringing about the nation's association with the EEC.

The tremendous initial support for Caramanlis has given way to a more critical attitude. The elections of November 1977 showed that a more balanced relationship among the various political parties in the country has begun to emerge. The distribution of parliamentary seats between the government and the opposition now measures much higher on Western yardsticks.

The Cyprus conflict is, and probably will remain for some time, the Achilles heel of Greco-Turkish politics. It also will continue to strain Greek economy which, in spite of all the difficulties, seems to have got going again.

The Largest Merchant Fleet in the World

While progress has been achieved in the last few years, Greece is not yet an industrial nation. Ten years ago, industry and agriculture accounted for roughly equal shares in the gross national product. Today, the industrial share is 150 percent of the agricultural share. This development has centered largely on the textile and foodstuffs industries and on shipbuilding, which has been so important in the last few years. A large part of the country's income of foreign currency is earned on the world's oceans. The Greek merchant fleet, a third of which sails under neutral flags, is the largest in the world, larger than even the Norwegian fleet. The illustrious names of shipping magnates Onassis and Niarchos are closely linked with it. The income from shipping and tourism and the money sent home by Greeks abroad balance the import figures and help to prevent what would otherwise be a chronic shortage of foreign exchange.

The chemical industry of the country accounts for about 15 percent of the entire industrial output, but this figure includes the oil-processing companies. Refinery capacity is 21 million tons a year. The largest company in this field is the Esso-Pappas petrochemical plant in Saloniki. For more than a decade, it was the only processor of any importance, but it now no longer meets the requirements of the market. As a result, a new petrochemical complex is to be completed by 1982 in the vicinity of Kavala in northern Greece. Its cost of US$650 million is to be financed by a state-controlled bank consortium. European chemical companies may participate in the project.

Ancient Culture and Young Wine

I have fond memories of a brief visit to Santorini, the island nearest Crete. I was especially intrigued by the famed excavations of Professor Marinatos, the discoverer of the ancient capital of Santorini. An earthquake in 1500 B.C. put an end to the Minoan cultural period on both Santorini and Crete.

During the Easter vacation of 1978, my wife and I took our children to Greece to provide them with their first impressions of both classic and modern Hellas. At the same time, we enriched our own historical and cultural knowledge of the country. The Greek soil, seas, and islands still hide many surprises. For example, the new excavation of Pella, the old Macedonian capital, has led to much new knowledge, especially through the well-preserved pre-Hellenistic mosaics. Even more sensational, however, was the discovery of the tombs near Vergina, a small village about 80 kilometers southwest of Salonika, where Professor Manolis Andronikos found two massive golden chests which have been identified as the tombs of Philip II, the father of Alexander the Great, and his wife Cleopatra. Our guide in the museum at Saloniki was surprised that we had been allowed to inspect these finds, some of which were stored in a safe and some in a laboratory.

During earlier trips I had seldom had the chance to leave the main roads. On this occasion, we toured unspoiled rural areas, which brought home to me the huge civilization gap that exists between the countryside and the few large cities and the reason why people are so eager to migrate into the city. Every third Greek inhabitant lives in Athens, which now has a population of about 3 million people.

11

The Chemical Industry on New Shores

The world port of Rotterdam, rendezvous of commerce and trade from all corners of the world: Looking across it today, it is hard to imagine the "modest, cozy and pretty little harbor town" of centuries past. When I visited it in the sixties, Rotterdam had become the world's largest port in volume of turnover. Manufactured goods, coal, ore, grain, and, increasingly, crude oil and oil products have together achieved a total turnover of more than 300 million tons a year, 215 million tons of oil alone. The Hoechst share in this turnover is 1.5 million tons.

It is clear that such a fascinating site on the North Sea coast offers tremendous attractions for industry to establish plants there. The giant port area offers some 4300 hectares for every kind of industry. The chemical industry is concentrated in the eastern part of Europoort and in the Botlek area.

The port's facilities can readily load and discharge ships of up to 250,000 tons. There is also adequate space for the supertankers constructed after the 1967 closing of the Suez Canal when transport of oil in small tankers around the South African Cape of Good Hope proved too expensive. As a result, Rotterdam is also the starting point of the northwestern Europe pipeline grid through which flow oil and petrochemical products such as ethylene. Hoechst is connected to this grid through its own pipeline between Cologne and the parent site in Frankfurt.

A Bastion of Large-Scale Chemistry

By the beginning of the 1960s it had become clear that both the Dutch and Belgian North Sea coasts would attract the chemical industry by their favorable geographical position. The large German chemical companies naturally followed this development very closely. During Hoechst board meetings, we often discussed whether the company should obtain a suitable site in one of these regions.

Once we agreed that we were interested in principle, John Brookhuis, then the head of our company in Holland and now responsible for our United States business, started to persuade us of the advantages of location in the Dutch coastal area. Our prime interest at the time was the establishment of a phosphorus plant. By 1966 we decided to drop anchor on the Dutch coast.

Rotterdam was not the only location we considered. Nearby was Antwerp, whose 600-year-old port had achieved a turnover two-thirds that of Rotterdam's. BASF had established itself in Antwerp as early as 1966. In the mid-sixties, the international chemical industry had begun to concentrate on the Antwerp area and to establish important bridgeheads there. BASF was soon followed by Bayer. Their investments at that time were around DM1 billion (US$250 million).

Further, the Petrochim subsidiary of the Dutch Petrofina oil concern and Phillips Petroleum of America built a cracking plant for ethylene. Union Carbide erected a plastics factory. For an investment of some DM2 billion each, those who provided themselves with an international rivit at Antwerp included Solvay, the largest Belgian chemical group; Monsanto, the American chemical giant; Rhône-Poulenc, the largest French chemical concern; Albatros of Belgium; Polymer Corporation of Canada; the AGFA-Gevaert photographic concern; and others.

One of the reasons why Antwerp became the "boomiest" city during that period was the firm initiative of the city itself, its general helpfulness, and the favorable conditions granted to the buyers of land. Initially, we looked at both Rotterdam and Antwerp, and were open to other possibilities. Toward this end, I traveled to Holland and Belgium at the end of August 1965 as a member of a group of board representatives headed by Professor Winnacker. We believed that the best solution would be found only after direct, on-site examination of the possibilities.

Our route included a helicopter trip over the estuary of the western Scheldt river and the small port of Flushing near its mouth. Flushing's position close to the open sea became more attractive when the local authorities assured us that we would be able to set up our own port facilities. Moreover, all the terms we were offered in Flushing were far more favorable than those of any other site that we had considered, especially with regard to electric energy supplies.

Holland and Germany—Links of Culture

Holland and Germany have always had much in common throughout the centuries, with their cultural and spiritual histories intertwined. The two countries share a common heritage of literature and art.

Visits to the world-famed Dutch art museums have given me rewarding, beautiful, and relaxing hours. For example, the Rijksmuseum, the Van Gogh Museum, and the nearby Stedelijk Museum in Amsterdam and the Mauritshuis in the Hague often exhibit some of the paintings that we reproduce in the Hoechst art calendar, the preparation of which is my most pleasant spare-time job.

Coming to Realities

Standing at the southern edge of the town of Flushing, one is able to see across the Scheldt almost to Flanders. Compared with Antwerp or Rotterdam, Flushing is a dwarf among giants. Its production is no more than a fiftieth of the volume of its large competitors. With 50,000 inhabitants, the town is one of the many seaside resorts that dot that coast, attracting a large number of vacationers in the summer. In the past, Flushing was also the Dutch terminal for ferries to England.

We Vote for Flushing

The small town had much in its favor as a Hoechst production site. First, industry was not nearly so densely concentrated as in Antwerp and Rotterdam. Nevertheless, Flushing offered favorable transport facilities for raw materials and finished products, whether by sea, by land, or on the rivers. We were even able to erect our own port installations for the handling of incoming bulk raw materials and outgoing products. Inland waterways could carry shipments between Flushing and the parent site, as well as to other works in the Rhine/-Main area.

The Flushing authorities gave unqualified support to our projects. For example, they ensured adequate supplies of nuclear energy and natural gas. A refinery already existed near the port, and it was not difficult to connect with an ethylene line nearby. As a result, this vital raw material for organic chemistry could be obtained from Dow, Shell, or other producers in Rotterdam and Antwerp. Finally, the Flushing region offered sufficient labor, even for the long term.

We had been offered an outstandingly suitable area directly on the coast, and 67 hectares were purchased for the first phase of our plans. At the same time, we had an option for an additional 250 hectares of immediately adjacent land.

Backed by Knapsack Technology

Our first operation in Flushing was a phosphorus plant. Hoechst thus drew on the long tradition of its Knapsack works near Cologne, the former Knapsack-Griesheim AG. Before the war, phosphorus had been produced by I. G. Farbenindustrie in Piesteritz on the Elbe in Saxony. The electrothermal process had been developed by Gustav Pistor, at that time professor in Frankfurt am Main, and a member of the I.G. board until 1937.

After the war, the highly productive installations at Piesteritz in the Eastern zone were no longer available for supplying West Germany. Production facilities had to be developed elsewhere. Hoechst started the first West German phosphorus factory in Knapsack in 1953, using an improved Pistor process. The first furnace in Knapsack had twice the Piesteritz capacity of 10,000 kilowatts. The second and third furnaces were rated at 50,000 kilowatts each. Soon, Knapsack had become the largest phosphorus producer in Europe.

Phosphorus is a raw material for many products, including water softeners, insecticides, and crop-protection agents. By far the largest amount is processed into phosphoric acid, one of the raw materials for fertilizers and a water softener in detergents. Over the years, requirements for phosphorus and phosphorus compounds increased rapidly.

Yet, there were limits to Knapsack's capacity. The production of phosphorus required imposing amounts of electrical energy and became both scarce and expensive in the highly industrialized Rhine and Ruhr areas. Additional current for Knapsack could not be obtained at a competitive cost.

Flushing's Promises

Conditions were far more favorable in Flushing. Electricity and natural gas were available at an economical price. The harbor facilities were another advantage, especially for phosphorus production. Thousands of tons of crude phosphate, required each year for the manufacture of phosphorus, arrive from Florida, the Soviet Union, and Morocco. In Flushing, they could be discharged directly at the company-owned harbor installations, thus yielding significant savings.

The Flushing plant went into operation in February 1968 with the production of sodium tripolyphosphate (NTPP), mostly for use in the detergent industry. By May 1968, the complete first-stage facilities went on stream. They included a phosphorus furnace with an annual capacity of 30,000 tons, and a further capacity for some 45,000 tons of phosphoric acid and 60,000 tons of sodium tripolyphosphate.

Energy supply figures vindicated our choice of Flushing. The first phosphorus furnace alone used some 400 million kilowatt hours of current per year. We were, and are, good customers of the local electric power company, which benefits in turn from the flexible operation of the phosphorus furnaces. Output can be reduced during peak periods and increased when the power station is underutilized, so that the electric power company is assured of good plant utilization at all times.

Prince Bernhard Remembers

During the short construction time of twenty-one months for the first Flushing phase, we invested some DM100 million (US$25 million). On September 13, 1968, the plant was opened with ceremonies and the Netherlands became a phosphorus producer.

At the inaugural event, Prince Bernhard of the Netherlands said: "The Western Scheldt is predestined to make an important contribution to the development of Dutch industry and that of our seaport." He noted that Hoechst's investments in this area were providing a strong stimulus. Prince Bernhard recalled personal memories that linked him with Hoechst and the German chemical industry. He disclosed that he had worked in Paris as an employee of I. G. Farben some thirty-three years earlier. "For this reason," he concluded, "it gives me great pleasure to perform the opening ceremony here today."

After the ceremonial addresses, the Prince dialed the number 440 on a special telephone. A picture of the works loading ramp appeared on a television monitor. Distribution chief J. A. Moerdijk pulled the lever of the loading installation and the first "official" consignment of sodium tripolyphosphate was loaded on a truck.

However, our Flushing activities by no means remained confined to phosphorus and its derivatives. We soon commenced the construction of plants for dimethyltherephthalate (DMT), the starting product of Trevira polyester fiber. After expansion of the plant, manufacture of polyester condensate chips was begun.

In 1977, a plant for the wet purification of phosphorus went on stream. A further plant for the production of alkanesulfonate, a new detergent raw material, is under construction and should be completed in the near future.

Environmental Protection

When we began construction in Flushing, Dutch action groups took up environmental issues. These problems had, of course, been identified and attacked by the chemical industry long before then. At the parent site in Hoechst, for example, working groups were formed to study the effects of production on the neighborhood shortly after the rebirth of the company in 1951. In 1961, a central department for the purification of water and air was formed. Its job is to ensure that harmful substances in waste air and effluents from the factories are kept far below the limits constituting a danger to health.

Like Hoechst, the majority of the other chemical companies also concerned themselves with environmental protection long before it became a daily topic for the public at large. The chemical industry itself developed reliable measuring methods needed to detect even minute amounts of contamination in air or water. At the same time, it developed purification processes to destroy harmful substances.

Today, no fewer than 350 people work at Hoechst in central environmental protection departments and working groups. No new production plant can be established without the agreement of these experts. In the meantime, Dutch authorities became highly adept at environmental protection at almost all levels of administration. As in other countries, laws and regulations were decreed and strictly applied in individual cases. In Germany, for example, a central notification and control agency has been established for the heavily polluted Rhine delta with its strong concentration of industrial installations. The agency investigates complaints, operates a smog-alert warning system, and has appointed a special representative for environmental questions affecting the region. The authorities have even decreed temporary partial closure of factories with inadequate environmental protection measures. In some cases, they have refused to approve construction of new installations with the potential for excessive environmental pollution.

While concerned about the environment, the Dutch realized that it would be foolish to throw out the baby with the bathwater. Authorities and trade unions agreed on the need to maintain existing jobs and to create as many new ones as possible. As a result, the Netherlands stands with those countries which seek to balance the need for effective environmental protection with those of an industrial civilization.

The Dutch know a great deal about a "clean" environment. Traveling across the flat Dutch landscape is almost like crossing a huge park. In spite of the dense population, everything shines fresh and clean. Even in the center of some of Holland's cities, filled with diverse industry, I sometimes feel as though I were wandering through an extensive garden. Holland's vast tulip fields, of incomparable beauty when in bloom, are a vivid expression of the Dutch passion for care of the landscape.

Wide experience in the field of environmental protection was of great benefit to us in Flushing. Some of the Dutch people initially regarded industrialization, and especially the chemical industry, with a degree of skepticism, particularly since Zeeland is well known for its pure air and Flushing is a popular seaside resort.

One local paper, the *Provinciale Zeeuwse Courant,* devoted a whole page to this subject when the Hoechst plant was opened. Reporters had visited the parent site at Hoechst to view Hoechst's extensive antipollution measures. Professor Winnacker made a point of personally showing them Hoechst's efforts in the field of environmental protection.

Such intimate concern with the neighborhood of new plants, especially those abroad, is no less important than studying the market and production potential and the finance possibilities. After all, it is the neighborhood that determines the climate in which we will have to live together as fair partners.

The upshot of these discussions is that the need for expert environmental protection is appreciated and, indeed, demanded in the Netherlands. The authorities take great care to ensure that industry strictly observes the law. At the same time, the Dutch have not allowed themselves to be overcome by environmental hysteria. They have balanced the need for cleaner air with the need for industrial progress, an attitude to which the trade unions have contributed much.

Expansion in the Netherlands

Flushing was our most important, but by no means our first or only, engagement in the Netherlands. Hoechst Holland NV, a sales company for our products, had been founded in the Netherlands as early as 1950.

In 1963, we acquired our first Dutch production plant, the Weerter Kunstoffabrieken, in the little town of Weert. We bought the plant from Mr. Krages, a Bremen timber merchant. At first, Phillips also had a share in the plant. Production at Weert comprises rigid film, plastic sheets, and polystyrene foam which, incidentally, brought us into close contact with the local furniture industry. Today, the plant has more than 800 employees.

In 1966, an opportunity arose for Hoechst to become involved in the production of polystyrene in Holland, the only important plastic then absent from the Hoechst line. We acquired a 50 percent interest in Polymeer Fabriken NV, in Breda, until then a wholly owned subsidiary of Foster Grant, an American company. Two years later, in 1968, Foster Grant withdrew entirely from this European offshoot, leaving Hoechst the sole owner.

Today, the Breda plant manufactures a large number of products, including normal and impact-resistant polystyrene as well as foam. Polystyrene foam is widely used as insulation against sound and thermal change, and as protective packaging for sensitive and fragile products. Breda polystyrene is sold under the name of Hostyren (normal and impact-resistant) and Hostapor (foam).

In 1972, we took over Wagemakers Lakfabrieken, another Breda factory. This company had rich experience in powder lacquers which allow low environmental-hazard paint application. However, this rather antiquated factory, designed on too small a scale, presented us with many economic problems.

In 1975, all these activities were merged into Hoechst Holland NV, now responsible for both the manufacture and sales of Hoechst products in the Netherlands.

Capitalizing on Geography

What about the prospects of the chemical industry in the Netherlands? The industry had its beginnings in salt, the precious mineral which has been produced in Holland since the thirteenth century by burning turf impregnated with seawater. By the early nineteenth century, there were some 130 saltworks in the Netherlands. The Dutch chemical industry broadened in 1835 with the opening of the first sulfuric acid factory. Production there was around 200 tons per year at a cost of 12 guilders per 100 pounds.

By World War I, Holland's chemical industry was still manufacturing only illuminating gas (with such by-products as coke, tar, and ammonia water), superphosphates, sulfuric acid, and various pigments for the dyestuffs industry. Zinc was the only metal produced. The processing of natural raw materials into alkaloids, essential oils, sugar, potato starch, oils and fats, rayon, and paper was an important activity.

However, Holland's chemical industry did not begin major growth until the 1930s. At that time, German patents in the ammonia synthesis field had just expired, enabling the Dutch to erect nitrogen synthesis plants alongside existing coke furnaces. After all, there were both an excess of coke oven gas and a great need for nitrogenous fertilizers. The rise of the inorganic chemical industry was initiated by gasification of coal in the province of Limburg, later supplemented by the production of ammonia and nitric acid, as well as salt-mining operations in Twente and development of salt electrolysis.

The availability of the two raw materials, coal and salt, played an important part in this growth. Caustic soda, chlorine, and hydrochloric acid were produced from common salt. Other new products were tin and cement.

Quinine Monopoly

Until World War II, the Netherlands held a key position in the world market for quinine. The island of Java in the former Dutch East Indies provided about 90 percent of all the cinchona bark in the world.

Yet, it was only the loss of these areas in 1948 that forced a reorientation of the Dutch chemical industry. A shortage of foreign currency forced import restrictions on products which could be produced domestically. Consequently, industry was able to launch new enterprises without much risk, and the postwar Dutch chemical industry had little to fear from its largely paralyzed German competition.

Industrial expansion was the motto of the Dutch government. Investment by foreign companies was intensively promoted, and the chemical industry of Holland recovered quickly from the war. Large new refineries in the Rotterdam and Amsterdam port areas promoted development of a petrochemical industry. In 1970, the total capacity of the oil refineries in the Rijnmond area near Rotterdam was 62.5 million tons; by 1975 it had reached 90 million tons. The 1960 discovery of natural gas deposits amounting to 2,000 billion cubic meters heralded the end of both Dutch coal production and the manufacture of coke for technical purposes. Ammonia synthesis was eventually based on natural gas.

Two Focal Points

Dutch industry is headed by two companies which are active in many fields. The number one position is held by Royal Dutch Shell, the product of a Dutch-British merger. The partners were the Royal Dutch Petrol Company, established in 1890 and originally operating only in Indonesia, and Shell Transport Trading Co. Ltd., formed in England in 1897. Since their merger in 1907, Royal Dutch and Shell have been joint owners of two holding companies which, in turn, hold interests in more than 500 companies. The Dutch own 60 percent and the British the other 40 percent of the two holding companies.

In 1975, Shell achieved a turnover of 78.6 billion Florins (US $29.1 billion). More than 80 percent was accounted for by oil and gas activities rather than by chemistry. The chemical field, in fact, contributed only 10 percent to the total Shell sales, with ethylene and propylene, in which Shell is the leader in Europe, responsible for most. The production range extends from solvents to agrochemicals and from various plastics to rubber.

Like other important chemical companies in the country, Shell is largely dependent on exports, which determine its starting position among international, and especially European, competition. Hoechst has maintained a close relationship with Shell since the early fifties, especially through a joint petrochemical company in Tarragona in which Explosivos Rio Tinto is a third partner.

Soap, Fish, and Margarine

Unilever is the second largest industrial company in Holland. Like Shell, it is another joint Dutch-British corporation with chemical products accounting for only a small part of its overall activities. Nevertheless, in 1977 chemical products accounted for Fl3.2 billion, or about 8 percent of the total sales of Fl43.7 billion.

Unilever has always understood how to keep close to consumers in the food, soap, and detergent markets. When established in 1927 through the merger of several Dutch and British companies, it was originally called Margarine Union. Renamed Unilever in 1952, it has more than a hundred subsidiaries. The United Kingdom companies belong to the British, and all the other international subsidiaries to the Dutch holding company of Unilever. Its chemical activities include fatty acid conversion, as well as paraformaldehyde, metasilicates, and synthetic resins production.

Other Chemical Groups

The leading Dutch company engaged mainly in chemical production is AKZO in Arnheim. A young and dynamic company, AKZO was formed in 1969 through the merger of the Allgemeinen Kunsteide Union (AKU) with the long-established Köninlichen Zout-Organon (KZO). The company's sales in 1975 were Fl9.7 billion. Thirty-five percent of this amount was accounted for by man-made and cellulose fiber production in which AKZO and its subsidiary, Enka-Glanzstoff, are leaders in Europe. The company accounts for approximately 7 percent of world sales in man-made fibers. AKZO has deeply suffered from the synthetic fibers misery of the last few years.

AKZO also has production plants for salts, heavy chemicals, chlorine, acetic acid, agrochemicals, formaldehyde and vinylchloride, and paints. Further, with the takeover of Organon, AKZO has become the most important Dutch producer of pharmaceuticals.

DSM, the second largest chemical producer in Holland, is less well known abroad than AKZO. DSM was originally founded in 1902 as Staatsmijnen, a coal-mining company. After the sale of its last coal mine in 1973, the company assumed the name of the Dutch State

Mines (DSM). Sales in 1975, amounting to Fl7.6 billion, were achieved almost entirely with chemical products. DSM is among the largest producers of fertilizers, ammonia, and caprolactam.

Among American chemical companies in Holland, Dow is strongly represented in Terneuzen. Although it sells the products of its United States parent, Dow has extensive production facilities south of Holland's Sloe area, including a cracker for light naphtha and plants for styrene and polystyrene. The ethylene obtained from the cracker is processed into polyethylene, ethylene oxide, and other derivatives.

The Start in Belgium

The geographical advantages which so greatly promoted Dutch chemical industry after World War II also accrued to neighboring Belgium. This youngest of West European states gained its independence in 1830, after the September Revolution in which the south seceded from the northern Netherlands. The new nation came to the not entirely disinterested attention of the major powers. France toyed with the idea of a close alliance with the new state, if not its total integration. Naturally, both the German princes and Great Britain strongly resisted these ambitions.

Belgium's independence was basically a compromise among the desires of the major powers. At the 1831 London conference, they agreed to the separation of the northern and southern Netherlands, at the same time guaranteeing the permanent neutrality of the young state.

Supplier of Royalty

Only the Netherlands attempted to block this development, and it took another eight years before Holland recognized Belgium's independence. A prince from the House of Coburg-Saxony, at that time the main supplier of European royalty, offered his services as national leader. He became Leopold I of Belgium.

Only by knowing the genesis of this country can we understand the recurrent stresses in this bridge between north and south in the heart of Europe. The strains express themselves in conflict between the Flemings of the north and the Walloons of the south. Frequently, these disputes are sparked by language issues. After the revolution that led to an independent Belgium, French became the official language. In the

1840s a Flemish movement was established; it gained increasing influence and has since succeeded in obtaining equal status for the Flemish language.

Yet, the differences between northern and southern Belgium are not confined to language. They are also reflected in the industrial structure. For some time, the sources of Belgian industry were the coal and ore deposits in the Walloon regions, where one of the centers of Belgian mining and an iron and steel industry were established. Between the two World Wars, this concentration sometimes produced social contrasts. For example, lack of labor in the Walloonian industrial areas frequently was resolved by immigrant Italian and Polish workers. But in Flanders, with its people living largely off the less profitable textile industry, unemployment was widespread.

The years after World War II to some extent balanced these contrasts. A significant role was played by the chemical industry. Dependent no longer on traditional coal but, instead, on oil and its fractions, it tended to be more attracted toward the coast.

Europe's "Gulf Coast"

What Rotterdam is to Holland, Antwerp became to Belgium. Both cities offer the same favorable conditions. In Antwerp, the most modern port installations are available to cope with the import of raw materials. Antwerp, like Rotterdam, has made available cheap land in the port area, though only for renting. A new canal between the Scheldt and the Rhine has greatly facilitated transport, especially into the German industrial areas. The American business magazine *Fortune* has compared the area between Rotterdam and Antwerp with the coast of Texas on the Gulf of Mexico, describing it as the "gulf coast of the old world."

Many of the best-known oil companies, including Exxon, Caltex, Gulf Oil, and Petrofina, also have erected refineries in Antwerp. This European gulf coast may extend even further. The Belgian government is busily engaged in constructing a new harbor in Zeebrugge as a complement to Antwerp.

Whenever I visit Belgium, I am impressed by the impact of the chemical and other industries on the industrial outlook of the country. The Flandrian provinces of the North Sea are flourishing, while the Walloon regions suffer from structural weaknesses. Sober facts confirm these impressions. In 1975, for example, more than three-quarters of the investments of the Belgian chemical industry flowed to the coastal areas of Flanders. The lion's share went to Antwerp and its surroundings, where more than a third of Belgian chemical output is generated. During the ten years between 1966 and 1975, chemical companies invested more than BF50 billion in the port area of Antwerp alone, creating some 8200 new jobs.

More evidence of the growth of the Belgian chemical industry is found in its domestic chemical sales, which increased about three times as quickly between 1969 and 1975 as did similar sales in the Federal Republic of Germany. However, problems accompany this rapid growth. The rise in labor costs is rendered even more serious by "indexing," the linking of wage rates to inflation. Many of my Belgian colleagues regard this development with special concern, because more than 70 percent of the country's chemical output is exported. Only the chemical industries of Holland and Switzerland have a higher export rate. These exports are vital to Belgium. A population of barely 10 million people forms too small a market for almost 700 Belgian chemical companies.

This high export rate is counterbalanced by very high imports of chemicals, 50 percent of the sales value of Belgian chemical industry. In the German chemical industry, by comparison, the import rate is only 20 percent.

The Legacy of Belgian Liberalism

This ratio is caused mainly by the many foreign companies operating in Belgium and importing large amounts of raw materials. These are then processed and the resulting materials reexported for further processing and manufacture of finished products. Such commercial practices can succeed only because free trade liberalism always has been the order of the day in Belgium, with protectionist barriers deliberately avoided. Foreign investors, especially in the chemical industry, have always found an open door, and economic discrimination toward foreign companies remains unlikely in the future.

Leading the list of chemical manufacturers, with sales of more than BFr80 billion (US$2 billion), is a Belgian company famous at the dawn of the European chemical industry: Solvay & Co. S.A. It shares with Hoechst not only its year of birth—1863—but also many milestones in the company's history. Solvay was established by the brothers Ernest and Alfred Solvay for the industrial exploitation of Ernest's invention for the manufacture of soda—sodium carbonate—from common salt, ammonia, and carbon dioxide. Ernest's discovery largely supplanted the earlier LeBlanc process. The Solvay company continues to be the largest soda producer in Europe, showing the head start gained in the chemical field as the result of Ernest Solvay's invention. At present, 67 percent of the soda produced in the European Economic Community comes from Solvay. The company also has a leading position in Europe in the manufacture of chlorine. Other important products in the company's line are salt, sodium compounds, calcium products, fertilizers, chlorine compounds such as vinyl chloride and polyvinyl chloride, and peroxide compounds.

Solvay is also engaged in the macromolecular field. The manufacture of plastics and finished goods from these materials accounts for an increasing share of the company's total sales, especially in polyvinyl chloride and polypropylene. Mainly because of its activities in the United States, Solvay is one of the world's largest producers in the field of low-pressure polyethylene, with an annual capacity of more than 400,000 tons. The company has more than 120 subsidiaries throughout the world, including Kali-Chemie in Germany, which also produces pharmaceuticals.

Solvay does not typify the structure of Belgian chemical industry in general. The industry is dominated, especially in the export field, by bulk products such as fertilizers, industrial chemicals, and plastics. Sophisticated chemical specialties play a surprisingly minor role. Two exceptions are pharmaceuticals and, thanks to the world-famous Agfa-Gevaert group, photochemistry.

Grand Seigneur *of Chemistry*

Although the company went public in 1967, Solvay resembles other family concerns, now so rare in large-scale chemistry. Some 80 percent of the shares are held by outsiders, but the family continues to be a major shareholder. Jacques Solvay, chairman of the administrative council and of the executive committee since 1972, represents the fourth generation to hold this position.

Jacques Solvay's predecessor, Count René Boël, also was a member of the family. I well remember this white-haired gentleman with his well-groomed moustache. The magnificent appearance of this *grand seigneur* of chemistry, combined with his entrepreneurial prowess and his remarkable gift for management, always made a deep impression on his partners.

Holding second place in the Belgian chemical industry, although well behind Solvay, is BASF, with sales of BF17.3 billion. BASF's production is mainly in plants in Antwerp. Union Chimique Belge (UCB) is of similar size. Established in its present form in 1961 through the merger of several companies, it produces mainly foils, films, and several mineral and organic chemicals. Another important UCB product area is pharmaceuticals, where UCB and Hoechst maintain a fruitful relationship through the Hoechst subsidiary, Cassella-Riedel. UCB and CRL have jointly developed a preparation to stimulate the cerebral nerve cells. In Belgium, the preparation is marketed under the trade name Nootropil; in Germany, it was successfully introduced in 1974 by Cassella-Riedel under the name Normabrain.

Neck and neck with UCB, in terms of sales, is Petrofina-Chemie, a subsidiary of Petrofina oil company. Petrofina-Chemie produces mainly plastics, partially with foreign partners.

Hoechst has no major production plants in Belgium. However, our sales activities are most encouraging in this economically hospitable country, which places no obstacles in the way of foreign companies.

Belgian Hoechst also looks after the company's interest in the Grand Duchy of Luxemburg, since the chemical market there is not nearly large enough to support its own sales organization. Further, the Netherlands, Belgium, and Luxemburg are increasingly merging into an economic union: Benelux.

Memories of Benelux

This first instance of supranational cooperation among European states was planned in the midst of World War II. On September 5, 1944, the exiled governments of Holland, Belgium, and Luxemburg, then in England, concluded a basic agreement for the future economic union of their countries. This agreement was confirmed after the war in a protocol signed in March 1947. Numerous obstacles delayed the eventual economic union of the Benelux countries until January 1960. As the European Economic Community grew, Benelux lost some of its original significance. Yet, it represents one of the first attempts of European unification beyond national boundaries.

As far as Belgium and the Netherlands are concerned, Benelux also expresses recognition of centuries-old affinities. The spiritual and cultural links between the Dutch and the Belgians are so close that "Belgian" art as such began only with the founding of the nation.

After 1830, Belgium made valuable contributions to European and international art. The graphic artist Félicien Rops was followed by the painter James Ensor, whose pictures of spectral masquerades and maliciously grinning larvae send shivers down the spines of the visitors to the Musée Royal des Beaux Arts in Antwerp. Among the Belgian sculptors, Constantin Meunier was one of the first to create realistic figures of workers, one of which can be seen on the Peace Bridge in Frankfurt. The slim figures by George Minne are derived from Jugendstil and expressionism and betray their intellectual affinity to the German Wilhelm Lehmbruck.

Churchill's Great Vision

In the world of commerce and politics, Belgium has a firmly structured and highly respected position within Europe. As the headquarters of NATO and the Commission of the European Economic Community, Brussels has become a symbol of the unity of the Continent.

In his speech in Zurich in September 1946, Winston Churchill, with prophetic vision, painted the ideal of a united Europe. However, he saw such development mainly as a way of bringing France and Germany together, with England staying on the sidelines. Churchill's idea was one of extreme boldness because these two nations had just been enemies for almost six years of bloody and bitter war. The miracle of this early French-German friendship would have been impossible without United States support. In West Germany, the idea of a united Europe elicited an echo that was as spontaneous as it was powerful. Most of my generation became enthusiastic Europeans. We viewed this route as a guarantee against a repetition of the events of the recent past and as a step toward building something new. Wherever there were signs of a new Europe, support of the "skeptical generation" was assured. This attitude was not without its egotistical motives. After all, Germany had lost the war. As Germans, we had nothing more to lose and could only gain as the result of European integration. Yet, in spite of such subconscious thoughts, our attitude stemmed from honest and unselfish enthusiasm for the vision of a united Europe in which national contrasts would become irrelevant.

In spite of Churchill's enthusiasm, my evaluation of the political and economic realities soon led me to recognize the many obstacles barring the road toward a European community. Therefore, my expectations were more modest than those of many of my contemporaries. Consequently, my disappointment was correspondingly less when the dream of a united Europe could not be achieved in a short sprint.

For many, the slow movement transformed enthusiasm into resignation and led to pessimistic and disdainful judgment of the European movement. I have remained a believer in the idea of European unity to this day. But, facing the facts, I must acknowledge that a great deal of patience will be needed before it is ideally realized. Despite such failures as the nonexistent common corporation law, inexorable progress is being achieved, as evidenced by the growing attraction of the European Economic Community for countries not yet affiliated. While England already has joined, Greece, Portugal, and Spain look forward to future membership. Surely, such attraction can exist only when a brighter future is seen beyond the present difficulties.

Hoechst AG, too has long tried to think in European dimensions. Coordination and harmonization of activities in the member countries of the EEC, and also in the other European countries, are of decisive importance. One of the means is our so-called Europa Conference, usually held at regular intervals. At this conference, the chiefs of the European agencies, together with the heads of the Hoechst organizations in America and Japan, meet to exchange their experiences and to try to chart Hoechst's road into the European future.

Of course, Hoechst has extensive industrial operations in many European countries. The ideal of European cooperation is being realized to a far greater extent in the economic area than in the political field. This discrepancy between the economic and organizational requirements of a multinational company, and the depressingly slow progress toward a true European community, will remain among the major problems for many years to come.

Parities change from year to year and market and price structures are subject to varying degrees of state intervention. Differences in fiscal law often prevent all the Hoechst subsidiaries in one and the same country from being merged into one group. Still, the persistent efforts of the technocrats in Brussels, frequently criticized unjustly, have achieved continual improvements.

While our Europa Conferences have taken place in Switzerland, Austria, Holland, Italy, Spain, and London, one of our most impressive gatherings was that in Brussels in 1969. Our site was the Atomium, the 110-meter-high structure erected for the World's Fair of 1958, representing an alpha-iron crystal magnified 150 billion times. This chemical symbol is most appropriate to a city that is the headquarters of the Conseil Européen des Fédérations de l'Industrie Chimique (CEFIC), the umbrella organization of Europe's national chemical industry associations, with whose progress I have been connected for many years. CEFIC's efforts clearly go toward the promotion of European unification, particularly in the chemical industry field. What was once said in another connection has become even more applicable to the idea of European unity during the last three decades: We have lost many illusions on the road to Europe, but never our dreams.

12

The Double Eagle Looks East and West

The celebrations to which Hoechst Austria invited us in May 1977 extended from a festive dinner to a rural lunch in Perchtoldsdorf Castle near Vienna. In Austria, one does not need much of an excuse to have a party. But this occasion provided ample reason. Hoechst Austria had at long last been able to move out of its overcrowded offices on the Seidengrund and into a modern administrative center. The site was appropriate for a company that is presently one of the largest private chemical producers in the country, with about 2400 staff.

The Austrian minister for commerce, trade, and industry, Dr. Staribacher, and many of our friends from Austrian industry had joined our colleagues of Hoechst Austria and the guests from Hoechst to celebrate the event. The dinner took place in the Palais Schwarzenberg, one of those numerous, history-laden Viennese palaces which, though long since turned to other purposes, still impart the atmosphere of past glory. The Palais Schwarzenberg was constructed under Duke and Field Marshal Philipp von Schwarzenberg, who is honored for his victory over Napoleon. I had lived in the hotel part of the palace in January 1969 when, after the big Europa Conference of 1966, we held a small follow-up conference. The date coincided with my fiftieth birthday. Our inventive Austrian hosts had engaged the Deutschmeister, one of Austria's most noted military bands, with a tradition going back to 1695, for a surprise birthday serenade. The band played the Radetzky March by Johann Strauss the elder, recalling the great popularity of the famous field marshal from Bohemia. The Radetzky March has always been part of Austria's history in both the good and the bad sense.

Under the Old Company Umbrella

My first professional encounter with Vienna took place in 1954 when, together with Walther Ludwigs, we tried to evolve our commercial concept on the spot. In the first postwar years, under American administration, small companies had been established which were conducting their business fairly independently.

Albert Creutzberg, a native of Frankfurt who had lived in Vienna for many years, had worked for I.G. in the pharmaceutical field before the war. He secured the company name Vedepha (*Ve*rtrieb *De*utscher *Pha*rmazeutika), the former pharmaceutical agency of I.G. It became the focal point of the new Hoechst organization in Austria, begun with only a few people. Our dyestuffs and chemicals business, partly taken over from an earlier joint I.G. agency, was also incorporated into the Vedepha organization. Dr. Hermann Kügler collaborated in these efforts from early 1950. During the war, as an I.G. representative in Rumania, he came to know Vienna well.

In those days, the future of the Austrian chemical industry looked rather bleak. The country never was one in which the industry really felt at home. Until 1850, its significance was very slight. Many starting products, such as nitrate, potash, sulfur, soda, phosphorus, and sulfuric acid, were in short supply; at that time, they were obtained or produced only by small companies with manual labor.

In contrast, there were large lignite and coal deposits in Bohemia, which had belonged to the Austrian monarchy. Since these were essential for the chemical industry as raw material and energy sources, Bohemia soon became a focal point for the industry. For example, the uranium deposit from which Madame Curie obtained radium was at

Joachimsthal in Bohemia. Not far away, nitrogen was produced. And the Galician sulfur deposits were large enough to support an important sulfuric acid production.

Because of its plentiful petroleum deposits, Galicia also had a sizable oil industry. In the Bukovine and the Austrian Alpine foothills, soda was produced. The Austrian heartland, on the other hand, was comparatively poor in raw materials. In 1919, when Austria lost Hungary and Bohemia as the result of the Treaties of Saint Germain and Trianon, it also lost the raw materials bases for its own chemical industry.

Austria as an Economic Area

The unavoidable structural changes affected the basic chemicals industry. Further expansion appeared almost impossible. In the next twenty years, Austrian industry, particularly the chemical industry, never quite recovered from the consequences of the Treaties of Saint Germain and Trianon. The entire economic structure was badly hit by the constriction of the heartland to a far smaller area. In 1938, following the *Anschluss* (the union of Austria and Germany), hopeful developments were interrupted once more.

Soon, the war economy took priority and all private initiative was constricted. Several large works were established during the war years in the *Ostmark* as Austria was then called. However, their huge size and one-sided production were to produce many difficulties in the postwar years.

At the nadir in 1945, the economic future of Austria was completely uncertain. It was essential to recreate a basic chemicals industry. Oversized plants, established during the period of German influence, had to be restructured. Moreover, it was necessary to overcome the consequences of dismantlement as quickly as possible.

However, this depressing balance sheet of the first postwar years bore one important asset: Austria had large hydroelectric capacities and a highly developed power industry. Both were essential for the gradual reestablishment of the electrochemical industry and its products.

The extensive salt deposits in Austria fostered the chemical industry. Salt is the starting material for the production of soda and its derivatives: alkalis, chlorine, and alkali products. The milestones in the development of Austria's postwar chemical industry were the erection of a polyvinyl chloride plant and a rayon and sulfuric acid factory.

As in all countries hard hit by the war, the construction industry in Austria was fully employed, another benefit to the suppliers in the chemical industry. The paint industry, in particular, experienced a considerable upsurge.

Penicillin V—An Austrian Invention

In the years before the war, pharmaceuticals were largely imported. Only fifteen years afterward, about 80 percent of domestic requirements were met by the domestic Austrian pharmaceutical industry. While the few factories that previously existed had done only packaging, Austrian research workers were now in a position to develop many preparations themselves. Penicillin V, the first oral penicillin, was an internationally acknowledged achievement of an Austrian research group. It became one of the major pharmaceutical exports of the country.

The élan with which the Austrians launched their industry after the war was remarkable. This achievement also gave great satisfaction to the Hoechst organization in Austria. Three tasks dominated the scene. First, we were very anxious to concentrate our diverse activities. Second, we wanted to examine the extent to which the Austrian presence could be utilized as a platform for our business with Czechoslovakia, Hungary, Rumania, and Yugoslavia in southeastern Europe. Third, we wanted to determine whether manufacturing facilities in Austria itself would provide a viable proposition for us.

We expected they would be in the case with polyvinyl acetate—the Hoechst Mowilith. Apart from pharmaceuticals, Mowilith production usually established our first foothold in foreign countries. A Mowilith plant went on stream in Austria as early as December 19, 1955, in Floridsdorf, an industrial suburb of Vienna. Many of the guests will never forget its inauguration, not least because one of the speakers, the plant manager, fell into a barrel of Mowilith during an excessively zealous tour of the factory.

The Captain of Kopenick *and Other Pleasures*

My leisure hours during visits to Vienna were always dominated by theaters and museums. The first play I saw there was Zuckmayer's *Captain of Kopenick,* with Werner Kraus in the lead. In addition to the Burgtheater, the Opera House had been restored in 1955 to the glory it had possessed in the days of the emperor. The Viennese love their opera above all else, and they made many sacrifices for its restoration. It has been claimed—perhaps with a hint of truth—that season tickets for the opera cannot be bought, but must be inherited. There is nothing unusual in queueing for days for a ticket.

In his *Wiener Knigge,* Jörg Mauthe tells a charming tale set in April 1945, when practically all life had retreated into the cellars and shooting was still going on across the Danube. A wretched old man stood in front of the loggia, the only part of the Opera House that had survived the fire. Holding a paintbrush in his hand, he was busy placing gold leaf on an ornamental stone tendril. Even if the story were only invented, Austrian reconstruction began with such minute and apparently meaningless acts and not with the construction of some vital secular building.

The Opera House is also the site of the Opera Ball, an outstanding social event attended by guests from all over the world. Anyone who has experienced such an evening (my wife and I have been present several times) simply cannot escape the impression of witnessing almost the last splendid example of a social culture without equal, except perhaps for the evenings at the Garniers Grand Opera or the Comédie Française.

The theater scene in Vienna is inexhaustible. There are at least two dozen large theaters and a number of small ones, some of them existing in cellars. In view of the Austrian preoccupation with everything connected with theater and opera, it is hardly surprising that issues relating to them, such as an appointment to the leading position at the Burgtheater or the State Opera, are debated with national passion. The Viennese are liable to mount the barricades if they hear, for example, that Herbert von Karajan is threatening to give up the baton in Vienna because some cultural bureaucrat proposes to cut the maestro's budget.

Important paintings and sculpture can be found in almost every one of Vienna's numerous churches and palaces. But the center is the Kunsthistorisches Museum on the Ring, whose art collection, one of the largest, is not just the result of acquisition, but has grown organically over the centuries. Its treasures are uniformly drawn from almost all cultural epochs. One of its unique features is the Breughel collection, the largest in the world. The Albertina, with its 40,000 drawings and watercolors and an infinite number of prints in a variety of techniques, probably has the most comprehensive print collection to be found anywhere.

World Politics in Vienna

In 1955 in Floridsdorf, we started our first Mowilith factory to be operated abroad after the war. At that moment, new political history was made in Vienna. Although the foreign ministers of the four occupation powers had not been able to reach agreement about the future of the country after more than 300 meetings, the Soviets in Moscow suddenly declared their readiness to accept the draft of the State Treaty. Incredibly, on March 5, 1955, the State Treaty so greatly desired by the whole of Austria was signed in Belvedere Castle. In April, a group headed by the former Federal Chancellor Julius Raab, and including Vice-Chancellor Schärf, Foreign Minister Leopold Figl, and State Secretary Bruno Kreisky traveled to Moscow.

The state agreement made Austria a unified economic area once more. Even those factories that had until then been under outside control were returned. The market, originally purely protectionist, was opened again to Western Europe. Most trade barriers were lifted and there was talk of fully restoring free enterprise. This concept had long been abandoned. Barely a year after the war, one of the so-called key laws of the Second Republic nationalized banks and industrial concerns.

No doubt, the emergency situation at that time was an essential reason for the passage of this measure by all parties represented in the Austrian parliament. Seventy percent of all the key industries had been destroyed. Gentle pressure by the occupation powers and the enthusiasm for nationalization prevailing everywhere in Europe at that time also played a part. Between 1946 and 1970, various Austrian authorities took turns administering the share rights of the seventy companies that had been nationalized. Finally, all activities were consolidated in the Österreichischen Industrieverwaltungs-Aktiengesellschaft (ÖIAG).

Would the Austrian economy have developed better without this nationalization? It is difficult to speculate. The first fifteen years are characterized by a considerable degree of reconstruction. In the nationalized sector of the economy, the balance between government and opposition was maintained even when there was a change in the ruling parties. Austria knew how to exploit booms and later how to live with recessions.

Many of my Austrian colleagues ask themselves today whether the helping hand of the state, which vigorously supported the economy during the years of reconstruction, will allow the kind of flexibility that enables companies to cope with such difficult periods as were experienced in the late seventies. Since 1976, the price and production indexes of the Austrian chemical industry have split apart. The causes include increasing competitive pressure, unused capacity, and reduced demand from domestic and foreign markets, as well as price controls.

Since 1971: Hoechst Austria

In 1971, the Vedepha company was finally converted into Hoechst Austria. Our plant in Floridsdorf had been enlarged to some 70,000 square meters. The factory was producing plastic dispersions, plasticizers, surfactants or tensides, auxiliaries, and cosmetics. It was also engaged in packaging agrochemicals and formulating pharmaceuticals. Plasma fractions have been produced since 1977 in a modern laboratory of Behring Diagnostics.

Through Austria Faserwerke, founded jointly in 1966 by Hoechst and state-owned Chemiefaser Lenzing, we established a strong position in the domestic man-made fiber market, weakened unfortunately by the later worldwide recession in synthetic fibers. Austrian per capita consumption in plastics is about 40 kilograms a year. This indicates a very rewarding market which we have exploited with our various Hostalen high-density polyethylene grades, especially Hostalen bottle crates, bottles, and strip. Lothar Arends, the present chief of Hoechst Austria, pays particular attention to this field—understandably, since he was formerly the successful head of plastics sales in Hoechst.

A Brilliant Career

Hoechst in Austria is prominent in the paint field. This development started in the Federal Republic with the acquisition of Chemische Werke Albert in Wiesbaden and Reichhold-Chemie AG in Hamburg, two closely linked companies.

Reichhold was originally headed by the brothers Otto and Henry who, after World War I, founded a resin factory in Vienna. They engaged the Austrian chemist, Dr. Herbert Hönel. He had once worked for Albert at Wiesbaden, but had then returned to his homestead. At Reichhold in Vienna, Hönel began developing synthetic resins.

Later, Henry Reichhold exported synthetic resins to the United States, first from Albert's production and later from his own plant in

Vienna. In 1927 he founded his own company in the United States to produce these resins directly in that country. When business appeared promising on both sides of the Atlantic, the Reichhold brothers built other synthetic resin factories in England, France, and Germany. Expanding its paint interest, Hoechst acquired the German plant in 1967.

Cooperation with Albert and Reichhold soon brought new international contacts. As a result, in 1968 we acquired a share in Vianova Kunstharz in Graz. This company also owes its existence and initial success to the same Reichhold chemist, Dr. Hönel. In those desolate years, he had founded Vianova in Graz with several partners. There he developed the water-soluble synthetic resins he had invented, but which had been ignored by the large European chemical companies. As late as the early fifties, automobile body primers were still diluted with alcoholic solvents, causing occasional explosions. Hönel therefore developed the first water-soluble primer.

The man who instantly realized the inherent possibilities in these water-soluble lacquers was Herbert Turnauer. He had come to Austria in 1946 from Prague, where his family had operated paint factories. He bought a small company in Mödling near Vienna—Peter Stoll oHG—which had been hit by the war just as hard as other chemical companies. A chemist, a forester who worked part-time, and Turnauer himself formed the entire original staff of the company.

One of those lucky coincidences, without which many an industrial career of the postwar period would have been unthinkable, ensured that Turnauer had adequate raw materials available immediately. While still in Prague, shortly before the end of the war he had been ordered to transport raw materials for paint manufacture from Czechoslovakia to Mödling. He simply followed the route of these raw materials after the war.

Turnauer acquired the manufacturing rights for the new water-soluble synthetic resins from Hönel and, together with Vianova, developed an industrial process for the paints produced from them. When he succeeded in winning Volkswagen as his first customer, he had no worries about the utilization of his plant for some time to come.

Several thousand kilograms of paint were shipped to Wolfsburg as the "premiere." Similar shipments were made week after week for many years. Soon, cars from Fiat, Lancia, Ford-Germany, BMW, Renault, Rolls Royce and even the Eastern bloc Moskwitsch and Skoda cars—millions of vehicles—were protected with Austrian paints. A close connection had been established between Vianova and Turnauer's company, which had meanwhile been renamed Stollack.

Nevertheless, Turnauer greatly surprised the industry at the end of the 1960s by selling his company to Hoechst. Vianova soon followed suit in conformity with industrial logic. However, Hoechst ownership did not alter the Austrian management structure of the companies. Today, Vianova owns a modern research center in Graz, headed by Professor Hänus, the *spiritus rector* of the company. Herbert Turnauer, now, as then, one of the great private entrepreneurs of the country, has remained a friend and adviser of Hoechst.

The Austrian Chemical Industry

As in other industrialized countries, the chemical industry in Austria was highly growth-conscious. Its output developed more rapidly than that of other industries. Only one-third of the total sales of the chemical industry come from companies existing in Austria before the war and nationalized in 1946.

The first place is held by Chemie Linz, which had a very difficult postwar start. Originally designed only for the production of nitrogenous fertilizer, the plant produced about two to three times as much as the country needed. The excess had to be sold in competitive export markets without an international sales force.

The most important task for Linz was to overcome its chemical monoculture and to build up new fields of production. At the beginning of the fifties, the company successfully added agrochemicals to its production and managed to penetrate into the pharmaceuticals field, partly through the acquisition of smaller companies.

Chemie Linz turned to plastics at an early date. Back in 1950, it began the production of plasticizers and phthalic anhydride. Important starting materials for resin production followed and, since 1967, the company has produced the synthetic resin raw material, melamine. With the start of the third melamine plant in 1976, Chemie Linz became one of the largest producers of this plastic product.

The Oil Field of Zistersdorf

Chemie Linz gained entrance into petrochemistry with the aid of Österreichische Mineralölverwaltung (ÖMV). This state-owned company has a monopoly in the prospecting for and production of oil and natural gas and the processing of domestic and imported crude oil and gas.

The history of Austrian oil goes back to the beginning of the 1930s.

Because of the economic crisis of that era, sufficient funds were unavailable for large-scale exploitation of the existing deposits.

During the war, production was increased from about 56,000 tons in 1938 to 1.2 million tons in 1944. Zisterdorf became the largest oil field in central Europe.

After the signing of the State Treaty, the Soviets returned oil production and processing to Austrian hands. The ÖMV was founded and new oil and gas deposits were opened up. A modern refinery, one of the largest in Europe, was established in Schwechat near Vienna. Today, it meets 80 percent of Austrian requirements.

At the end of 1961, Danubia Petrochemie AG in Schwechat took up the production of polypropylene. Stickstoffwerke Linz held a 40 percent share in this company, as did Montecatini, whose process was being used; the remaining 20 percent was owned by the Austrian state. In 1963, after plastics sales had developed unsatisfactorily, Montecatini withdrew from the company and the Österreichische Stickstoffwerke acquired its share.

From the outset, ÖMV and Linz realized that the production of polypropylene alone was not enough. The manufacture of polyethylene was considered, with Hoechst as a first negotiating partner. Primarily because we could not reach agreement concerning sales, Danubia Olefin GmbH was founded in 1967. ÖMV and Chemie Linz, through Petrochemie Schwechat, each have 25 percent of the shares and BASF has 50 percent. Current production capacity is 66,000 tons per year of high-pressure polyethylene. The necessary ethylene is obtained from a steam cracker which ÖMV erected in 1960. In addition, plants for the production of oxo-alcohols and phthalic anhydride have been established in Schwechat. In 1977, annual sales of Chemie Linz AG were 7.6 million Schillings (US$500 million), half of which represents exports. Thus, Chemie Linz alone accounts for a quarter of the export yield of Austria's chemical industry.

I have described the development of Chemie Linz in some detail because it is a good illustration of how a dynamically managed company can succeed in difficult diversification. It is always an intellectual pleasure for me to negotiate with its chief, Dr. Buchner.

However, the intense activity of the European chemical industry in Austria was due not only to the formation of economic blocks in Europe, but also to its special geographical advantage as a bridge to the East. During my many trips to Vienna, which began in the 1960s, I observed with admiration that the Austrians had succeeded in maintaining a special relationship with the countries of the old monarchy, in spite of historical and political divergencies. Even if Vienna has not fulfilled all expectations as a mediator for southeast Europe, it is without doubt an excellent environment for human and commercial encounters.

Foreign Sales Only from Exports

The remaining 40 percent of Austrian chemical sales is achieved by the many small chemical companies manufacturing special products mainly for the domestic market. These domestic producers have hardly any foreign production facilities. All their foreign sales are achieved from exports. As a result, the export quotas of the large nationalized companies are far in excess of 50 percent of the total sales.

In the past, the nationalized chemical companies of Austria allowed business with the Eastern bloc to run down. Considerable new capacities have been established in these countries, enabling them to meet their own requirements to a large extent, as well as to export relatively cheaply. In addition, foreign exchange in the countries of the Eastern bloc is in such short supply that imports are strictly controlled.

This development has meant that Austrian industry, and in particular the chemical industry, is looking more often toward the West. Such trade is all the more appealing because business does not contend with commercial tariff barriers in the Western nations. At the present time, a fifth of Austria's chemical exports are destined for the Council of Mutual Economic Aid (COMECON) countries and three-fifths for countries belonging to the Organization of Economic Cooperation and Development (OECD). The largest trade partner of the Austrian chemical industry is the Federal Republic of Germany, which in 1978 provided half of all Austria's imports and absorbed some 20 percent of its exports.

Jobs from Chemicals

Since 1970, the production value of the Austrian chemical industry has grown by about 50 percent, although unevenly in the various sectors. Plastics have recorded the greatest growth rate, followed by the pharmaceutical industry and then by man-made fibers and fertilizers. The organic chemical sector, based largely on the starting materials and derivatives of oil and natural gas, is in the top position.

It seems remarkable to me that the number of chemical plants in Austria has increased by about half within the last twenty years, new companies accounting for only 36 percent of the total. Such remarkable growth is reflected in rising employment. In 1972, for example, the chemical industry employed some 63,000 people, indicating a 70 percent increase in twenty years. Compared with the 20 percent gain in employment in the rest of Austrian industry, the chemical sector is clearly forging ahead in job formation.

Looking to the EEC

The share of the chemical industry in the total output of Austrian industry is well above 10 percent, placing it in the top group. But even this performance will have to be improved in the future, especially through involvement in the EEC and its markets.

The comparatively large amount of foreign capital investment in Austria also causes problems. If foreign investments in domestic companies are made only for momentary advantage, then such cooperation will not be successful in the long run. On the other hand, developments show—and I believe Hoechst Austria can be cited in this context—that long-term cooperation between domestic and foreign companies is of considerable benefit to both partners.

An Island of the Blessed

"The leading Hoechst men like to come to Vienna," wrote Professor Walter Heinrich of the University for World Trade in Vienna in a flattering 1957 commentary on our "House in the Seidengrund." I believe not much has changed since. On the contrary, Vienna provides new fascinations beyond the attractions that the historic city of "Vienna Imperialis" has long held for visitors.

There is another decisive factor: an old admirer of the city once meditated, "He who cannot love Vienna must be an enemy of man." It is probably the many "human genialities" in which the Viennese, according to Alfred Polgar, love to engage that make them so attractive. In this sense, Vienna is surely still somewhat of an "island of the blessed."

Vienna is not all of Austria. To do justice to the country, one has to enlarge one's geographical image of the "island." This starts with the "villages under the heavens," as Grinzing, Sievering, and the other towns on the edge of the Vienna Woods are lovingly called. It continues in the sometimes still dreamy spots of the Burgenland. A trip through this wonderful countryside is an introduction to the true heart of the Austrians. On the slopes of the isle of Vienna, perhaps more truly blessed people are to be found than in the capital city. Yet, the business visitor has to leave the island to meet the ups and downs of hard competition in which there is little time for human sociabilities.

13

Where States and Plans Steer the Economy

Unfortunately I do not speak Russian, although I had ample opportunity to learn the language. My mother hails from the Baltic, and my father had lived in Russia for more than twenty years. They spoke Russian with each other quite often, especially when we children were not supposed to understand what they were talking about. Russian, therefore, had a kind of secret attraction for my brother Carl and myself. That is why I tried, for quite some time, to grasp at least the Cyrillic letters and the complicated syntax. A fresh attempt after the war also failed because at that time English and French were of special importance to me for professional reasons. Nonetheless, I did follow the developments in the Eastern countries with considerable interest. At least in theory, I was fully informed about the political and economic events that governed the fortunes of the countries of the Eastern bloc.

In 1959, I took my first trip to an East European country: Poland. My destination was Warsaw. I boarded the train in East Berlin on a cold March day. Crossing the snowy expanses of the country, the compartment became colder and colder until the heat failed altogether. Drinking the vodka that was offered on the train was the only way of warming up.

I have two distinct memories of this visit: the generous hospitality of our Polish discussion partners, which helped overcome many practical difficulties, and the successful reconstruction of the old city of Warsaw, which had been devastated in the war. Warsaw contrasts very favorably with many German cities that were reconstructed without maintaining their cultural and architechtural heritage. In succeeding years, the first visit to Poland was followed by many other trips to Eastern Europe, where it was my primary objective to familiarize myself with the economic situation of our neighbors there.

The Counterpart to OEEC

The Council of Mutual Economic Aid (COMECON) was formed in Moscow in 1949. It was the East's answer to the establishment of the Organization of European Economic Cooperation (OEEC) by the West in 1948. The members of COMECON are the U.S.S.R., Bulgaria, Hungary, Poland, Rumania, Albania (inactive since 1961), Czechoslovakia, the German Democratic Republic, the People's Republic of Mongolia, Cuba, and Vietnam (since July 1978). Although Yugoslavia is a partially associated member, it occupies a special position; because of its particular governmental and economic system, it therefore must be regarded, from a commercial point of view, as an nonaligned state. The People's Republic of China, too, has an ideologically and politically differing status.

Together, the countries composing COMECON have more than 426 million inhabitants, about 10 percent of the world's population, while a mere 259 million people live in the European Economic Community. In contrast to the EEC, COMECON is theoretically only an advisory body. The governments of the COMECON states can accept or reject its advice without violating the statutes of the organization. But political reality is different.

The common ideology and the dominating position of the U.S.S.R. make it perfectly legitimate, from the Western point of view, to talk of COMECON as a political entity. COMECON is not concerned with the liberalization of exports among the countries involved, or with a customs union, a common agricultural policy, or the free movement of capital and labor. Its sole aim is coordination: the designation of focal points for national investment and production in various sectors. This goal calls for economic plans, usually for a five-year period and carefully integrated before they are finally passed by the national bodies.

As far as can be judged, these coordination efforts have so far succeeded only in part, since they repeatedly came to grief through the autarchy inherent in the economic systems of the individual COMECON members. Nevertheless, considerable progress has been made in various sectors, and these advances go beyond the mere coordination of economic plans. Good examples are the installed and the planned pipelines for natural gas, oil, and ethylene and the resultant close supply relationships within COMECON industry, the creation of supranational industrial organizations such as Interchem (for the improvement of cooperation between the chemical industries of the COMECON countries), and finally, the realization of joint investment projects as part of the "complex" program.

Even if it is repeatedly disputed, there is no doubt that the Soviet Union is trying, through these programs, to consolidate integration within COMECON and to pursue its own economic aims. The most striking example of this is the Soviet policy regarding the supply and price of oil. Through this policy, every COMECON country that is poorer in raw materials than the U.S.S.R. will become largely dependent on the Soviet Union.

Commerce Is Growing

The total foreign-trade volume of the COMECON countries is estimated at more than DM450 billion (US$195 billion). Barter trade between the COMECON countries amounted to about DM260 billion in 1977 (US$110 billion). The corresponding figure within EEC is about DM890 billion (US$385 billion). This comparison alone shows the difference in function between the commerce of East and West. Insufficient flexibility in the barter trade within COMECON is the real reason why this organization does not even approach the trade volume found within the EEC.

Barter trade between the EEC and COMECON amounted to DM75 billion (US$32 billion) in 1977. Of this, 46 percent was the share of the Federal Republic of Germany, the East's biggest trade partner. Even compared with other regions, this volume is by no means negligible.

East-West Trade Has Its Problems

These figures cannot hide the fact that the development of trade with Eastern Europe has long been fraught with problems for Western companies. For one thing, the export goods scheduled to provide the East with foreign currency for its necessary imports are agricultural and thus come up against the protectionist agricultural policies of the EEC. Also, the range of industrial goods that the East could or would export has so far been rather limited.

It has therefore always been difficult to attain a balance of payments. The two years 1976 and 1977 have shown how hard the individual COMECON countries are trying to improve their negative trade balances with the West. More than 65 percent of German exports to the East are made up of capital and consumer goods, while almost 50 percent of imports consist of foodstuffs and energy.

The chemical industry enjoys a high priority with its economic partners in the East. Unfortunately, there are no official figures on chemical plant exports. It can be assumed, however, that a considerable part of the machinery and plant supplied by West Germany is allotted to chemical production. We also gather from their economic plans that all the COMECON countries plan greater production increases for the chemical industry than for industry as a whole. Moreover, the chemical industry, with its 16 percent share of total exports from the Federal Republic, is the biggest sector after capital and consumer goods.

Another peculiarity is that since the mid-sixties, the East has tried to buy from Western countries the necessary technologies for the development of its own industries. Similar procedures are followed on a large scale by Western concerns, which usually have the assistance of engineering companies to erect the appropriate plants on the selected sites, sometimes to the turnkey stage.

Cashless Payment with Question Marks

The tricky issue in all transactions involving plant is always the question of payment. Since COMECON countries suffer from chronic shortages of foreign currency, there is a constant search for new ways of settling accounts. However, cooperation has been more talked about than practiced. When our Eastern neighbors talk about technical and economic cooperation, they have the idea that a plant should not only be erected with the aid of bank credits, but that it should also be largely refinanced with products from the same plant.

Western companies have learned to appreciate the difficulties of such procedures. They will not readily agree to be paid for plant with their finished products. Who can tell in advance whether, and to what extent, Western markets will be able to absorb those finished products when the supplies arrive at a much later date? But in each case, attempts should be made to find just solutions for both partners. This calls for a certain measure of mutual understanding, responsibility, and flexibility. At any rate, I remain confident that the contacts developed since the beginning of East-West trade will increasingly facilitate the solution of problems.

Since January 1975, the mandate for commercial agreements with COMECON countries has passed into the hands of the EEC authorities in Brussels. Only agreements for economic and technical cooperation remain within the competence of national authorities. The Federal Republic of Germany viewed the development of trade with the East with much sympathy even before the new Eastern policy was adopted. But it rejects all attempts to subsidize this trade. The Federal Republic offers German exporters only state guarantees that secure export credits for business with the COMECON countries, the transactions being conducted by the state-owned Hermes insurance company. Other nations, such as France and Italy, go much further in this respect. In view of the high indebtedness of the COMECON countries to the West, especially since the beginning of the seventies, it would be desirable for the industrialized countries of the West to agree to a coordinated policy on credit and interest in trading with the East.

Is the East in Hock?

I do not consider it wise to dramatize the indebtedness of the East, as is done at times in public discussions. Even though the absolute level of the COMECON debt at the end of 1977 was estimated at approximately US$50 billion, the amount should not give rise to undue concern, considering the economic potential and raw material resources of the two main debtor nations, the U.S.S.R. and Poland. Furthermore, estimates concerning the indebtedness of the COMECON countries must be regarded with reservation. Our Eastern neighbors would gain from a little more candor in this respect.

The Name of the Obstacle Is Bureaucracy

Let us now turn from generalities to Hoechst's Eastern bloc affairs. The enormous market of the Soviet Union with its 260 million people— roughly equivalent to the total population of the EEC—is, of course, of interest to us. Naturally, the foreign exchange situation and the increasing commercial interlocking of the COMECON countries seriously restrict normal business. In addition, there is the opacity, the parallel and conflicting jurisdictions within COMECON's political and economic administration.

Conducting business with the East takes a great deal of time and patience. Many transactions fail because the necessary cooperation among the various government offices in Moscow is not achieved, or not achieved in time. For example, we have been trying for many years, in vain, to set off our regular, extensive purchases of Kola crude phosphate for our works in Knapsack and Flushing against the products exported to the Soviet Union.

Hoechst also faces the problem of how best to conduct business when such severe restrictions are still imposed on personal contact between the producer or supplier and the consuming industry. The establishment of our advisory office in Moscow in 1971, the first by any company from Federal Germany, was a positive step. The function of this office is to advise and look after visitors from Germany, to follow up contacts, and to negotiate with Soviet authorities.

The facilities that we had rented originally in the Hotel Ukraina soon proved inadequate. In February 1975 we were able to move into another office, where we have about a dozen people today. Five are from the Federal Republic, the rest are Russian. We hope that this link between Hoechst and the U.S.S.R. will contribute to the consolidation and expansion of our promising business relationship with that country.

The organization of the Hoechst agencies is not the same in all the Eastern bloc countries. For instance, after corresponding laws were passed, we opened our own technical advisory office in Poland in October 1977. In Rumania we have had such an office for many years, with a completely Rumanian staff. But in most cases, the official representation of a Western concern is handled by one of the state-owned companies established specifically for this purpose. In the instance of a large corporation such as Hoechst, a special department is set up and deals only with that company's business. Our people, especially our expert technicians, can visit these state-owned agencies, which aid them in discussions with export companies, factories, and institutes.

In Yugoslavia, we can send permanent delegations to these companies. Our people in those delegations have official work permits enabling them to act as advisers and to establish closer links with all the various departments of the parent company. In a large industrial concern, this is often both difficult and crucial. These delegates also establish communication between Hoechst visitors and the offices of the company's business partners, as is sometimes essential for the transaction of business or its follow-up.

In political discussions, we sometimes ask ourselves whether it makes sense to promote the sale of plant as part of our trade with the Eastern bloc, when this can be done with the help of large Western credits. Our company's policy has always been to realize to the full any opportunities to trade with our Eastern neighbors. Often, as in the period after the currency reform, this approach called for personal sacrifices by those concerned with the promotion of such business. But I believe it was well worth the effort. Long before our foreign policy resulted in a rapport with the East, we had made our contribution through economic cooperation to what I believe to be reasonable, neighborly relations.

In the Shadows of the Kremlin

In the spring of 1969, the first sizable Hoechst delegation traveled to Moscow at the invitation of the State Committee for Science and Technology. Professor Winnacker had been invited to give a lecture at one of Moscow's universities on the subject of present and future chemical problems. For this occasion a large audience, as interested as it was sympathetic, had come from plants and offices all over the Soviet.

During this stay, our delegation visited the ministries with whom we already were acquainted. First among them was the chemicals ministry, which had long been headed by Leonid A. Kostandow, a man of international note. We had a lengthy and interesting talk with him about the extent to which the Federal Republic, and in particular Hoechst, could participate in the U.S.S.R.'s expansion plans for its chemical industry. We found a whole panorama of possibilities.

A conversation with Jermen Gvishiani, the vice-president of the State Committee for Science and Technology, was also productive, and we later concluded a cooperation agreement with the committee. It led to an active exchange of scientific views between delegations from the U.S.S.R. and from Hoechst. It also enabled us to provide technical advice to some of our customers in the Soviet Union and to those who processed the products we supplied to the U.S.S.R.

Since then, the Hoechst staff has regularly visited factories in Moscow, Leningrad, Kiev, and other cities. Together with Russian scientists and plant managers, we carry out practical experiments and evaluate the results jointly. This, by the way, is the only successful method of introducing new products to the Soviet market.

Side Trips into a Large Country

The casual visitor to the Soviet Union often knows solely Moscow and Leningrad. Other perspectives of this huge country are opened up only when its vast interior can be explored.

We traveled on board a twin-jet Ilyuschin of Aeroflot, the largest commercial air fleet in the world. Our group, accompanied by two representatives of the chemical ministry, journeyed around for about a week. Our main stops on this trip were Kiev, Tschimkent, where Uhde had erected a large chemical plant that was just being expanded, Samarkand with its relics of Islamic culture reminiscent of Isfahan, and Tiflis.

From Kiev, we traveled by hydrofoil up the Dnieper to visit a large chemical plant in Tscherkassy. On our return journey, the minister who accompanied us invited us to a meal in a guest house of the Ukrainian government. During our lively conversation, he suddenly became very thoughtful and said, "In the area around here, thousands of Russian and German soldiers died in the encirclement battles of the last war. That must never happen again between our two peoples."

One central fact should be fully recognized when considering the Soviet Union. Within sixty years, that nation has developed from the agrarian country of the Tsarist empire to the second largest world power of our time. That is, without question, an extraordinary achievement.

Agriculture still accounts for a fairly high share of the nation's output, about 18 percent of gross national product, compared with 3 percent in the United States and West Germany. The opening up of Siberia continues to be one of the most important projects of the Soviet Union.

Whoever visits some of the new industrial complexes that have been developed over the last few years with the technical and financial help of Western industry gets an idea of the U.S.S.R.'s enormous effort to industrialize its agrarian economy. It simply cannot be denied that visible progress has also been made in supplying consumer goods. Also remarkable is the growing number of modern skyscrapers, which are giving an increasingly metropolitan character to Moscow's silhouette.

One of the Soviet Union's central and lasting problems is how to establish a highly developed industry and at the same time satisfy rising consumer demands in the absence of a market economy and under strict control from the top.

A Wealth of Raw Materials

The chemical industry reached significant proportions at the beginning of the 1960s. Today, the fertilizer sector in particular has gigantic industrial capacity, and in this field, the U.S.S.R. has become one of the world's leading producers. Other giant plants for starting and intermediate organic products, for plastics and man-made fibers, are either planned, under construction, or completed.

The strength of the Soviet economy lies in its wealth of raw materials. It is among the world's largest producers of many starting materials for petrochemicals and the chemical industry. Lignite and coal reserves alone are estimated at 60 to 70 percent of world totals. Oil and natural gas deposits are probably around 11 billion tons, or 23,000 billion cubic meters. This is about 12 percent of presently known oil reserves and 36 percent of natural gas reserve. Most of this wealth is located east of the Ural mountains. Production and processing therefore require large investments in a comprehensive infrastructure. The natural gas and oil currently produced is needed primarily in the country itself. This again is an opportunity for long-term development of the barter trade between East and West.

The last stop on our trip was Leningrad, a city of waterways like Stockholm or Venice. It is without doubt the Soviet city with the greatest attraction for visitors from the West. This appeal is due not only to the cultural wealth of the art collections, such as in the Hermitage Museum, the envy of many a Western museum. It is also the atmosphere that radiates from this city, rebuilt in its former late baroque splendor. Leningrad gives an impression of openness, and for many visitors it is the gateway to Russia.

The Machinery Specialists of COMECON

I visited Czechoslovakia, our nearest Eastern neighbor, before I went to Russia. The Czechoslovak republic is among the most highly industrialized countries of Eastern Europe. Its considerable potential, however, lies not in its raw materials, which are largely imported from its neighbors—crude oil for example, from the U.S.S.R. Rather, it is based on a well-developed infrastructure and a hardworking, technically gifted population that can draw on the cumulative industrial experience of many decades. It is hardly surprising, therefore, that Czechoslovakia's exports reached a considerable volume soon after the war.

Czechoslovakia is the machinery specialist of COMECON. But its chemical industry still has to make up much ground. Before the war, the chemical plants in Aussig produced a respectable range of organic and inorganic products as well as dyestuffs. At that time, they worked closely and cordially with the I.G.

In recent years, Czechoslovakia has become a tourist country again, impressing West European visitors with a wealth of monuments that are part of the history and the artistic development of central Europe. Prague, the Golden City, the City of a Thousand Spires, is particularly rich in art and history: St. Veit's Cathedral, the Karls Bridge, one of the oldest universities in Europe, the Hradjin. Together with well-preserved streets and lanes, all this combines to give the impression of a medieval city that has hardly suffered from the war and has a beauty that is rarely matched.

Elemental Sulfur from Poland

The resentment of the Polish population toward the Germans can easily be understood. However, it has not prevented the establishment of close economic ties between the two peoples, not even during the first postwar years. Poland has made giant efforts to promote its industrialization. This also applies to the chemical industry, which produces inorganic chemicals and fertilizers, plastics, organic intermediates, and dyestuffs. Poland is an important supplier of these products within COMECON.

In addition, elemental sulfur has become of utmost importance for the Polish export economy. While only ten years ago Poland's share of world production was only 4 percent, its present-day share of about 5 million tons, or 27 percent, has made it one of the leading world producers.

Not so long ago, coal reserves, in the order of about 40 billion tons, were discovered in eastern Poland in the vicinity of Lublin. With energy costs soaring, this is a particularly important resource. It will no doubt be exploited by the chemical industry, particularly through coal gasification.

The Poznań Fair has long been an important meeting place for the international chemical industry. Like the Leipzig Fair, it offers a showcase for presenting new products and developments from East and West to a wide range of consumers. To Hoechst's technicians and experts, Poznań therefore offers welcome opportunities for advice and service to customers.

Poland, too, faces the problem that her exports do not balance her import requirements. As a result, there is a high foreign debt, which demands great flexibility on the part of the West if mutual trade is to be developed or maintained.

Rumania's Oil Riches

Even before the war, Rumania was a rich oil country, and its refinery in Ploesti was one of the oldest in Europe. In the meantime, the chemical industry of the country has so developed that Rumania can meet only two-thirds of its crude oil requirements from domestic production. The rest has been purchased from Arab and Iranian sources.

Links between Hoechst and Rumania were reestablished in the early fifties. We have learned to appreciate the Rumanians as efficient business people and inventive industrialists but have also found them to be tough businessmen who defend their interests with great ingenuity.

Rumanian plans for the chemical industry during the last decades were perhaps too ambitious. Only a portion could be realized. Nevertheless, in cooperation with Western companies, numerous modern industrial complexes have been established. Our subsidiary, Uhde, has constructed several ammonia-urea installations, an oxo-synthesis plant, and a polyester fiber complex in Iasi, in the northeastern part of the country. One of the major stages went on stream in 1970. That event was celebrated at Bucharest in the presence of the chemicals industry minister, Mr. Florescu, and many representatives from technology and science.

In the mid-fifties, I traveled to Bucharest as a member of a Hoechst delegation that included Drs. Winnacker, Gaebel, and Rossow. The subject of our discussion was industry cooperation between Rumania and the Federal Republic. On a more recent trip in 1974, this time with Dr. Sammet, we were able to note with satisfaction how well this cooperation had developed in almost twenty years.

In the course of those years, there have been many talks with Rumanian politicians about joint projects, including the feasibility of financial participation by Western industries in state-owned enterprises. But I know of only one case in which a West German company was actually involved in this type of cooperative project.

When we gained control of Stollack and Vianova as the result of our acquisition of the Turnauer group in Austria, both companies had considerable holdings in Eastern Europe. They had, for example, constructed a paint and synthetic resin factory in Rumania which was obviously operating satisfactorily. This venture highlights the problems that attend the construction of new industries in former user countries. Stollack lost one of its important markets for finished paints because Rumania now buys only those special products she can not yet manufacture herself. There will be additional difficulties when the finished products from the newly erected Rumanian plants are exported to neighboring markets.

Hungary's Wealth: A Many-Splendored Thing

Tradition and an eventful historical past find their many-faceted expression in Hungary's capital, which offers visitors one of the most beautiful city panoramas of central Europe. Budapest came into being in the middle of the last century when the independent cities of Buda, Pest, and Obuda were merged into one. Of Hungary's 10 million inhabitants, more than 2 million currently live in Budapest. Because of its numerous architectural and historical monuments, this city exerts a particular attraction for tourists. Art and culture of the past survive in such impressive buildings as the church of Saint Matthew, the parliament building, Budaer Castle, and many other historic structures.

Although Hungary is not rich in raw materials, it has famous wines which help redress the balance. Tokay is known the world over, as are the excellent dry white wines that constitute about 60 percent of Hungary's wine production.

The chemical industry accounts for more than 15 percent of Hungary's total industrial output. The manufacture of pharmaceuticals is a tradition that earned Hungary fame as the "pharmacy of the Eastern bloc." The textile and clothing industry, too, occupies an important position, both in relation to the national economy and to COMECON as a whole.

There are long-established relationships between Hungarian and German chemical industries. Hungary is a customer of Germany's man-made fiber industry, importing knitted and woven garments made from synthetic materials. Of countries outside the European Community, Hungary is one of the largest exporters to the Federal Republic. Cooperation has increased recently between West German and Hungarian factories, research institutes, and authorities, nurturing the hope that even closer relationships will be developed in the future.

Bulgaria: Attar and Industry

Bulgaria regards Russia as its liberator from Turkish rule during 1877 and 1878, and from the German occupation in this century. Such historical experiences are understandably the ties which bind Bulgaria to her large and powerful neighbor in economics as well as politics.

The world knows Bulgaria mainly as the country of attar, a fragrant oil usually made from rose petals. This precious product is an important export, and it is considered pure and unadulterated only if it carries the state's official seal of quality.

Bulgaria is about to implement an extensive industrialization program, designed in accordance with socialist ideology, to complement the country's agrarian economy. Examples of this are the industrial-agricultural complexes. The two largest, in Burgas and Dewnja, employ 15,000 people and cultivate up to 50,000 hectares of land. Production begins with crop cultivation and livestock fattening, and it ends with the canning of vegetables and meat. There are eight such collective farm units, a type of vertical organization.

Burgas also houses the largest refinery in Bulgaria. From our point of view, cooperation with the Bulgarian chemical and agrochemical industry is beneficial. Furthermore, the government's attitude to industrialization is more realistic here than in many neighboring states.

The Other Germany

World War II and its consequences severed innumerable human and family relationships between central and western Germany, as well as between industrial associates and their connections. But on both sides, people who had formerly worked together tried to reestablish old economic contacts, even when prospects of normalization still seemed very remote. In the first years after the war, barter was written in capital letters.

With the consolidation of economic and political conditions, an interzonal trade agreement, concluded in 1951, became the basis of intra-German trade. The "swing," an interest-free overdraft granted by Bonn to businesses in East Germany, and the duty-free transit of goods were of great help in getting this barter trade started. It has since grown to considerable proportions.

There were years in which personal contact with the users, the production plants in the German Democratic Republic, was greatly restricted. The reason was not only the construction of the Berlin Wall in 1961. There was also an export monopoly so strictly enforced in East Berlin that it was virtually impossible to discuss mutual business anywhere except at the Leipzig Fair. Not much has changed in this respect, although the autumn fair in Leipzig has become an important international meeting place for the chemical industry of the East and the West.

Within this framework, new developments in intra-German trade are initiated from time to time. In the recent past, extensive transactions, especially in the industrial plant business, have been completed in this way. Hoechst and its engineering subsidiary, Uhde, have been very active in this field. In 1975, for example, an agreement was concluded for the construction of a chlorine-alkali electrolysis plant with subsequent vinyl chloride and polyvinyl chloride production. A polyester filament plant went on stream in Guben as early as 1973.

The chemical industry of central Germany has always been important. In the former chemical triangle of Leuna-Buna-Bitterfeld, large new lignite conversion plants were constructed even during the war, with synthetic gasoline facilities being produced at Leuna and Buna rubber at Skopau. Since that area has no raw materials except lignite and potassium, it was hardly surprising that acetylene chemistry became very important there. Based on the close links with the COMECON countries and on the pent-up demand of this economic area, increasing efforts are being made by the German Democratic Republic (East Germany) to effect cooperation with companies from Western Europe and Japan. Apparently, this is necessary for the country's industrial expansion and modernization. It can only be hoped that such growing economic cooperation will also contribute to the normalization of human and political relationships between the two Germanies.

Yugoslavia: Balancing between East and West

In 1918, a new state was created from a conglomerate of peoples in the southeastern part of the European Continent. On November 29, 1945, this state was renamed the Socialist Federal Republic of Yugoslavia. The Yugoslav policies of nonalignment and labor autonomy began only after the country's ideological break with the Soviet Union in 1948. Society, politics, and the nation's economy have since borne the stamp of its distinctive system.

It is well known that Yugoslavia is eager to have good relations with all nations without aligning herself politically with any particular group. This balancing act has obviously been successful. The country has, in fact, become a member both of COMECON and an associate member of the Organization for Economic Cooperation and Development (OECD). Yugoslavia does not wish to be associated with the EEC, but aims at maintaining good contacts with the West. This special position between East and West affords Yugoslavia a significant and interesting role in global politics.

Labor autonomy is the key to the country's internal policy. In accordance with the constitution of 1974, the working people are to administer, or help to administer, all aspects of society. This tenet applies to the economy, social services, and communal policies. Already, it is primarily the workers in the factories whose decisions determine investments, wages, production programs, company management, and many other issues. They also bear the risks of their decisions.

The state intervenes as little as possible in this autonomous system of administration. The same applies to the framing of the nation's economic plans. Unlike such plans in the COMECON countries, the social plan in Yugoslavia is conceived in very general terms and might be described as a summary of guidelines.

Whether the system of labor autonomy has stood up to international comparison, and will continue to do so, is a question as yet without an answer. It is a fact, however, that industrial development has surged ahead in recent years. The population, at any rate in the northern regions, has achieved a remarkable standard of living. This was evidenced by the comprehensive display of consumer goods that I recently saw while strolling through Belgrade's shopping center. One establishment deserving mention is the Beogradanka, Belgrade's tallest and most modern building, whose lower floors accommodate a store with goods ranging from slivowitz (potent plum brandy) to automobile parts and plows. I had a quiet chuckle when in some shops I discovered shirts bearing the Trevira label.

With respect to barter trade with the West Germany, Yugoslavia is in third place among the Eastern European countries, preceded by the U.S.S.R. and the German Democratic Republic. West Germany is Yugoslavia's most important Western trading partner. In exports, Yugoslavia is oriented more toward the West than the East, the ratio being something like 60 to 40.

The chemical industry occupies a leading position in exports as well as in the medium-term planning of the country. Although the majority of chemical products has still to be imported, it is expected that by 1980 large capacities, especially in plastics and man-made fibers, will be created in Yugoslavia itself. To promote this development, the state has supplemented the principle of labor autonomy with a law for foreign investors, under which foreign capital can be invested in domestic production companies.

Hoechst is one of the pioneers of such joint ventures. Many years of cooperation have shown that the country gives foreign partners a fair chance, despite certain legal limitations. The positive development in relationships between Hoechst and the Yugoslav market is largely the work of Yugohemija, the Belgrade company that looks after the interests of Hoechst. It celebrated its twenty-fifth anniversary in November 1977. Since its first agency agreement with Hoechst is almost as old, this memorable celebration represented a milestone for Hoechst as well. Five German staff members from Hoechst work in Yugohemija, a sign of mutual friendly relationships.

Cooperation and hospitality are attributes that also apply to our partners in the joint ventures. There is, for example, Iplas in Koper, a small town on the Istrian coast not far from the Italian border. Together with Iplas and Yugohemija we founded Polisinteza, a company that produces Hoechst synthetic resin dispersions. A short time ago, Koper was simply a fishing village. Today, the town has one of the largest harbors in Yugoslavia and is about to become the most important industrial center of Slovenia.

Zrenjanin in the Vojvodina region, known as Yugoslavia's granary, is also a rapidly developing town. It is the site of the agricultural combine Servo Mihalj, with whom we have cooperated for more than twenty years. But the outstanding achievement to date is the establishment of a plant for Hoechst pharmaceuticals through another joint venture, Yugoremedija. In this way, pharmaceuticals have become an important facet of the Zrenjanin mosaic.

The combine there not only engages in agricultural and livestock farming but also processes its products. Its showpiece is the modern slaughterhouse, which exports a large part of its preserved meat products to the United States. The Yugoslavs have always been very hospitable, but the local produce that Zrenjanin offers its visitors, including the slivowitz from its own distillery, is outstanding and memorable.

First Stop in China

It was the autumn of 1972 when the Air France Boeing 707 came to a halt in front of the modern arrivals building of Shanghai Airport. Our small delegation soon shook off the fatigue of the long flight from Rangoon to Shanghai. The formalities were simple and courteous. There were no crowds. In China, flying is a privilege, and the few aircraft, usually of Russian origin, are always fully booked. We were received by a committee of the appropriate state company. The reception was as friendly as at all the other stops on our trip.

In 1972, diplomatic relations between China and the Federal Republic of Germany had just been reestablished. They were therefore a timely subject in the discussions with our Chinese partners. The hope was expressed often that the economic relationships of the past would now be revived. It was considered taboo to talk about Chinese-Russian relationships, and the topic of personal life in China, such as family and school, was also rarely raised.

Chinese policy originally aimed at building the economy through the country's own efforts and at remaining independent of the world abroad. This was the reason for the modest volume of China's export trade. For a long time, China had a fifth of the world's population and only 1 percent of world trade.

In recent years, this policy has changed considerably, especially at the expense of the COMECON countries. In the 1950s, about three-quarters of China's imports came from socialist countries. By now, the industrialized countries of the West have taken over most of this share. West Germany's share in these imports is high. Today, it is the second largest trading partner of the People's Republic of China after Japan.

Hoechst returned to the Chinese market quite early and established connections with export organizations there. Uhde has already built a number of chemical plants. The first, an acetaldehyde installation in Shanghai, went on stream at the end of 1976. A vinyl chloride plant followed at the end of 1977.

According to our present knowledge of the new five-year plan, China is going to make a major effort to create a largely independent industry and economy by the year 2000. However, this growth will hardly be possible without further imports of Western technologies and products. China must therefore mobilize its huge raw-materials potential over the next few years and export a great deal more in order to earn the foreign currency needed for its own imports. At the same time, the basic principle of self-reliance is not likely to be totally abandoned in China's foreign trade policy.

The Attractions of Chinese Cuisine

It is impossible to write about a visit to China without mentioning the perfection and variety the Chinese cuisine offers to the traveler. The hospitality of Hoechst's Chinese partners was simply overpowering. There was not a single lunch or dinner at which we were not introduced to some new regional specialities.

We were pleased that our hosts accepted our return invitations, which we had carefully arranged with the help of experts in Chinese cuisine. This, incidentally, was not very difficult, because there is truly no shortage of gastronomic possibilities in the large cities that we visited—Shanghai, Peking, and Canton. We had the impression that good cuisine is one of the things that the Chinese value even today. It is offered at prices that even the modestly paid working population can afford.

Business discussions recessed, our hosts gave us some insight into the history and cultural riches of their country as well as the life of modern China. They did this with the same kindness and care that had gone into drawing up the program of our visit.

Two Examples of a School System

At the time of our visit, it was not easy to convince our hosts that we wanted to visit a Chinese university. Professor Winnacker's fervent wish was fulfilled in Peking, but our impressions were rather mixed. The university buildings obviously went back to the thirties and had been inspired by the campus system of American universities. But teaching and research buildings, as well as the students' dormitories, were largely empty. We were told that China's university system was being subjected to a thorough overhaul. In the future, university places would be reserved primarily for the children of the working population. It was also emphasized that undergraduate education would be kept brief, lasting four years at the most. Thereafter, the students, whatever their field, would be discharged to gain practical experience. Only after the students had proven themselves in their chosen fields would they be allowed to return to the university to complete their studies. We had a most interesting discussion about this approach with a number of professors. Although quite outside our Western traditions and experience, it seemed to us, at any rate at first sight, that such a program of study, related directly to practical conditions—especially in the sciences—would produce an academic generation unlike its counterpart in Germany. Certainly it would not need a long period of time after entering industry to jettison its excessive academic ballast.

Dental Treatment under Acupuncture

In Peking, we had an opportunity to join a tour of one of the large clinics. After an explanation by the chief physician, we witnessed five demonstrations of how acupuncture is used in anesthesia prior to surgery. First we saw some teeth removed under acupuncture anesthesia with no problem at all. Then we witnessed an operation in which a woman's gallbladder was removed. In this case, too, acupuncture appeared to afford almost complete anesthesia. We did, notice, however, that with other patients, suggestive persuasion by the doctors and a change in the needle arrangements were necessary to render the operations entirely painless. As a matter of fact, during our discussion, the doctors freely admitted that up to that time there was no scientifically based explanation of the mechanism of acupuncture, that only practical experience guided its application. They explained to us that although acupuncture was successful as an anesthetic, one had to allow sufficient time for preparing the patient psychologically for the operation and for this type of anesthesia. In the meantime, the research institutes of the country are working intensively to discover the scientific basis of acupuncture.

A China Showcase

We concluded our trip with a visit to the Canton Fair, where we encountered a world totally different from Shanghai or Peking. The exhibition halls offered space for the exhibition booths of foreign traders and customers, but predominantly, they displayed the exports China had to offer. These were largely products of the Chinese crafts and textile industries. There was a tremendous selection of ivory carvings, wood and bamboo work, and carpets of traditional and modern design. The Canton Fair is regularly visited by Western countries, which also exhibit there, usually through branches in Hong Kong. The Hoechst Hong Kong company too has a special China department, which is staffed by Chinese and Europeans versed in the language. They maintain contact with their Chinese discussion partners not only during, but also before and after, the Fair. This is essential to concluding contracts during the period of the Fair.

The tactics employed by the Chinese buyers put quite a strain on the staff members who spend several weeks in Canton. Quotes are invited, but the qualities of the products are generally not clearly defined. After the offer has been made, the Chinese often wait until shortly before the Fair closes before they accept or reject the offer.

In Canton, as in Hong Kong, one has one's finger on the pulse of China's economic life. For those who feel at home in this world, respect for Chinese custom is almost a passion. Even if great reserve is shown on both sides, some human contact is established, which may well be important at a later date. Naturally, in this environment our Chinese people from Hong Kong play a crucial role. The European who participates in the Canton Fair is restricted in function to keeping in touch with the parent company in Europe, and to ensuring that acceptable offers are made within the framework of the possible. This is by no means easy, because centralized China exploits its privileged buying position thoroughly, both in Canton and elsewhere.

Pivot for the China Trade

Hong Kong, where the Bank of China and branches of the import and export authorities can operate unfettered, is the great distribution center for the China trade. This spatially limited city-state with more than 4.5 million inhabitants has grown rapidly through heavy migration from the Chinese mainland. Theoretically, Hong Kong will be a British crown colony until 1997, but its future is already governed largely by China. It seems that Peking will be content for some time to come to allow the city, with its active commerce and industry, to retain its present status.

It is an enormous achievement of the autonomous Hong Kong administration that it has been able to create housing and, to a large extent, also jobs for the many people who have come from the mainland. The backbone of its industry, which employs far more people than commerce or communications do, is textile and plastics processing. Hong Kong is no longer one of the regions with the lowest wages. Large trading concerns that have more or less permanent headquarters in Hong Kong now divert their large orders to countries with even cheaper labor, such as Korea or Taiwan.

Hong Kong has so far adapted well to these changes. It cannot be denied, however, that the economy of this busy center is very dependent upon international economic conditions. Indeed, it would be no exaggeration to claim that the textile industry of Hong Kong has become a kind of barometer for trade prospects in many parts of this field.

Eastern Trade and Public Opinion

What can we learn from these considerations of the economic relations between East and West? First of all, there is a lack of market research data. Not only is direct business contact with end users limited in the countries of the East, but the economic statistics of this part of the world are not clear enough to provide real information.

Eastern trade, and consequently cooperation with the East, cannot be divorced from political ambivalence—certainly not in the Federal Republic. In West Germany, there are many conflicting jugments. On the one hand, the Federal Republic, like the German Reich before it, is regarded as the traditional bridge between Eastern and Western Europe. A leading position in trade with the East is thus presumed, sometimes a little too readily. On the other hand, there is concern that large-volume trading with the Eastern countries might one day result in an alarming dependence on them. Occasionally, other concerns are voiced as well. We are supplying more and more plants and machines to the East. Is there not a danger that competition from the products manufactured in those plants will eventually present Europe with greater problems than it may care to have?

Although West Germany's bridge function is confined to economic matters, such economic relationships are bound to result in a certain degree of dependence on both sides. But there is also one other fundamental consideration: In 1975, when East-West trade reached its highest figures, the Federal Republic's exports to the Soviet Union were only 3 percent of its total exports. Exports to all the Eastern bloc countries combined amounted to less than 8 percent of the total exports of West Germany.

These figures do not, to my mind, reflect any significant dependence. Moreover, growth rates like those experienced from 1972 to 1975 are hardly likely in the future. Indeed, in 1976 and 1977, German exports to the East were already beginning to fall off slightly. One must remember that the Eastern bloc continues to demand our highly developed technology. Surely this fact puts the recipient under a greater obligation than to merely to fulfill the supplier's desire to obtain additional raw materials in return for this technology.

However, as far as the German chemical industry is concerned, it has never fought shy of fair competition with imports from the East, nor will it do so in the future. Naturally, the industry will have to protect itself against the kind of dumping that we have witnessed occasionally.

The East European countries, on the other hand, should remember that compensation and barter deals of the kind increasingly demanded are not a suitable way of expanding trade between East and West. Especially in question are "buy-back" arrangements in which chemical plants are paid for through their own chemical products. In this connection, an appeal to reason must be made, not only to our Eastern trading partners, but also to companies in the West.

It is highly desirable that the governments of the industrialized Western countries reach agreement among themselves about their economic policy toward the East, that they create as uniform a framework as possible for international competition in trade with the East. It should be recalled that the final KSZE document prepared at the Helsinki Conference of 1973 represents a jointly developed basis for all future forms of cooperation between East and West.

The fear that we are facilitating the speedy rearmament of communist countries because our credits partially relieve them from providing consumer goods has never seemed a valid argument to me. Surely, we have witnessed often enough that the private sector in those countries has to subordinate its requirements to the dictates of policies. Does anyone seriously believe that we could exert the slightest influence on this policy, let alone on the speed of rearmament, by restricting cooperation and trade? I, for one, do not think so.

14

New Power Regions in the Far East

An Opel Rekord was the only office car that Hoechst had available in Japan in 1954. It was in this car that Dr. Gustav Bunge, the first Hoechst delegate to Japan, collected me at the Tokyo Airport. The postwar boom was then in full swing, and this spirit was reflected in the Ginza, the entertainment quarter of Tokyo. Thousands of colored neon signs in strange, ornament-like letters lit up the district. One could not fail to notice that full employment had created a high standard of living for everybody, in welcome contrast to most of Japan's neighbors. A strong, independent economy had been established, an economy which was to prove itself resistant to successive crises.

While Dr. Bunge was driving me to the Imperial Hotel, we tried to sum up the position that Hoechst had attained in Japan in 1954. In figures, it looked fairly respectable. At that time, we had sales of DM2.1 million (US$525,000). But this figure bore no relationship to the purchasing power of Japan's 100 million people. Also, we did not yet have a properly developed organization, only some modest beginnings.

Japan is not only the second biggest market among the industrialized countries of the world; it is also by far the most difficult. This can be appreciated only by gaining some perspective on the earlier economic history of the country.

In 1854, the American fleet forced Japan to open itself to the outside world after 300 years of isolation. Treaties of trade and friendship, signed first with the United States and then also with other countries, soon led to closer contact between Japan and the rest of the world. But the real industrialization of Japan started only after the collapse of the Tokugawa shogunate at the beginning of Emperor Meiji's reign in 1868.

Then Japan's scholars and rising young men were dispatched to America and Europe to acquaint themselves with the legislation and industry of various other countries. In addition, foreign experts were invited to Japan to help it accomplish in decades what other nations had achieved in the course of centuries: the country's transformation into a modern, industrialized state.

The success of this venture can be gauged from the realization that by 1918 most of Japan's industries were standing on their own legs. It took the chemical industry somewhat longer to achieve this, but its progress was nonetheless remarkable in view of the fact that Japan owns virtually no sources of raw materials. Only lime and sulfur are present in sufficient quantities. There is some, but not much, coal, which is of such low quality that only part of it can be turned into coke.

In this situation, the Japanese chemical industry profited from political developments. Japan turned its attention to neighboring Korea and China, including Manchuria, which had all the necessary raw materials. Following the Japanese-Chinese conflict in 1894–1895, Japan secured interests in Korea, which it started to administer in 1906. In 1910, Korea was integrated into the Japanese Empire. After World War I, the Treaty of Versailles transferred the Sino-German leasing rights over Tsingtao to Japan. In 1932, when Manchuria was declared an "independent" state under Japanese control, Japan fulfilled its old desire of securing China as both a source of raw materials and an economic market.

At the beginning of the century, the first hydroelectric installations were constructed. With their aid, carbides, calcium cyanamide, and caustic soda were produced. In 1909, for example, the Nippon Chisso Hiryo company, founded in 1906, took up the production of calcium cyanamide in Minamata under a license granted by Frank Caro in Berlin. This company's name, incidentally, as well as the township of Minamata, became notorious in the 1960s in connection with problems of environmental protection. The coke production plants especially set up for the steel industry, which also blossomed at the beginning of the century, yielded tar products and thus provided a basis for the dyestuffs industry.

When foreign and especially German supplies of chemical products were cut off during World War I, Japan set up plants for sulfuric acid, alkali and electrolysis, dyestuffs and celluloid. The first experiments to produce synthetic ammonia were performed in 1921, again by Nippon Chisso Hiryo in Minamata. In 1928, the Sumitomo group, active in copper mining, succeeded in breaking into the chemical industry. Under license from Nitrogen Engineering, U.S.A., it was able to begin the manufacture of this ammonia. The oldest chemical company in Japan is Nissan Chemicals, formed in 1887.

The rayon industry was established in the mid-thirties. Shortly before World War II, it was joined by a broadly based nitrogen industry. The former I. G. Farbenindustie provided the technology and start-up technology for five large nitrogen plants alone. Acetylene derivatives, such as acetic acid, acetic anhydride, and ethyl acetate, were also produced for the first time before World War II. Large titanium white plants also sprang up around this time.

Stormy Postwar Development

Dr. Karl Braus, a chemist sent to Japan by I.G. before World War II, played a decisive role in the establishment of the Japanese nitrogen industry. I got to know him in Egypt early in the fifties when he was managing a nitrogen factory near Suez. Some years later, after he had returned to Europe, Braus visited us in Hoechst. I asked him whether he might be interested in going to Japan as our technical liaison representative. We had reestablished many old connections and created new ones. It seemed advisable to us, therefore, to acquire the services of an expert with knowledge of both the language and the country.

Karl Braus took on the job with much pleasure and executed it with great skill for a number of years. He profited, of course, from his close relationships with Japanese industrialists. He also laid the foundation for the technical department that was later created within Hoechst Japan. This department deals mainly with the exchange of technology between the two countries.

After Japan's surrender in August 1945, Japanese industry seemed to have come to the end of the road. It had been enormously strained during the war and had suffered vast destruction. The large cities had been bombed to rubble and the industrial sites were in ruins. At that time, neither Japan nor the rest of the world thought that such a crushed economy could recover quickly. The situation was not dissimilar to that in Germany. Both countries experienced the famous "economic miracle" which provided them with the highest growth rates in many years.

Japan's development passed through several phases. The first, lasting until about 1950, was a period of integration into a smaller world. The country also had to get over the shock of a lost war. The first manufacturing activities got underway during 1947 and 1948. The government's first priority was to ensure food supplies. But even in 1947, Japan began to direct some of its attention to iron and coal. Those industries were to become the foundation of a broadly based development. To stabilize growth, the government introduced a vigorous deflationary policy in 1949, which created confidence in the currency. People started to save again. Efficiency and cost-consciousness became the main mottoes of Japanese companies.

With the outbreak of the Korean war in 1950, Japan became the most important staging area for the United States armed forces. The resulting boom benefited the chemical industry. In 1956, the first petrochemical processes were introduced. Considerable capacities for methanol, polyvinyl chloride, melamine, and polyethylene were developed. The large petrochemical centers of Iwakuni (Mitsui Petrochemical), Yokkaichi (Mitsubishi Petrochemical), and Niihama (Sumitoma Chemical) were formed in 1957 and 1958.

The names of these leading chemical companies are well known. After World War II, the Japanese conglomerates, called *zaibatsu*, were dismembered by the Americans. Their individual companies were made autonomous, an action somewhat parallel to the dismemberment of I.G. Today, those companies have once again merged into large groups, called *zaikai*. In contrast with earlier times, however, they are no longer controlled by a central office and by the trading companies. Today, those groups are integrated more loosely but still very effectively through the banks and through personal links and coordination.

The production of nylon 6 started at the beginning of the 1950s and was soon followed by acrylonitril and polyester. The technology for those developments was acquired from abroad, especially from the United States.

Even today, many sectors of Japanese industry have to rely on foreign technology; the nation's chemical industry was late in engaging in research. Research expenditure has now increased to about 25 percent of sales cost. The initial policy of buying modern processes worldwide and using the limited facilities within the country for improving them has clearly paid off. Considerable development potential was established through these acquired processes and the country's own research, so that Japan is now increasingly exporting its own technologies.

Our Far Eastern Interests

West Germany has an essential share in the processes and patents acquired by Japan. After World War II, for example, Hoechst concluded sixty agreements for licenses, patents, and know-how with Japan. The Hoechst engineering subsidiary Uhde designed and erected fifty-seven chemical plants, including eleven acetaldehyde plants (569,000 tons per year), four acetone plants (66,000 tons per year), and three Oxo plants (193,000 tons per year), as well as facilities for sulfur recovery, synthesis gas/ammonia, and other products.

One focal point of the Japanese chemical industry is fermentation of aminoacids, enzymes, antibiotics, and the mature technologies of the large-volume products, especially the derivatives of petrochemistry. Petrochemical centers with large plants facilitate a favorable cost structure, but they are particularly susceptible in times of general recession. The last recession pinpointed this structural problem of the chemical industry, especially among the monocultures of petrochemistry and man-made fibers. Attempts are therefore being made to achieve better risk distribution by striving for new, comparatively small product volumes with a high degree of refinement.

The Japanese economy has experienced an upsurge that is almost more impressive than the German economic miracle. In addition to state aid, the personal characteristics of the Japanese are to a large extent responsible for this success. The Japanese are intelligent and diligent, tenacious and adaptable. The industrialists, too, have made outstanding contributions. A final factor, one that should not be underestimated, is the pronounced feeling of solidarity that distinguishes the Japanese. More than any other people, they are public-spirited and ready to make sacrifices if the situation demands them.

Industry and Government

Life in Japan is so different from that in the Western world that a visitor needs some time for adjustment. On many occasions, I have been deeply impressed by the respect paid to older people. In Japanese companies, as well as in our own establishments there, the seniority principle is honored to a striking extent. The position of employees in the hierarchy and their emoluments are not normally governed by their individual performance, but by their length of service. Although this standard may sound absurd to us, it has the advantage of promoting teamwork.

The tendency to take both performance and seniority into account when deciding salaries and wages is of recent origin. However, as is always the case when deeply rooted tradition is being questioned, the subject has to be approached circumspectly. Another discovery that surprised me on my first visit to Japan was the decisive importance of banks and commerce for industry. Traditionally, government and industry work well together. This has proved to be so even in times of crisis. The forms of this cooperation are varied, and they operate on a variety of levels.

The most influential institutions for industry are the *Keidanren* (Federation of Business Organizations) and the *Nikkeiren* (Japanese Federation of Employers' Associations). The government is represented by the MITI (Ministry for International Trade and Industry). MITI coordinates the expansion of industry through careful examination of intended projects. The aims of MITI's economic policy are recognized. Discussions of investment policy and coordination through MITI take place in public. Large and medium-sized corporations are increasingly and voluntarily coordinating their production plans and the dimensions of new projects. Examples include the cooperation of coke-oven plants with chemical companies and combines for the production of petrochemical products or for PVC manufacture.

The initiative always lies with the individual companies, MITI acting as a coordinator. The government will make no decision against the declared intentions of industry. Both parties find it easy to understand one another because the senior officials of the ministries and the top industrial managers have come from the same élitist Tokyo University. They know one another, and they feel they are among colleagues.

The exchange of people between government and industry is promoted deliberately. Industrial recruits work in government offices for a time. It is also quite common for senior officials, when they have reached the retirement age of fifty-five, to go into industry, where they can provide valuable services because of their governmental experience.

The Banks

A distinction is made between city banks, trust banks, and institutions dealing with securities. All-service banks like those in the West are not known. The city banks are fairly liquid, since the Japanese, because of their underdeveloped old-age pension system, have one of the world's highest levels of personal savings. In addition, the central banks pursue a very generous monetary policy.

With a level of 15 percent, compared with 55 percent in the United States and 35 percent in Europe, the equity of Japanese industry is extremely low. Since neither the issue of shares nor of bonds plays a large role, industry has to rely on the banks for financing. Every large concern has its own personal bank, which usually provides the major part of the external financing.

Compared with the great number of companies, there is only a small circle of financially powerful city and trust banks. In practice, therefore, a few institutions are able to exert a controlling and coordinating function. The most important city banks are the Daiichi Kangyo, Fuji, Sumitomo, Mitsubishi, Sanwa, and Mitsui banks. Among the trust banks, Mitsubishi, Mitsui, and Sumitomo are the leaders.

To a far greater extent than in other countries, projects and commercial policy are discussed between company management and the banks. Real company collapses have remained a rarity.

Environmental Protection and the Future

The problem of water and air pollution by industry has been a major topic of public discussion in Japan over the last ten years. Japanese chemical industry was assailed in the foreign press even more vigorously than its counterparts in the United States and Germany. It is necessary, however, to recall the situation at the beginning.

The imposing economic upsurge of Japan after World War II led to fundamental changes in the sociological, economic, and ecological areas. While in Europe similar changes were not so geographically confined, Japan was hit much harder by the concomitants of expansive industrialization. The populations of Tokyo, Osaka, and Nagoya increased by leaps and bounds. Many factories were established in the suburbs of the cities, so as to be close to the available labor, the markets, and the consumers. In 1955, these three cities contained more than 30 million people, or almost 35 percent of the country's population. Within fifteen years this figure rose to 45.6 million, or 44 percent of the population. The close proximity between the production facilities and the markets has, without doubt, contributed greatly to the rapid economic upswing.

Under such circumstances, the side effects of an economy directed solely toward growth were not long in manifesting themselves. Only 30 percent of Japan's total area of 37.8 million hectares is habitable or arable. The country's 110 million people create the second largest gross national product in the free world.

Unfortunately, not enough attention was paid to city and plant planning. The result was extensive environmental pollution in many fields. In the large cities, exhaust fumes from cars and heating installations, as well as domestic sewage, were the main culprits. In the sixties, however, the public became concerned about this problem. Astonishing progress has been made since then.

The alarm signals for stricter laws were triggered by Minimata disease (mercury poisoning), Itai-Itai disease (cadmium poisoning) and asthma in Yokkaichi resulting from sulfur dioxide air pollution. The first comprehensive law to set standards and provide for preventive measures, the basic law for environmental pollution control passed in 1967, has been followed by many other laws that have induced communities as well as industry to intensify their efforts. Japan has become something of a pioneer in the environmental protection field because of the extraordinary success of its tremendous moral and financial efforts.

However, the current environmental regulations present formidable problems to industry, especially to the chemical industry. Many Japanese firms fear that countries not subject to the same stringent regulations may for this reason become more competitive internationally. Without question, Japan has proportionately the highest investments for environmental protection among the industrial nations. In 1975, the chemical industry in Japan spent 21 percent of sales revenue for this purpose, compared with 12 to 15 percent in West Germany.

Investments in Japan

In my opinion, Japan will continue to be among the larger growth markets. However, foreigners have always found it hard to gain a foothold in that country. In most industrialized nations today, investments abroad are more or less balanced by the foreign capital invested in their own country. In Germany, capital exports and capital imports are almost identical. In Japan, however, capital imports amount to only about 20 percent of capital exports. An essential factor in this development was Japan's concern to protect her domestic industry and to avoid excessive losses of foreign currency through the transfer of profits by foreign firms.

Foreign investment is still governed by the current laws of 1949 and 1950. In the meantime, as a full member of the Organization for Economic Cooperation and Development (OECD) since 1964 and a member of IMF since 1973, Japan has moved closer to the free economy system. The yen was made convertible and transfer of profits was allowed. It seems to me that two criteria by which Japanese authorities evaluate foreign investments are of particular significance: overall economic usefulness, and effect on the balance of payments.

Let me quote a recent example. For some years now, Dow has attempted to obtain approval for a caustic soda plant. Although in theory this sector is entirely open to foreign investors, Dow has had no luck so far. The project was obviously opposed by the domestic caustic soda industry, whose own existence might thereby be endangered. Despite liberalization, many difficulties are due to tedious bureaucratic procedures: usually, several ministries are responsible for a single matter. This again shows the close cooperation between government and domestic industry.

Upon joining OECD in 1964, Japan accepted the liberalization code, but was able to push through a number of exemptions. Under pressure of growing foreign criticism and in view of Japan's increasing dependence on exports, the abolition of the remaining obstacles to foreign trade was eventually recognized as essential for further growth. Gradually, most of those obstacles were indeed removed.

In principle, 100 percent of foreign capital investments are now also approved. But case-by-case decisions are still made concerning new companies in the fishing industry, agriculture and forestry, refining, leather, and retail chains. Foreign investors may nonetheless prefer joint ventures together with a Japanese partner. Such a partner will be familiar with the pecularities of a complicated language, with the national mentality, the workers, and trade unions, and with the merchandising systems. On the other hand, joint ventures with a foreign minority holding, which always used to be preferred by the government, are likely to become rarer, while fifty/fifty joint ventures and those with a foreign capital majority will increase.

Effects on Neighboring Countries

Since Japan has no sizable mineral wealth, she is even more dependent on imports than Germany. More than 95 percent of all raw materials (petroleum, phosphate, salt) and approximately 80 percent of the energy carriers (oil, natural gas, coal, nuclear fuel) must be imported. For this reason, Japanese industry has made a considerable effort, with the aid of government subsidies, to obtain raw material supplies from all over the world. The focal materials are oil and natural gas from the Middle East and Indonesia, and salts from Australia, India, and Mexico. But it needed the dollar crisis of 1971, and even more, the oil shock of 1973, to convince Japanese industry that its companies had to become truly international.

Exports alone will not suffice in the future. The Ministry for International Trade and Industry (MITI) pointed out in a study that, compared with American and European concerns, Japanese companies have hardly invested abroad, in spite of equal technological standards. They were urged in the study to increase their investment in consumer countries. MITI reported that European and American countries earned between 20 and 50 percent of their profits from foreign subsidiaries. The corresponding figure for Japanese companies was about 2 percent.

In the meantime, Japanese concerns have begun to establish subsidiaries and associate companies in Southeast Asia and the Middle East, and even in Europe, the United States, and South America. The Japanese chemical industry is also looking for new markets in several developing countries. Chemical complexes are—or were—planned in Iran (by Mitsui), Saudi Arabia (by Mitsubishi), Singapore (by Sumitomo), and Indonesia (by Mitsui). In order to maintain and expand their market, the Japanese are also investing in Europe. Petrochemical plants are being exported, frequently in cooperation with the large trading concerns providing the credits. Tokyo is also placing high hopes on future cooperation with China and the opening up of the Chinese markets for Japanese products.

During recent conversations with Japanese industrialists, I had the impression that the exaggerated original expectations from the Chinese market have in the meantime been trimmed back to more realistic dimensions. No doubt, the Japanese too have realized that the economy would develop only very slowly, step by step, if it had continued to adhere to the principle of self-sufficiency now relaxed. Without fundamental political changes, no other course would be thinkable. Exports and the plant business with China may now develop more slowly. As I found when I visited the Canton Fair, it cannot be ignored that the Japanese are pursuing the Chinese market far more intently than other countries. That, of course, is not difficult to understand.

First Dyestuffs, Then Pharmaceuticals

From 1895 until the end of World War II, the Johann Grodtmann company represented Hoechst dyestuffs, and later, as a subagency, various divisions of the I. G. Farben, in China and Japan. It was not surprising, therefore, that Johann Grodtmann, whom I got to know during his retirement in Vevey, together with his son Carsten, founded new companies in Japan, Hong Kong, and China after 1945 to rebuild their business in East Asia. However, the attempt to resume business with China, even on a small scale, was condemned to failure after the founding of the People's Republic of China.

Hoechst's activities in Japan, which had greatly modernized and rationalized its textile industry, also began with dyestuffs. Together with Grodtmann and another European partner, we founded the Dyestuffs & Chemicals Trading Co., Ltd. Its Japanese headquarters were in Osaka, near the large textile centers around Kobe and Nagoya. Supported by the technical resources of the parent company and a well-equipped advisory laboratory, the company prospered. Some years later, it was merged with the newly founded Hoechst Japan Ltd.

It was not easy to reintroduce Hoechst pharmaceuticals in Japan. In this case, it was not a matter of regaining the trademarks. Rather, we had to reestablish ourselves in a market in which we had not been active for twenty years. Eventually, however, Gustav Bunge was able to announce the first successes. He had assumed responsibility for the entire Hoechst business in Japan during the mid-fifties. In 1956, Nippon Hoechst, in which Kofuku Sangyo had a 60 percent interest, was founded with headquarters in Nagoya. Our partners were the Miwa family, originally active in the textile field. In the favorable climate of the boom years, Miwa had established a medium-sized corporation, including a large trading house, textile factories, an optical plant, and also pharmaceutical manufacturing facilities.

The establishment of this company had to be celebrated, of course. Miwa senior invited our wives and us to a geisha party. According to the seniority principle, the guest of honor had the privilege of being served by the highest ranking geisha. The classic Japanese cuisine is rather strange to Western tastes. Tempura and sukiyaki dishes are the exceptions. But, since these are not part of the classic cuisine, they are not served at geisha parties. Such parties are quite harmless. Much rice wine is drunk, and there is a lot of joking and banter. At the end, there is always some dancing or some kind of musical offering.

A New Partner

Although our association with the Miwa family had begun in great harmony, it was not destined to last. Hoechst had too little opportunity to influence business policy. This lack applied especially to the marketing and promotion of new products. In 1967, after long but friendly negotiations, we took over the shares of our partner, now trading under the name of Kowa, and integrated the pharmaceutical business into Hoechst Japan, Ltd. It had just reached a volume of DM37 million (US$9.3 million) per year.

With this step, all sales were brought under one roof, in conformity with company policy. However, at that time, a manufacturing license for pharmaceuticals could be obtained only with a Japanese partner. Alfred Dienst, who had taken over the Hoechst management in Japan in 1966, succeeded in gaining Mitsui Petrochemicals Industries Ltd., with whom we had had friendly relations for many years, as a 50 percent partner in a new manufacturing company named Nippon Hoechst.

In 1967 a sizable pharmaceutical plant was built in Kawagoe, about an hour's drive west of Tokyo. Now with its own research establishment and staffed entirely by Japanese, this operation strengthened our position in the market and contributed to our worldwide innovative potential. A few years later, the manufacture of cosmetics, mainly the Schwarzkopf range, was also begun in Kawagoe.

We have maintained a long relationship of trust with the leading people of Mitsui Petrochemicals: Takeshi, Ishida, Yasuji Torii, and

Shozo Nagami. It had grown at the beginning of the sixties, when we undertook a fruitful exchange of technologies for polyolefins.

In Japan, too, the management style of the large concerns is greatly influenced by the man or men at the top. I sensed this very clearly in 1966 and 1967, when we negotiated on a joint venture with Mitsubishi for the manufacture of reactive dyestuffs. Mitsubishi's President Shinojima received our Hoechst delegation in Kaitokaku in a well-tended town palace belonging to the company's founder, Baron Iwasaki. This palace now serves as guest house for the Mitsubishi group. Ladies in lovely kimonos served drinks and helped the guests to find their way around the Japanese cold buffet.

A New Joint Venture

Kasei Hoechst is a fifty/fifty company, operating a now flourishing dyestuffs plant in Ohamamachi, a small town about two hours' drive from the textile center of Nagoya. The acquisition of the Berger group in England provided Hoechst with a 50 percent interest in Kobé Paints Ltd., producers of marine paint in Japan. Our partner is the Kawasaki group, active in steel and shipbuilding. This has provided a link to Hoechst Gosei which, as a partner company, produces plastics dispersions made by Hoechst processes and those of its partners. It does this in conjunction with Nippon Gosei, in which Mitsubishi Kasei has a minority interest. Guided by Mr. Okado, its president, the company is one of the top groups in this field.

The Role of the Trade Unions

As the only French pharmaceutical company in Japan, Roussel-Uclaf had the courage to establish itself under its own banner. In accordance with our agreements, Roussel-Uclaf has retained complete independence. Our French colleagues entered into partnership with Chugai, a Japanese pharmaceutical company. This arrangement proved particularly helpful to Roussel when the labor union struck and closed the plant down for several months.

There is much talk in the West about the alleged lack of aggressiveness in Japanese trade unions. But that is only partly true. Certainly, in well-managed companies there is a solidarity among the employees that far exceeds Western expectation. But this closeness requires much managerial skill and adaptability. For this reason, and especially in plants where foreigners have leading positions, it is essential for personnel and social problems to be entrusted to experienced Japanese managers. A sensible balance in top and in middle management between nationals and foreigners is even more important in Japan than anywhere else. The complicating factor here is that qualified Japanese prefer to work for domestic companies, whose reputation lends them prestige. One cannot compensate for this feeling with material advantages, but only with careful concern for the work environment and opportunities for training and promotion.

Our cooperation with Roussel-Uclaf in Japan is not confined to the occasional dinner at Tokyo's Maxim's. Hoechst and Roussel help each other in many respects in a pharmaceutical market in which the indigenous industry has a promising research potential and dominates the field. This mutual help concerns problems of manufacture, development, and product approvals. Our sales organizations are separate. After the United States, Japan is the largest pharmaceutical market in the world, so a company that claims world renown must have an appropriate presence in that country. The results will be beneficial for both parties, especially where new developments are involved.

Future Plans and Their Limitations

When I talked with my colleagues in the new administration building on Gaien-Higashi Avenue in Tokyo about the developments that our business is to pursue over the next ten years in that part of the world, we all accepted one fact. Our exports to Japan will not continue to grow by leaps and bounds. Moreover, with the exception of pharmaceuticals and agrochemicals, we are not likely to have many products that offer qualitative superiority over their Japanese counterparts. At best, we will have a time lead in one instance or another.

The Hoechst group, with more than 2000 employees, is an important factor in the market. It has gained market shares that must be maintained and, indeed, expanded. There is a sales force that should also benefit from new products. But new products need new manufacturing plants. In my view, it is sensible not to enter the field of mass products in such a competitive country, with large capacities already available and able to handle a growing export trade. It is better for us to concentrate on specialities and novelties. Our immediate plans, therefore, cover organic pigments and surfactants, with which new terrain can still be opened up. This will once again be effected in partnership with local partners. The principle of not becoming wedded to only one company in Japan so far has proven a sound one.

Japanese Technology for Europe

Some years ago Kalle obtained the agreement of the parent company to enter the field of office copiers other than engineering copiers and printing plates. This development focused attention on Japan. Japan's precision engineering and optical industry has occupied a leading world position for many years. A complicated agreement was concluded with Ricoh, one of the market leaders. This also included cooperation with an expert group in the United States.

The Infotec business of Kalle is conducted with equipment that is produced in the very modern Ricoh factory at Atsugi, near Tokyo. In addition to its modern office copiers, Ricoh's facsimile transmission equipment is outstanding. Within a few minutes, facsimile copies of originals can be sent via telephone to any desired destination equipped with the corresponding unit. Whether close technical cooperation with our Japanese partner will eventually lead to joint production in Western Europe is still an open question.

Another good example of the flow-back process of know-how from Japan to Western Europe involves Takeda, one of the leading pharmaceutical producers in Japan. Takeda, with whom I.G. had collaborated closely, has formed a joint venture with Roussel-Uclaf for the manufacture and marketing of its new pharmaceuticals in France. I hope that this marks the beginning of a relationship that will be as fruitful as the development in Japan. The research-intensive Japanese pharmaceutical companies are clearly scoring great successes, and they are as unlikely as Western companies to be satisfied in the long run with merely handing out licenses.

Perspectives

The hangover that one encountered among Japanese industrialists immediately after the oil shock of 1973 has meanwhile given way to a more reasoned attitude. Although expectations of unlimited growth have been scaled down considerably, on my most recent visits I sensed that the Japanese again have faith in the future of the chemical industry. And not only within the boundaries of the Japanese nation but also, and especially, in the Southeast Asian area. Doubtless, Australia, America, and Western Europe are also markets for Japan's exports and industrial activities, especially for its more highly developed products.

China's future, whether expressly mentioned or not, colors all discussions of Japan's economic significance in East Asia. All political dreams of a vast pan-Asian area have been abandoned, just as Germany has had to give up similar visions. But a certain political and industrial sense of mission has remained, if this is not describing this attitude too dramatically.

Japan is no longer a low-cost country. Comparable salaries have reached Western levels, at any rate as far as large-scale and medium-sized corporations are concerned. Doubtless, there still is a considerable gradient difference in scale, and thus a competitive advantage due to subcontracting companies, which are usually small family firms. But I am sure that this advance will not last for very long.

One fundamental difference persists, however. The older and the younger generation both perceive their plant or office as a kind of second family. Western visitors will note with delight how in industry, public service, and especially in the service sectors, people make every effort to do their work as well as possible. This attitude will, no doubt, ensure that Japan, so little favored by nature, will maintain and expand her place among the large industrial nations. Such a position will be of great political importance for the free world. As a democratic and economically strong power center in East Asia, Japan is an indispensable factor in politics and economics, especially in a crucial part of the globe with many unsolved problems.

South Korea: The Little Neighbor Becomes a Big Competitor

The Japanese, preoccupied first with survival and then with their own economic miracle after World War II, had a big surprise: their little neighbor, South Korea, was duplicating their miracle almost unobserved. In 1977, South Korea achieved a small excess in its trade balance for the first time with 10.3 percent real growth of gross national product and exports worth more than US$10 billion. Tokyo realized with mixed feelings that a second Far Eastern miracle was taking place on its own doorstep. In the world market, Japan's former small neighbor has become a serious rival.

I first visited Seoul in the mid-fifties. It seemed almost an American garrison town, not unlike those in Germany immediately after the war.

Even then, this region south of the 38th parallel was full of life and initiative. Partition gave South Korea the worst of the deal. Except for textile factories, all industry and all raw material sources were in North Korea. But alliance with the United States soon had a beneficial effect on the economy. Moreover, the refugees from the north, seeking new means of existence, provided fresh momentum.

The Japanese ruled the country from 1910 to 1945, but had shown little interest in its industrialization. They were more concerned with transferring Korean labor to Manchuria, their conquest and heavy industry base.

Times were difficult in 1953. The textile industry then operated under fairly primitive conditions, manufacturing cheap articles and paying very low wages. Our first contacts with the new South Korean state were made through a Sino-German who had been working for I.G. Ernst Bochow had established FOHAG, the Fernost-Handels AG, as an import business in Korea. But this was only one of his activities. Another and far more lucrative enterprise was to supply the American troops there with articles of everyday use, including such items as beer from Germany and Hong Kong.

Our prospects for continuous development improved considerably when we made contact with S. K. Kim and his partners. As a modest beginning, they had in 1954 founded the Union Drugs Company, which in 1957 introduced the first Hoechst pharmaceuticals into South Korea. S. K. Kim was an extraordinarily enterprising man. His slight command of English and our equal ignorance of Korean did not preclude communication, although they did complicate matters at times.

As early as 1958, S. K. Kim and his partners founded the Han-Dok Remedia Industrial Company, with whom Hoechst signed a technical assistance agreement in 1959. This was, incidentally, the first agreement of its type concluded in South Korea with a foreign partner. The starting point was Han-Dok's existing pharmaceutical plant, which now began manufacturing Hoechst products. We soon found one of our technicians prepared to go to Seoul. This pleasant young man, W. V. Janecek, went to Korea as a bachelor and established contact with the Korean people in a most natural manner: he married a very charming Korean girl. The two of them remained in Seoul for some years and then returned to Europe.

Initially, our cooperation with Han-Dok Remedia was confined to the technical assistance agreement. After some years, however, we also became financial partners, although Hoechst held a minority share. The joint venture did well, and for a number of years ranked second among the many pharmaceutical companies set up in South Korea by then.

In 1968, Han-Dok Remedia took over the industrial business abandoned by FOHAG when that organization withdrew from Korea as business with the American troops began to dwindle. Its industrial operations were largely confined to dyestuffs and other products for the textile industry.

At that time, we remained largely aloof from the fast-growing chemical industry. In part, this was because our Korean partners were pharmaceutically oriented. But another decisive factor was the difficulty of interesting qualified people from the parent company in a prolonged stay in Seoul, given the modest living conditions of those days.

All these conditions have, of course, changed greatly since then. The tremendous diligence and toughness of the Koreans soon led the country into a wave of industrialization. Favorable working conditions tempted processing industries that in earlier years would have settled in Hong Kong, Taiwan, or Singapore. Nowadays, South Korea is described as the Ruhr of the Pacific. But this may be as much an exaggeration as the suggestion that Korea, with its 37 million inhabitants, is on the way to becoming a second Japan.

A tremendous amount has been accomplished in the industrial development of South Korea. The government, which has been in the charge of the authoritarian General Park Chung Hee since 1961, created the political conditions under which economy can develop in a climate of internal calm. President Park is following a clear and hard anticommunist course, not only internally but also externally. President Carter has projected a gradual withdrawal of the American forces in South Korea. It is, however, scarcely imaginable that the United States would contemplate a South Vietnam-style departure: Indeed, the withdrawal timetable has been suspended. Korea has become too important—strategically, politically, and economically—for such a departure.

Nonetheless, the South Korean government has formed its conclusions from the partial withdrawal of United States troops from the country. Its 37 million people have been marshalled into a nation under arms, and General Park is firmly resolved to offer the strongest resistance to any invasion from the north. Although the demarcation line has usually been hermetically sealed, there are tentative gestures from time to time toward a political understanding with the North Koreans.

Pony Carts for Western Europe

In the meantime, Seoul has become a modern capital in which business firms from the United States and Europe operate on an open-door basis. A steel industry has been developed, electronics have found their place, machinery and automotive industries have been built up. Since the spring of 1978, even pony carts from South Korea have been offered in Western Europe.

South Korea has ambitious plans for petrochemistry. Some production plans are already in existence, others are at the stage of concrete projects. The leading partners of the Koreans are the United States, closely followed by Japan. The natural resentment against Japan that existed for many years—after all, the Japanese occupation lasted thirty-five years, from 1910 to 1945—has largely been overcome. At any rate, it no longer surfaces when economic matters are involved.

I remember a visit to Seoul about ten years ago. We spent the evening in the private home of our partner, S. K. Kim. His house was furnished tastefully in Eastern and Western elegance. There were about a dozen guests, and the lady of the house offered us a meal which, although totally different, could easily hold its own against the best Chinese cuisine. We drank rice wine in ample amounts, as is the custom in Korea, partly for climatic reasons. One of the highlights of that evening was the performance of Kim's thirteen-year-old daughter, who played the piano in the most accomplished manner. She has since become a music student at the University of Freiburg.

The partnership between the Kim group and Hoechst has meanwhile become even closer, and it now extends also to pharmaceutical raw materials. Cooperation in research was initiated by KIST, the Korean Institute for Science and Technology. KIST was planned and its construction supervised by the Battelle Institute by order of the United States government. It is now a national institution which continues to receive strong support from the United States. Today it exercises considerable influence over the development of domestic technologies in Korea. Despite this advantage, Han-Dok Remedia is unfortunately no longer the number two company. But it continues to be one of the large, modern, and progressive pharmaceutical companies in the country, and we are certain that, together with our partners, we shall be able to maintain a sizable market share.

Our industrial agency has since been thoroughly reorganized. We shall not abandon the attempt to catch up with the stormy development of industry and chemistry in this rising Asian country by extending our activities beyond the normal business. Fortunately, conditions in Seoul for Western staff are thoroughly acceptable these days. There is a different mentality, of course, and the language barrier. But that can be overcome, although Korean is said to be even more difficult to learn than Japanese.

Taiwan: Far Eastern Contrast Program

When discussing power centers in the Far East, it is impossible to leave out one country that, back in the 1950s, was only a geographical concept even in Germany: Taiwan, the young industrial center on the edge of China. My first visit to the country in 1957 was at a time when Taiwan was showing signs of assuming an important economic role, and not only in the Far East.

From the airplane, which lands in the middle of the built-up area of Taipei in a relatively steep descent, the first impression is of an inner city designed in checkerboard fashion with almost total traffic chaos. From the other side of the plane, the imposing Grand Hotel, constructed on a hill in the style of a giant Chinese shrine, is also visible. Both aspects of Taiwan show that the visitor is entering, not some entirely modern city, but a city where progress and tradition go hand in hand. From the ground, the hotel, with its magnificent Chinese colors, does indeed look quite different from hotels in any other country. Every effort is made to show the visitor that Western comfort can be readily combined with Eastern tradition.

Apart from the main street, which looks like that of any metropolis, the town itself gave the impression of a huge building site. Anything on wheels is used for transport. I was fascinated, for example, by the number of people and things that a simple motorcycle can transport. Not infrequently, one can see a whole family, up to four to five persons with their belongings, rushing by on a motorcycle.

The indigenous population of Taipei clearly takes a great deal of pleasure when the unprepared visitor, on the first evening, ventures into the old city of Taipei where there is not only the main temple of the city but also the so-called snake-market. Incomprehensibly for the newcomer, live poisonous snakes are ripped open and blood and gore pressed into a glass of rice wine which is then offered for sale to visitors. My companion told me on this occasion that about 200 years ago, the original population of Taiwan made human sacrifices to its gods and that in Chiayi, in the center of Taiwan, there is a temple to commemorate the last human sacrifice, Wu Feng.

During a trip through Taiwan, the stranger will be surprised by the persistence with which the population has maintained its original language and, indeed, much of its independence generally. Influence from the mainland has never been particularly great. It is probably typical of the Chinese way of thinking that cultural confrontations are allowed to mature. This is the reason why the indigenous Taiwanese and the mainland Chinese who came after the end of the Chinese civil war live together without any kind of difficulty. The differences in language remain but the differences in mentality and educational standards are gradually becoming blurred. Land reform was initiated immediately after the Chinese had taken over. Because the unpopular Japanese tutelage had come to an end at the same time, the Taiwanese welcomed the mainland Chinese and showed their eagerness to cooperate with them. Large landowners were compensated with shares in Japanese industrial undertakings, thus providing the basis for both an intensive and a socially balanced agriculture as well as for an indigenous industry.

Taiwan is a country of contrasts. In the north of the island, in the vicinity of Taipei, and southward along the entire western coast to Kachsiung, another city with a million inhabitants, one industrial complex follows another. The east and south, on the other hand, can boast of great natural beauty, with romantic rocky coasts interchanging with wonderful bathing beaches. A journey through the mountainous center with the unique Tarokko gorge and scenery that reaches up to 4000 meters is no less unforgettable.

On our way to Asia, we were told in Honolulu that pineapples, for a long time an important export item, could no longer compete with the exports of Taiwan. When traveling through Taiwan, the reason can be readily identified: relatively small, well-managed fertile plots on which sugarcane, pineapples, bananas, rice, and, in winter, vegetables are grown. These are characteristic of the intensive Taiwan agriculture.

Taiwanese industry is in a similar situation to that of neighboring Japan: There is a complete absence of mineral wealth. Taiwan has no oil and only a few natural gas deposits in the center of the island where an ethane cracker was constructed. Two further naphtha-cracking installations were erected in the south. They provided the basis for the establishment of a petrochemical industry which is today producing the entire range of modern plastics and man-made fibers.

The new Hoechst building in Tokyo houses administration as well as facilities for the production of pharmaceuticals, pigments, and plastics.

The traditional artist's brush and modern technology blend well in Japan. Hoechst pigments fit into the Japanese tradition.

Overleaf: The bridge across the Bosporous, connecting Europe and Asia, is protected by anticorrosive material manufactured by the Hoechst U.K. subsidiary BJN.

Dr. C Ishibashi, President of the International Medical Society of Japan, meets with Kurt Lanz.

T. Fukuda (right), former Japanese Prime Minister, discusses Hoechst's planned activities in Japan with Kurt Lanz and former German Ambassador to Japan, Professor W.G. Grewe (second from left).

Hotel Taj-Mahal in Bombay.

In Mulund, India, Hoechst produces pharmaceuticals and conducts research on antibotics.

Kurt Lanz and Dieter Laengenfelder discuss India's pharmaceutical market problems with Indian colleagues.

...and traditions of host countries are duly respected by Hoechst worldwide. Here, Kurt ...icipates in a ceremony in which a red dye rubbed on the forehead means welcome.

Rice is the basic nutrition for billions of people. Thiodan, a Hoechst insecticide, is used successfully to fight insects threatening the harvests.

Hoechst established the first pharmaceutical facility in Kabul, Afghanistan, and today 300 people are employed there.

...des the main location in Melbourne and the large division in Sydney, Hoechst also has branches in ...ane, Adelaide, and Perth.

On a visit to Peking, Professor Winnacker (center) and Kurt Lanz (right) view the Great wall of China.

The capacities far exceed the amounts that can be utilized in the country. For example, in the polyester fiber field, about 10 percent of world capacity is located in Taiwan. There is consequently a great dependence on exports. The excess amounts can be unloaded to only a limited extent in adjacent neighboring markets, so some of them inevitably reach overseas markets where they greatly interfere with the market structure. There cannot be much pleasure in Taiwan over the profits achieved from this kind of business, not even among the partners who come mainly from the United States.

Apart from the textile and plastics industries, the electrical and the electronics industries play a large part in Taiwan. It can be expected that in the next few years, Taiwan will attempt to achieve an important position in heavy industry, steel, and ship construction—at any rate in the Far Eastern regions.

Taiwan reacted with remarkable flexibility to the economic crisis of the last few years, which incidentally started with the textile industry in the Far East. By means of shorter work hours, sliding wage scales, and similar measures, unemployment was kept within tolerable limits.

The contrast between progress and the Taiwanese consciousness of tradition became clear to me when, on a single day, I visited both a supermodern textile factory and one of the main sites of the Far East, the National Museum of Chinese Art in Taipei. In the morning, we met a management group which had ideas about future intentions. The Taiwanese industrialist, however, is not the type who weighs matters too carefully. His strategy starts with the target. Factors like market size, infrastructure, and training levels of the workers are analyzed only later.

On the afternoon of the same day, an elderly professor who spoke fluent German led us through the National Museum of Chinese Art. It houses art treasures thousands of years old, the like of which can be found only in the vicinity of Peking. For the Western visitor in particular, it is an experience of a special kind to observe the devotional admiration with which groups of young people, usually pupils and students, absorb these testimonials of the unique artistry of their predecessors.

The Chinese honor their past and believe in their future, but the present, the shortest span of time, is not necessarily the most important. Perhaps this philosophy tells us more about the people and possibilities in the Far East than the visitor from the West can gain from complex analysis.

15
The Third World and Fifth Continent

Burma: Buddhism and Planned Economy

One of the most beautiful of Buddhist sacred places in Southeast Asia is the Golden Pagoda in Rangoon. Visitors are impressed not only by its massive and beautiful structure, but also by the great variety of memorials and symbols. We were no less fascinated by the people kneeling in prayer before the altars, offering their flower gifts, and then hurrying along to the next memorial.

I saw the Golden Pagoda for the first time in 1956. Although my wife and I had taken a little honeymoon trip in 1955 from France to Spain and Portugal, we wanted a longer one before the desired offspring might make this difficult. As a result, my wife and I traveled through Southeast Asia, visiting Burma, Sri Lanka, Thailand, Manila, India, Pakistan, Singapore, and Malaysia.

Burma is a unique mixture of Buddhist tradition and an ambitious planned economy. Buddhist tradition manifests itself in the innumerable pagodas. The most beautiful of them rises in the center of Rangoon. Bearing the name of Schwedagon, it is the heart of the city. Three tons of gold cover its enormous domed roof, and according to the high priests of the temple, its crown is adorned with 50,000 jewels which gleam in the sun in indescribable colors.

Burma was not very friendly toward strangers in 1956. Tourism was no longer promoted, and a visitor's stay was limited to 24 hours. Hostelries were correspondingly primitive. But these conditions were more than offset by Burma's cultural wealth and the charm of its inhabitants.

Following earlier socialist experiments, Burma has had, since 1962, a military dictatorship of a special kind. Its aim was to establish a Burmese variety of socialism. That experiment is now more than fifteen years old. U Nu was followed by General U. Ne Win, who rules in tandem with the only real power in the land, the military.

The socialist planned economy has resulted in complete impoverishment. Revisited in 1976 after twenty years, Rangoon gives the impression of a capital in a state of degeneration. It has no traffic, no attractive shops, and only primitive hotels. To visit a country under such conditions needs weighty reasons indeed.

The state-managed economy of the country did not allow us to establish an agency for Hoechst in Burma. We have, however, continued in the tradition of I. G. Farben and are selling dyestuffs, pharmaceuticals, and other products, working together in an unusual arrangement with the other two I.G. successors and with a Burmese agent in Rangoon. He, of course, is not in a position to do any proper trading. His activities are confined to offering technical advice to the state purchasing monopolies that regularly invite offers.

Further expert support for such business is provided by our neighboring friends in Bangkok. This assistance has enabled us to maintain a volume of business that is quite respectable, although it shows few possibilities of growth. Ultimately, our progress there will depend largely on the industrial and agricultural development of the country. But regrettably little has happened in this respect over the last two decades.

There was a phase in which Hoechst, even if indirectly, had been assigned a remarkable role in plans for establishing a petrochemical plant in Burma. The assignment came some years ago and caused much discussion at Hoechst headquarters. General U Ne Win's government planned to build a proposed large-scale refinery, followed by a manufacturing plant for high-density polyethylene. That would have been fairly risky, mainly because Burma has no existing processing industry of any significance, making the project very dependent upon neighboring markets. But, in any case, the plan came to nothing because of financing difficulties, and there are no signs that it will ever be realized. Nonetheless, I nurse the hope that Burma will not give up after some timid attempts at liberalization and that it, too, will participate in the vital development of Southeast Asia's active nations.

Sri Lanka: Radiant Country

The next stop on our trip in 1956 was Colombo, the capital of Sri Lanka, the former British colony of Ceylon.

> If you leave the island of Angaman and sail a thousand nautical miles southwest by east, you reach the island of Ceylon. Both in size and otherwise, it is the best island in the world. It extends for 240 miles; it used to be larger still, but northern storms that rage with giant force have made the mountains brittle so that they have collapsed in some places and drowned in the sea. The inhabitants are idolaters and not obliged to pay tribute to any other power. Men and women walk about almost naked. The people are not at all warlike; if they need soldiers, they hire them in adjoining countries.

This is an extract from an enlightening account written by Marco Polo 700 years ago. It is probably one of the first declarations of love for that part of the earth which now calls itself "the shining, radiant country—Sri Lanka."

In 1956 Colombo, the country's only large city, still had the charm of a past epoch. There were reasonably comfortable hotels, though their concern with gastronomy and tourism had already faded. At that time, Sri Lanka was in a transitional phase. It had regained its independence, but the many European owners of tea plantations had remained, although they knew that their almost paradisiac existence in the hills of Ceylon would soon come to an end. And, indeed, they did not have long to wait. In 1956, Prime Minister Bandaranaike, who aspired to national independence, took over the government. After his death, his wife continued to rule until 1965. She was elected as Prime Minister in 1970.

The Buddhist Singhalese, who account for less than 80 percent of the nation's inhabitants, came into even greater conflict with the Tamils, whose ancestors had come from southern India centuries ago, and with half-castes who had been brought from India in the nineteenth century. The result was racial discrimination which seems, however, to have been attenuated in the succeeding years. The centralized state control of the economy has led to the impoverishment of the population because it made the rich poor, but it did not make the poor rich. A giant administrative bureaucracy grew up with its concomitant corruption and office patronage. Although the prerequisites for a successful tourist industry existed, an increasingly hostile attitude developed toward foreigners. It nullified the modest beginnings formerly made to develop tourism.

A relatively short time ago, it was realized that this was bad policy, and that the tropical garden of Southeast Asia was becoming impoverished. Attempts are now being made to reverse that course. There can be little doubt that Sri Lanka, given the attractions of its landscape and the wealth of its centuries-old Buddhist culture, will succeed in attracting tourists. However, facilities expected by modern travelers must be created. It would appear that the new generation of Singhalese politicians is beginning to come to terms with realities in this sphere as well.

Hoechst was no more successful than other Western companies in establishing itself in Sri Lanka. When we thought the right moment had come, the trend toward state socialism was already so firmly established that we had to be content with makeshift arrangements.

We were able to enlist local agencies for the sale of both pharmaceutical and industrial products, which have, alas, only a relatively modest market in Sri Lanka. We support these agencies, of course, by visits from experts either from neighboring India or from Hoechst. In this way we have been able to develop a modest business.

If economic controls are eased, if the purchasing power of the population increases, and if industry becomes more flexible, Sri Lanka may well become a growing market for Hoechst products again.

Ten years later, on a second visit, I found the country almost unchanged. Nothing of the old had been destroyed, but the paint on the houses was flaking off, the hotels were shabby and uninviting. Business was stagnating, and private initiative hardly had any opportunity to expand. The country had become even poorer.

Bangkok: In the Last Kingdom of Southeast Asia

It was humid and hot when we landed in Bangkok, capital city of Thailand, in the spring of 1967 on our way back to Germany from Tokyo. Our friends awaited us at the airport with the welcome news that an audience in the Royal Palace had been arranged for the next morning. Driving along the highway into the center, we got into heavy traffic, indicating that industrialization in the vicinity of Bangkok was proceeding rapidly.

Actually, tourism had become the main business of Bangkok. New luxury hotels, unmatched by any existing elsewhere in Southeast Asia, had risen everywhere. Bangkok had taken on the role of a recreation and rest center for the American troops in Southeast Asia and particularly Vietnam, with all the benefits and disadvantages that might be expected.

At about 11 o'clock the next morning, we entered the royal precinct, the park which surrounds the Royal Palace. Courteous and friendly court officials conducted us to the master of ceremonies, who took us to the room, looking out on the extensive gardens, where the royal couple received us.

Queen Sirikit was as beautiful as we knew her to be from the illustrated magazines. She also proved to be an intelligent woman who spoke English and French fluently and who devoted her interest to the welfare problems of Thailand. Her concern was to have more doctors and medical supplies for the hospitals, and to organize medical stations in the country, together with all the ancillary measures that such provisions required. Since she knew that we represented a large pharmaceutical company, the Queen's questions and general conversation centered on those subjects. King Phumiphon, however, was more interested in agricultural problems.

At the time of my first visit in 1956, Bangkok was still a provincial town. Its narrow streets had not yet been choked by traffic, and the hotels had been modest. The rooms in the Oriental Hotel, still one of the best, were furnished mainly in the old colonial style. Ceiling fans provided a cooling system of remarkable efficiency. There was also a new wing with a dozen air-conditioned rooms in which we were fortunate enough to be quartered. From those rooms, we had a dramatic view of the huge Menam Chao Phayo River, which is one of the most important transport routes in the country. Junks moved ceaselessly up and down the river. Timber, rice, corn, and soy beans were their main cargoes.

Hoechst came back to Thailand at a fairly early date, the first products to be marketed being pharmaceuticals. Mr. Rebhahn, a druggist who had acted for Bayer and Hoechst in Vietnam, had settled in Bangkok after the war, marrying a Thai girl who was as efficient as she was attractive. Rebhahn's business partners were the Swiss trading house of Berli Jucker & Co. This was, in fact, our first pharmaceutical agency, which did fairly well with our traditional product line.

In the long term, however, the business clearly needed fresh investments to introduce new products. However, these were beyond both the resources and the intentions of our Swiss partners. Subsequently, through a friendly and successful partnership, we gradually formed a Hoechst Thai company. Officially established in 1959, it took over the dyestuffs and the industrial business. The latter had been revived by a Chinese trading company headed by U Chu Liang, who had secured the services of Europeans from former I.G. organizations.

I clearly remember my first visit to U Chu Liang. He had installed his office on the extensive top floor of his warehouse, directly on the Menam River. This warehouse was filled with products from almost every country of the world. U Chu Liang was a sincere and hardworking partner, although he conducted business in a style acquired during his bazaar days, which no longer really suited the new economy of Thailand. We secured U Chu Liang as partner for Hoechst Thai, but later dissolved this partnership in all friendship. Since his death, we have maintained business relationships with his son, who took over the business.

Factories on Bangkok's Periphery

In the period of rapid economic growth during the Vietnam war, our business in Thailand developed by leaps and bounds. Ten years after the founding of Hoechst Thai, we had more than 200 employees, mainly Chinese and Thais who for the most part spoke a second language, usually English.

It soon became evident that we could not confine our activities to trading, and that it was necessary to formulate pharmaceuticals in the country itself and to produce plastics dispersions and textile auxiliaries. We acquired space in an industrial area on the periphery of Bangkok. Today, this area is completely surrounded by housing, so that we can operate only with the existing facilities. New facilities will have to be located further away from the city, because, understandably, no new industry will be allowed in residential areas.

Before the oil crisis of 1973, many plans were made in Bangkok, one of them calling for the establishment of a petrochemicals industry. The Japanese were the leading protagonists. They were looking in many areas in Southeast Asia for locations that offered a secure source of raw materials.

These petrochemical plans have long since been shelved. Thailand has no oil of its own, and—a decisive factor, to my mind—in spite of its growth, it will remain a relatively small market, not justifying large industrial installations for a long time to come. For this reason, the large-scale industries that have been established, mainly under Japanese control (for example, the polyester factories of Teijin), are now struggling to remain competitive in the export business.

In the meantime, Thailand has carried out its first experiment with a government in the style of Western democracy. It appears, however, that the country is not yet ready for such a form of government. The good intentions of the new government after the revolution of 1973 have largely evaporated because the educational level, at least of the older generation, is not adequate for shared political responsibility.

As a result, a military government was back in power by 1978. It faces difficult tasks, because Thailand's political environment has not changed for the better. Fortunately, the American domino theory did not prove to be valid. The departure of the American troops from South Vietnam and the tragic events of Laos and Cambodia have had no direct consequences for Thailand and the other Southeast Asian states. But the American troops have also been withdrawn from Thailand at the behest of the Thai government, thus depriving its economy of a powerful flow of foreign exchange.

Relations between the United States and Thailand, temporarily strained, have been restored, and Thailand continues to feel part of the Western world to which it looks for the economic support to smooth its path into the future. Such help is essential, not only to raise the standard of living, but also to cope with the unending stream of refugees from surrounding countries.

Though flooded by tourists, Bangkok has remained attractive. There is hardly another place in Asia that can compete with its hotels and amusements. But more important are the many cultural attractions of the city and its surroundings. The Buddhist temples in the Thai style, with their carved gables and staggered roofs covered with colored tiles, are among the world's great wonders.

Twenty years ago, Thailand was regarded as the land of the free peasants. But harvest failures and low income from agricultural products impoverished the rural population. Many farmers had to sell their land to large landowners from the city. It is asserted that about 60 percent of formerly independent peasants have now become tenants. Large landowners and agents control most of the trade in rice, fertilizers, and agricultural chemicals.

Thailand: 700 Years of Tradition

In solving its difficult political problems, Thailand will no doubt succeed by adapting to the changes in neighboring countries. There is no other country in Southeast Asia that has managed as Thailand has to escape foreign domination for more than 700 years of its history. Even today, the traditional forces of the royal family, the predominant Buddhist religion, and the freedom-conscious population are exerting a great influence on the affairs of the country. In the power play between the Americans, the Chinese, and the Russians, Thailand performs an important role: to maintain the equilibrium. It is therefore enjoying the protection of the United States and China, a guarantee that the country will continue to deserve its name, which, in translation, means "land of the free."

Twice in the early years of our activity in Thailand, I had an opportunity of making an excursion to Angkorvat. We flew in a rickety DC-3 to the capital of Cambodia, Phnom Penh, which architecturally shows many traces of the French colonial period. From Phnom Penh it is only a short distance by car to the temples of Angkorvat. In size and artistic design they surpass even the magnificent temples of southern India. The French rescued these thousand-year old sites from the primeval forests which had overgrown them during past centuries. On my first visits, I saw temple ruins whose masonry had been penetrated by tree trunks and roots. The erection of these temples in but a few decades must have demanded a vast effort from the people and greatly sapped their strength. Perhaps this is why there was no significant Khmer culture after the abandonment of Angkorvat following a wartime defeat in the fourteenth century.

Manila: Requiem for Eric Stern

With his wife and small son, Eric Stern had left Vienna in the mid-thirties just in time. He went to Manila where he found a new home. When I met him for the first time in 1955, he was the manager of the chemicals department of Menzi & Co. This company had cooperated with I.G. before World War II. Afterward, it also looked after Hoechst's industrial business. But Eric Stern's scope was limited because

Menzi & Co. had no sizable funds for its chemical business. Menzi junior, a Swiss citizen with Philippine nationality, was an influential man with many connections in commerce, industry, and the political circles of the Philippines.

I tried at that time to establish an independent Hoechst company in partnership with Menzi and his company. But these efforts were unsuccessful. So we decided to go our own way in conjunction with Philippine financial partners. Eric Stern was happy when we entrusted him with the management of Hoechst Marsman & Co., the small company we founded in 1958. Our partner, Mary A. Marsman, was an American citizen who, with her husband, had established a trading company supplying the mining companies of the country. Apart from agriculture, ore deposits are one of the major sources of the country's income.

In 1960 we established the first pharmaceutical plant in the buildings of the Marsman company. Mrs. Marsman died shortly afterward, and the small company became independent. After being renamed Hoechst Philippines Incorporated, the company acquired 10,000 square meters of land in 1965 in Mandaluyong for the erection of a pharmaceuticals plant. To this were later added an agrochemicals plant and a small surfactant production.

We were thus well established in markets very important to Hoechst and were able to exploit the country's potential. In those early years, the only significant industry was the textile sector. But that suffered from the illegal imports of finished textiles, so business was rather unstable.

After 300 years of Spanish colonial rule, 50 years of American colonial administration, and 4 years of Japanese occupation, the country became independent in 1946 in compliance with its 1936 treaty with the United States. The wife of our friend Eric Stern had been killed in 1945 during the brutal rearguard actions of the Japanese. It was a tragic loss for our friend. After that, he regarded the Philippines with a mixture of love and hate. On the one hand, the country had given him and his family refuge; on the other, he had lost his wife there under the most harrowing circumstances. Finally, the administration refused him a Philippine passport to which he was surely entitled after living so many years in that country.

Somehow, Stern was always a tragic figure. His son had emigrated to Australia to study and remained there. Thus, old Eric Stern was left alone to eke out a modest existence from his work with Menzi & Co. This changed only when he took over the management of the Hoechst company. At that time, the Philippines were at a low ebb both politically and economically. The country was not prepared for the sovereignty granted by the Americans. There were troubles and internal insecurity. Corruption affected almost every facet of life.

The situation changed in 1965 when Ferdinand Edralin Marcos was elected president. He favored an authoritarian regime from the beginning, and later gained official approval of that regime. After the administrative apparatus had been thoroughly purged, the country changed visibly. Once a dirty town suffocating in traffic chaos and threatened by general insecurity, Manila has become a clean and orderly city with 1.6 million inhabitants. The city center seems very modern. Many new hotels have been erected along the beautiful boulevard on Manila Bay, and processing industries are expanding on the periphery of the city.

Marcos, supported by his politically ambitious wife Irmelda, has initiated an agrarian reform program. Rice growing has been modernized and agricultural workers without land are to become small land holders. This experiment seems also to be blessed by success. The International Rice Institute, located in Los Banos in the vicinity of Manila, has become the rice-growing authority throughout Southeast Asia.

Rice constitutes about one-half the agricultural yield and is the staple food of 46 million inhabitants distributed over more than 7000 islands. Two-thirds of the population live on the main islands of Luzon and Mindanao.

The other part of the Marcos program is the promotion of industry. This applies especially to woodworking, which is intended to replace the export of timber, but it also encompasses mining and the textile industry. The chemical industry has so far been confined to a few basic chemicals. However, the first plastics production plants have meanwhile been established. They, in turn, have been followed by the appropriate plastics processing plants. Refineries and petrochemical industries are also planned, although they are not likely to be started until the beginning of the 1980s. In the meantime, off-shore oil has been found off the island of Palawan, which should help to relieve the rather strained balance of payments.

There have always been foreign investments in the Philippines. Before the Americans granted sovereignty to the country, they concluded the Laurel-Langley agreement, which secured for the United States the right of military, political, and above all, economic intervention. That agreement has now expired, so American capital no longer enjoys preferential treatment. Moreover, foreign investments must now involve Philippine partnership. But it is not difficult to find reliable financial partners, as we did for our own company.

Eric Stern had surrounded himself with a group of German and Philippine colleagues whom he managed in a pàtriarchal manner. He always took a great interest in promising new recruits. One of his protégés was Dieter W. Dopheide, for quite some years now Hoechst's top man in Athens. Eric Stern died rather suddenly in 1973. A heavily built man, he had long been suffering from ill health. The many years of uncertainty and insecurity had left their mark. In 1975, I visited his simple grave in the Jewish section of the main cemetery of Manila.

Supplying 7000 Islands

The citadel of Manila, which is maintained as a national monument, is near this cemetery. The only monument that survived the extensive destruction of the city in World War II, it reminds the viewer that Manila had been founded as the headquarters of the Spanish colonial government in the middle of the sixteenth century. Apart from Tagalog, the country's language is English, spoken by almost everybody. Spanish is confined to high society, although the names of people reveal the long Spanish rule.

With more than 7000 islands, the country is not easy to manage economically. The industries are concentrated mainly on the two major islands. The pharmaceuticals, on the other hand, have to be distributed over the entire country. This was a task that exceeded the resources of the small Hoechst company. We therefore enlisted the help of a Philippine distributor who works for Hoechst as well as other companies, covering the country down to the last corner with an army of sales representatives.

In agrochemicals, indispensable for the modernization of agriculture, we are trying to stay independent, particularly in providing advisory services. Unfortunately, we suffered a serious reversal in this field some years ago when we found that the head of our agricultural department had committed extensive frauds and had made arrangements that harmed our business. Luckily, this was one of the very rare cases in which people in leading positions betrayed our trust.

Luitpold Schneider played a vital part in the establishment of our organization in Manila. Schneider had been an employee of I.G. in India, where he was interned during the war. On his release, he worked for Naphthol-Chemie in Offenbach until it was merged with Hoechst. Later, he was appointed regional director for Southeast Asia in Bangkok. From there, he traveled regularly throughout Southeast Asia for many years, using his long experience to offer advice and guidance in commercial and human relations to the young people whom we had sent into these markets with little foreign experience.

In time, the function of regional directors became obsolete. However valuable the function may have been in the early years, the managers of the Hoechst companies abroad became eager to have direct contact with the parent company rather than through the mediation of an interposed authority.

The work begun by Eric Stern is now being continued by younger people. It looks as if internal security in the Philippines will be maintained under Marcos. The rapidly growing population, increasing at approximately 2 percent a year, will have every opportunity to develop its agriculture and industry. The country is likely to occupy an important position in Southeast Asia not only in political and strategic respects but also in economic considerations. Following the change in power in Vietnam, Marcos has established relations with the People's Republic of China and certain East European countries. The Philippines aim at pursuing a balanced foreign policy.

A young team consisting of Philippinos and Germans has the task of conducting our business in the Philippines and will contribute to the nation's industrial and agricultural development.

India: Subcontinent of Contrasts

Who has not been fascinated by India in one's youth? I devoured the romantic *Jungle Book* of Rudyard Kipling. The film *Bengali* is one of the outstanding memories of my younger years. Later, I tried to penetrate India's intellectual world and the Buddhist creed. I was especially fascinated by the story of Siddhartha Gautama, the son of a noble family, who lived almost half a century before Christ. He sacrificed glory and wealth to search in solitude for the meaning of existence, until after seven years he really became the Buddha, "the enlightened one."

Today, Buddhism is no longer widely practiced in India. Most Indians confess to Hinduism. But throughout other countries of East Asia, the various schools of the Buddha's disciples are still encountered.

Rabindranath Tagore, an Indian philosopher of this century and much read in the West in the thirties, engaged my attention for a time. Tagore, who came from an old family of scholars in Calcutta, made great efforts to promote understanding between the Indian and Western cultures. He founded a school in Schantiniketan in Bengal to spread his idealistic teaching of a new Indian inwardness.

During my school days, I also heard for the first time of Gandhi, whom his followers called the Mahatma, "the great soul." He was India's leader in the campaign of passive resistance against British colonial rule. In the thirties, his teaching had a special relevance elsewhere for people who saw the violence meted out, as the order of the day, to dissidents.

During our honeymoon trip to India in 1956, we were on the road for almost six weeks. At first, we needed some time to adjust to the stark contrasts between poverty, misery, and dirt on the one hand and the cultural and historical treasures on the other. We closely observed the experiment of Indian democracy and we saw how questionable its success may be. We also noticed the Indians' many reservations about foreigners. But my wife and I probably fell a little in love with India during this first extended trip there. For this reason, we have returned often and with pleasure.

Work and Food for 850 Million?

The graceful charm of Indian babies, carried on their mother's hip, can easily let one forget that population growth is the biggest of all Indian problems. India's population today is estimated at more than 600 million. This means that every seventh person in this world is an Indian. There will be 850 million Indians in the year 2000. Where are they to find food and work? In the countryside, the joint family system is regarded as a kind of substitute for social insurance, nonexistent at present. Some time will be needed to convince the ordinary people that the joint family system is not an acceptable solution in the long term.

Since gaining independence, India, like all young nations, has strongly promoted industrialization. The state continues to exert great influence on economic development. Jawaharlal Nehru, largely molded by English culture and civilization, elevated the socialist pattern of society to a principle of Indian policy.

Nothing changed in this regard after Nehru's death. The planning bureaucracy has in many ways become the worst enemy of rapid and efficient industrialization. It has suppressed all private initiative in favor of state enterprises that have been established only slowly and rather inefficiently. Foreign investors, prepared to create new production facilities in this vast market, were increasingly discouraged by the bureaucracy. Later, laws even prevented the establishment of foreign companies.

Nevertheless, some successes have been achieved in industrialization. In my view, this is largely thanks to a substantial number of large families that, with strongly nationalistic feelings, did not allow their ambitions to be curbed by the unrealistic restraints of the bureaucracy. The enormous industriousness, intelligence, and skill of the Indians were essential factors in the upsurge.

Foreign development aid has helped to promote industry and agriculture more than the Indian public has. Some years ago, we visited a steel mill in Rourkela that had been constructed by a German consortium. For a time, the mill had a bad reputation because of the inexperience of its local managers. Today, this steel mill is one of the pillars of Indian industrialization.

Unrestricted development aid was also employed to compensate for many years' neglect of Indian agricultural policy. Some fairly impressive examples demonstrate how agricultural yields can be substantially increased through good organization and relatively simple methods of cultivation and irrigation. Since food supplies are so critical for a country like India, agriculture may share emphasis with industrialization in national development plans.

Many Restrictions

In the autumn of 1968, we accompanied a large group of European journalists to attend the inauguration of Polyolefin Industries, Ltd. (PIL) in Bombay. In two-and-a-half years, a factory had been constructed for the annual production of 30,000 tons of high-density polyethylene, a plastic which had been available in India only in small, imported quantities.

The company, established after years of negotiations, was headed by the Mafatlal group, which maintained friendly relations with Hoechst. The ethylene raw material was supplied by the neighboring Nocil (National Organic Chemical Industries, Ltd.), in which Shell, Mafatlal, and Public each had a 33 percent share. The PIL shareholders were Mafatlal, Hoechst, and Nocil. The plant has operated well since then and has contributed to the development of the plastic processing industries in India, thus creating new jobs.

We have limited possibilities for expansion, despite growing market demand, because Nocil, the raw material supplier, could not obtain permission from the authorities to expand its capacities for base materials. I am convinced, however, that the industry established there will overcome its problems and develop further. The economy of India urgently needs the product. Even the bureaucracy in Delhi will not be able to ignore this reality in the long run.

After the war, Hoechst resumed its activities in India as soon as possible. With its large textile industry, the country was one of the most important dyestuffs customers for I.G., and we were soon able to build on this tradition. Daji Chendwankar, an Indian official of the former I.G. agency, organized a small trading company and collected some of his former staff around him. He then skillfully reestablished the business in dyestuffs, auxiliaries, and chemicals. Money was provided by a rich Islamic family that had made its fortune in the bazaars. But that business was on the wane because the textile industry, having come of age in the meantime, wanted to obtain raw materials directly from the manufacturer.

Unfortunately, our friend Daji, with whom I spent many pleasant hours in both India and Germany, died at a very young age. For a time, his colleagues continued the business to the best of their ability, but it had to be abandoned eventually.

Fortunately, we had meanwhile renewed our connection with the Mafatlal family, important in the Indian textile industry and, even before the war, a customer of German chemical industry. In partnership with this family, headed by Arvind Mafatlal, the oldest brother, we founded Hoechst Dyes and Chemicals Ltd. This company has developed successfully. In addition to its activities with PIL, it constructed its manufacture of plastics dispersions. It has not been able, however, to develop any further production facilities because of the restrictive licensing policy pursued by Delhi.

At the same time, a new pharmaceutical organization was formed by Hoechst. Vittal Mallya, an enterprising private industrialist whose success had been mainly in the brewing industry, had offered Hoechst a partnership. This led to the establishment of a large modern pharmaceutical plant in Mulund, an industrial suburb of Bombay. The company operates as Hoechst Pharmaceutical Ltd. (HPL). It employs more than 2000 people, more than half of them in the manufacturing operation. In line with our downphasing policy, a large part of the required base material is manufactured in the country.

Some years ago, we also established a research center to develop pharmaceuticals specifically designed for India. This center, employing two dozen natural scientists of Indian nationality, has achieved its first successes in the search for new antibiotics.

The pharmaceutical industry in India does not have an easy time, especially when it is partially owned by foreigners. Pharmaceutical prices and profits are again the subjects of political discussion, although there is as yet no medical insurance system of the Western variety. Following parliamentary inquiries, it was proposed that pharmaceutical patents should be limited and that foreign ownership should be curtailed. I hope that the Indian authorities will realize in time that both research and the pharmaceutical industry will require close relationships with the Western world for a long time to come. There is hardly another field in which international contact and exchange are so important.

With respect to dyestuffs, German chemical industry did not find it so easy to prepare in good time for domestic manufacture in India. Indian industrialists, acquiring know-how from Italy (Indian Dyestuffs Industries and Amar Dye Chem) and from England (Atul), had started up their own small plants. Together with Indian partners, Bayer founded Colour-Chem Ltd. in Thana near Bombay, a company which Hoechst joined in its establishment phase in 1956. This undertaking thrived and is today one of the largest Indian dyestuffs manufacturers, although it has to contend with powerful domestic competition.

Hoechst Dyes & Chemicals Ltd., founded with the Mafatlal family, was able to supplement its decreasing import volume with Colour-Chem products. In addition, there was the Hostalen from PIL and from its own production, as well as Mafron, a fluorinated hydrocarbon from Navin Fluorine Chemicals. The volume of imported chemicals, however, has lessened from year to year.

Sunrise over Mount Kanchenjunga

In the autumn of 1960, some colleagues and I went on a round trip of the country in a small plane chartered from Indian Airlines. This trip started with a direct flight from Bombay to Benares. In Benares, the holy of holies on the Ganges river, the Hindu religion and the life-style of its believers are most dramatically seen by European eyes.

Memories of Darjeeling fascinate me even today. Early one morning, we made an excursion to Tiger Hill to see an unforgettable sunrise above Mount Kanchenjunga. In Kerala, at the southern tip of the continent, we found definite traces of Portuguese missionary activity. Goa, a Portuguese colony for a long time, had been incorporated into India in 1961 by Nehru, in this case not quite so concerned with the nonviolence he was fond of preaching.

Our trip also took us to Calcutta, a city of 6 million people and, to me, a nightmare of dirt and misery. But Calcutta has extensive industry. Most of it was founded during the British era, as were the well-run country clubs that ease life for the foreigner.

Some years previously, we had established a pharmaceutical plant in Chittagong, which now belongs to Bangladesh, and we paid a visit to this locality. The mouth of the Ganges river had been devastated by violent storms, and nearly the whole of West Bengal was under water. The people, already the poorest of the poor, were thus robbed of even their last possessions.

Proceeding north, the climate became more pleasant. Besides the industrial visits, we found time to enjoy some of the cultural riches of the country. The splendid Taj Mahal had been built by Emperor Shah Jahan as a tomb for his wife. In the moonlight, this marble symbol of undying love appears almost supernatural. I have seen even the most unemotional characters moved by its beauty.

In the government center of New Delhi, we had an audience with the President, Dr. Radhakrishna. We talked about economic problems in an informal and friendly way. On leaving, Professor Winnacker handed the President a document promising three scientific scholarships. "Only three?" asked the aged President with a smile. "Six would have been better."

The final stop on our trip was Madras. Industrially, this part of India is little developed. But, as always where there are many people, the city and its catchment area was a large market for Hoechst pharmaceuticals. For this reason, we had for some years maintained a branch office there that dealt with hospitals and doctors.

Our office had arranged a tour of the General Hospital of Madras. The hospital has more than 1000 beds. Even more impressive, however, was the clinic, visited every morning by hundreds of sick people seeking treatment. It was most inspiring to see the devotion with which the Indian doctors and attendants looked after them, even though the available means for clinical treatment were fairly limited. During our visit, we donated a significant quantity of pharmaceuticals.

Birthday in Southeast Asia

In Madras at the end of January, we celebrated my birthday together with our friends. The wine was supplied by the liquor store in the hotel. Afterward, we went to the adjoining cinema, probably the only available amusement in Madras, at that time populated by about a million people. The cinema featured *Carmen Jones,* a movie version of the classic opera *Carmen.* Today, very few foreign films can be seen in India, and they are usually American hits with Indian soundtracks. The indigenous film industry is well on the way to replacing foreign films altogether.

Because of its enormous population alone, India has tremendous reserves of energy. To garner them is not an easy task for the government. There is no shortage of trained and trainable people, but opportunities for work and work training are inadequate. If India were to deal more freely and less dogmatically with the Western world, the existing potential could be much better utilized.

Afghanistan and Pakistan

In 1977, I revisited Kabul for the first time in eight years. Between my visits, the capital of Afghanistan had changed—greatly to its advantage. In the suburbs, we visited the Hoechst pharmaceutical factory which Professor Winnacker and I had discussed when visiting Afghanistan in 1969. It was, incidentally, the first pharmaceutical factory in the country. Under the direction of Dr. Schaden, we had succeeded within a remarkably short time in turning this expensively equipped factory into a paying proposition. The company has some 300 employees, and it is producing drugs under license for other pharmaceutical companies as well.

For thousands of years the area of the present Afghanistan was the crossroads of the overland trade between India, China, and Europe. Whenever this trade remained unaffected by war, Afghanistan flourished. At various times, it has been the center of large empires, which included both the present Afghanistan and large areas of Central Asia and India.

It is generally said that present-day Afghanistan dates from 1747. At that time, Ahmad Shah Durrani established the area as an independent state under Persian hegemony. During its wars against the British in 1841–1842 and 1878–1889, Afghanistan defended its independence successfully, but it had to give up some of its territory in the East. Eventually, it drifted into a kind of protectorate relationship with British India.

Although Afghanistan remained independent in matters of internal politics, its foreign policy was determined by Britain. This status greatly affected the country's economic development. Never a colony, Afghanistan did not profit from the investments of the colonial powers. On the other hand, the Afghans were cut off from contacts with other major nations. This isolated geographical situation kept the country at a fairly low state of development during the nineteenth century. Afghanistan achieved full independence within its present-day borders only in 1919, in a third war against the British.

Five Years, Two Revolutions

In 1973, there was a revolution in Afghanistan, the king was deposed, a republic was proclaimed under Mohammad Doud, its new president and prime minister, and plans were made to speed up the nation's development. Germany, which has had excellent relations with Afghanistan since World War II, played a major role in these plans.

Both Doud's political career and his life came to a sudden end when the country erupted in revolution in the spring of 1978. As the military broke into the government palace, he was shot and killed. It remains to be seen what policy the new team under Noor Mohammad Taraki intends to follow, and whether it will be able to escape Moscow's tutelage.

Afghanistan still lives primarily by agriculture. Its best-known products are probably the skins of its karakul sheep, carpets, and lapis lazuli, which comes from a mine opened up 5000 years ago. The color of those stones is unique. Lapis lazuli from Afghanistan has been found even in the tombs of the Pharaohs of the Egyptian dynasties. Raisins, dried fruit, skins, and pelts are exported. Recently, Afghanistan has also produced considerable quantities of natural gas, most of it piped to the Soviet Union.

On a sunny February morning in 1977, our small column of cars set out for the Khyber Pass on Afghanistan's frontier with Pakistan. The

Khyber Pass was important both in antiquity and in the frontier wars that raged only a few decades ago between the British and the Pathans. Today, this area enjoys absolute peace. We viewed the beautiful landscape and made the descent into the plain around Peshawar, the capital of the North-West Frontier province. This area has become famous through the excavations of the Gandhara culture, which shows both Greek and Buddhist influences.

Pakistan

Our next stop was Karachi, the most important industrial and commercial city and port in Pakistan and formerly its capital. But for strategic and psychological reasons, this honor was transferred to Islamabad, a new city in the north of the country. Together with its twin city, Rawalpindi, Islamabad has some 100,000 inhabitants, while Karachi numbers some 3.5 million people.

On the long flight to Karachi in the fully booked DC-9 of Pakistan International Air Lines, I recalled some of the history of the Hoechst developments after the last Indo-Pakistan war. I also thought about the development of the new states of West Pakistan and Bangladesh, formerly Pakistan with an eastern and a western part.

During my first visit to Karachi more than twenty years earlier, my wife and I had stayed in a hotel on the outskirts of the city. At that time, Karachi had the appearance of a small harbor town, which of course it had been during the British colonial era. The hotel was called Beach Luxury, and people used to joke about its name, saying, "No beach, no luxury." Since then, the ambience has changed. Karachi has become a large modern city. There are decent hotels, the best being the Intercontinental, an oasis for the globe-trotter. The slum areas, formed during the original immigration period from 1947 to 1948 by refugees from India, have been replaced by standard housing tracts.

An Artificial Creation

The founding of the Islamic Republic of Pakistan in August 1947, shortly after India had won its freedom, was a violent operation. Its initiator was Muhammad Ali Jinnah, who had risked much and did not flinch from even the bitterest conflict. Much blood flowed before Pakistan emerged as an independent state from the historic process that ended with India's departure from the British colonial empire.

Many years later, it was clear that this state was an artificial creation. Inevitably, separation came in 1971 after the East Pakistan candidate, Sheik Mujibur Rahman, had obtained an electoral majority. General Yahya Khan opposed the result. Civil war in East Pakistan was soon to lead to military conflict between India and Pakistan in both the east and west of the country. The Indian army soon overran the Pakistanis. East Pakistan gained its independence and became the People's Republic of Bangladesh.

West Pakistan was thought not to have much chance for survival. However, it recovered from defeat by India and separation from the geographically smaller but more populated eastern part of the country. Indeed, in West Pakistan the standard of living increased more quickly than if the new state had had to bear, as before, the burden of underdevelopment in Bangladesh. I believe it was always an illusion to imagine that the eastern part of the country, 2400 kilometers away from the capital, could be governed from the west.

Hoechst, too, was greatly affected by the division of the country. Far-seeing people in the parent company thought of locating the required pharmaceutical plant in the eastern area. Therefore, Chittagong, the second largest city and East Pakistan's export center, was chosen as headquarters. Hoechst Pharmaceuticals Company Ltd. started production there in 1960.

Then came the division of the country, which no one had expected so quickly. Until then, there had been no sign of the advantages which Hoechst had expected when it decided in favor of the industrially underdeveloped East Pakistan. These included fewer import restrictions on raw materials and more favorable working conditions. But now, our seeming strategic error proved to be a blessing in disguise. The pharmaceutical plant in Chittagong became one of the few large pharmaceutical producers in the new state of Bangladesh. It therefore received preferential treatment from the government, which was concerned with the prevention of epidemics and disease. To this day, we are most satisfied with our business in Bangladesh.

In West Pakistan, on the other hand, the import of finished specialties had become more difficult, if not impossible. However, we had foreseen this development and had started to plan a new pharmaceutical factory as early as 1970. With our two partners, Pir Ali Gohar and Syed Barbar Ali, we therefore founded a new company, Hoechst Pakistan Ltd., and began construction of a pharmaceutical plant on the outskirts of Karachi. This very modern, but very expensive, plant went on stream in 1972, and several other manufacturing facilities were soon added. Foreign investment in the chemical industry is now largely at a standstill. Between 1971 and 1977, under the government of Ali Bhutto, since overthrown by the military, state socialism was the official policy, and the country suffered a wave of nationalizations.

Chemical Industry with 15,000 Employees

The main artery of the country is the Indus River and its tributaries and canal system. As a result, most economic activity takes place along the axis of Peshawar–Lahore–Hyderabad–Karachi.

The agricultural center of Pakistan is the Punjab. Its largest cities, Lahore and Lyallpur, are also the largest industrial centers after Karachi. More than two-thirds of the labor force are employed on the land, where the main crop is cotton. Pakistan is one of the largest cotton producers in the world. Not surprisingly, then, the textile industry is one of the oldest and also one of the most progressive. The remaining industries, too, are concentrated in the agricultural centers of the Punjab and Sind.

Karachi is the commercial and administrative center. The chemical industry employs some 15,000 people, many in fertilizer factories producing mostly nitrogenous fertilizer. Some 65 percent of the country's requirements can be produced domestically. The state-owned National Fertilizer Corporation is the central body for the expansion of the industry. It was both a great honor and a great burden for Babar Ali when the government entrusted him with the management of this company. He held that position until 1977.

I should also mention a national polyester fibers factory with a capacity of about 10,000 tons per year. We had an opportunity to participate in this venture, but doubted its profitability. Other industrialization plans are afoot to exploit the important natural gas deposits. The gas is pumped from the north of the country down to Karachi. Oil production, though limited, helps to reduce the import requirements of this country, chronically deficient in foreign exchange.

Twenty years ago, we and Bayer allowed ourselves to be talked into investing 15 percent each in Pak Dyes, a national dyestuffs project, located in a development area in the north. Neither we nor the users nor the state derived much satisfaction from this enterprise in Iskanderabad. It still exists, but I doubt whether it has any chance of developing further. In 1970, our acquisition of Berger provided us with a well-established paint factory in Karachi.

As everywhere, there were enterprising businessmen in Pakistan whose services we were able to enlist after the war. In Karachi it was Richard Rauer, an I.G. color technician from China, and F. W. Staeber, who later pioneered our business in Afghanistan. Staeber belonged to a group of former I.G. people who were interned in Dehra Dun in India at the beginning of World War II. The same happened to Karl Josef Seyfried, who was in charge of our dyestuffs business for many years and to whom the dyestuffs division owes much gratitude. Also interned was Rolf Magener, who has been the financial chief of BASF in Ludwigshafen for many years. He eventually escaped from a British internment camp with the mountaineer and expert on Tibet, Heinrich Harrer. He described his adventurous escape in the widely read book *The Chances Were Zero.*

In many respects Pakistan is a difficult country. But the great vitality of its people may one day provide it, like India, with an important role in the Middle East. Even today, the Persian Gulf states employ thousands of Pakistani technicians and skilled workers who have learned much in industry and at the universities of their native country. Many received their vocational training in Europe. For the time being, however, they are selling their know-how abroad where it is fetching more money than in their own country. A decisive factor in the future development of Pakistan is, of course, the long-term relationship with India. That was worsened by the war of 1971 and especially by the still unsolved Kashmir problem. But the leaders of the Indian subcontinent will surely realize one day that peaceful coexistence is indispensable for the prosperity of all.

16

The South Pacific: A Potpourri of Islands

When I paid my first visit to Singapore in the mid-fifties, the Raffles Hotel was still the center of social life and the preferred hostelry of the foreigners. The hotel was built in the typical British colonial style, with large fans on the ceilings instead of an air conditioning system. The entire city is set in ample tropical surroundings, but its mixture of nineteenth-century European and Asiatic architecture, designed for the requirements of the tropics, lacked distinction.

In 1956, Singapore, the "city of the lion," was still a British crown colony, as were the adjoining Malaysian territories. It gained limited independence three years later. At the behest of the government of Malaya, which had attained independence in 1957, it joined Brunei, Sabah, and Sarawak in northern Borneo in 1961 in the Federation of Malaya. Singapore became an independent republic in 1965. The racial contrasts between urban Singapore, with a population that was 75 percent Chinese, and the fertile Malaysian region were simply too great. Economic activity was controlled largely by the Chinese.

At that time, Hoechst was still represented by Behn, Meyer & Co., the Singapore office of an old and important Hamburg trading house. In its spacious offices, bounded by vast warehouses, there was a corner with a few desks behind which a few people were tending the affairs of the three I.G. successor companies and other German chemical firms as well.

It was clear to me at once that this arrangement would not be adequate in the long run to take care of the potential of Singapore and the adjoining country. But it took until 1967 before we finally established Hoechst Singapore to take over from our Hamburg friends.

The company started with a German boss and half a dozen local employees. That the agency developed quickly and successfully was hardly surprising in view of Singapore's rapid economic growth in those years. Since then, I have visited Singapore several times on the way to or from the Far East, and I have been able to observe its progress *en passant*, as it were. In 1977, however, I was able to stay longer, talk to our people there, and make closer contact with them.

I found that the city had undergone a fundamental change. The mostly single-story colonial buildings along the main thoroughfares had been replaced by skyscrapers with offices and spacious stores. In but a few years, Singapore had assumed great importance in Southeast Asia as a supermarket for goods from all over the world. In quality and price, it holds its own with Hong Kong. Luxury hotels have been built whose equals are found only in Bangkok and Hong Kong.

Lee Kuan Yew, undisputed ruler since 1965, is also the head of the People's Action Party. The party holds all sixty-nine parliamentary seats, governing through a democratic single-party system. Fortunately, however, this monopoly of power does not imply any restriction in personal freedoms.

Singapore's economy soon revealed the hand of efficient Prime Minister Lee. He has created a growth climate for both domestic and foreign industries in the confined space of Singapore, smaller in land area than Hamburg.

In contrast with most other developing countries, Singapore courts foreign capital. Investors enjoy substantial tax privileges for the first years. Shipping and the oil, iron and steel industries, as well as metalworking and precision-engineering companies, dominate the industrial scene. In addition, there are textile processing plants, a remarkable building materials industry and other small and medium-sized enterprises.

In the last few years, no new labor-intensive industries have been set up in Singapore. Even existing companies are moving to the neighboring countries of Malaysia and Indonesia, where wage levels are still considerably lower. This, too, is in line with the economic policy of the Singapore government.

Socialist but Anticommunist

Singapore's foreign policy has become extremely flexible. The government describes itself as socialist, which it certainly is. But it is just as certainly anticommunist, and does everything in its power to prevent the penetration of communist influence.

Although Singapore has no oil, there are a number of important refineries on offshore islands. One of their functions is to provide the starting material for the ambitious petrochemical developments which the city-state has been planning since the oil crisis of 1973. The plans attracted investment by a Japanese consortium headed by Sumimoto. However, it is now no longer clear what advantages Singapore offers the petrochemical industry. It lacks its own raw materials, and even for Singapore itself, the petrochemical industry would provide few new jobs.

Singapore, in terms of growth and prosperity, towers over nearby countries. Whether this condition will endure depends on the future flow of foreign capital. I have no doubt that investment will continue, particularly in view of the Japanese expansionist urge, which can be felt throughout Southeast Asia. On the other hand, much Chinese capital too has found its way into Singapore. It may have been guided by the thought that Singapore might one day assume the role of Hong Kong. To be honest, I hope this will not happen soon.

Malaysia: Rubber and Oil

In the spring of 1957 on the way back from Indonesia, my first stop was Kuala Lumpur. I was with a small group, visiting a number of countries in East Asia. We wanted to see how rubber latex was obtained in Malaysia, and our friends took us to one of the large plantations. The country is the largest latex producer in the world, supplying more than half its total requirements.

Very briefly, latex production goes like this: The bark of caoutchouc trees, which can be up to 20 meters high, is scored—usually in a V-shape. The latex which has accumulated underneath the bark will then ooze out of this incision for several hours. It is collected in beakers and poured into buckets through filters. The latex coagulates under the influence of formic or oxalic acid. This viscous substance is put through a calendar in which it is pressed into sheets 1.5 to 3 millimeters thick. The sheets are then dried, rolled into bales, and shipped.

As representatives of the chemical industry, we asked ourselves whether such an apparently primitive operation could compete with synthetic rubber in the long run. But the plantation owner remained quite confident. He convinced us that cultivation and yields were only just at the beginning of their potential development. Indeed, much has happened since then with the aid of state institutes. The yield per hectare has increased considerably, and "Standard Malaysian Rubber" has acquired an excellent reputation.

The years since then have proved that rubber plantations can be justified since their product can stay competitive with synthetic rubber, although the latter has become much cheaper in the meantime. In practice, a blend of both materials is often used and offers a number of advantages. Rubber will therefore continue to be a rich source of foreign exchange for Malaysia. The wealth of the country is rooted in its agriculture, which is greatly favored by the tropical climate. In addition, there are natural resources like tin, of which Malaysia is the world's largest producer. There also are plentiful oil reserves on the coast of the South China Sea. Thus, Malaysia, which gained its independence in 1957, has a very healthy economic basis, which has been further strengthened by federation with Sabah and Sarawak, by an infrastructure established during the British era, and by a highly efficient administration.

Racial diversity is probably Malaysia's most precarious problem. The Malays themselves constitute almost 50 percent of the more than 12 million inhabitants, but the efficient, industrious Chinese are the economically decisive factor, apart from the large foreign companies. This situation persists in spite of the preference given to the Malays, the Bumiputras, or "sons of the earth." The Chinese play the leading roles in the economy, in commerce, and in trade. But this state of affairs is now to be changed by law. Malaysian participation and—in the long term—Malaysian majority ownership are to be the rule.

When Malaysia became independent, we felt it was necessary to establish Hoechst Malaysia in Kuala Lumpur. This was done in 1968 with a small group of Europeans and about three dozen Malaysian staff, most of them of Chinese origin.

This company did not confine its activities to imports, which soon flourished. Textile auxiliaries for the fairly well developed textile industry were produced in the country itself, as were emulsifiers for crop protection agents. The acquisition of the Berger group in England provided us, quite unexpectedly, with a paint factory in Malaysia, with a branch company in Singapore. Both are flourishing and cooperating well with the Hoechst companies.

We also produce plastic pipes from Hostalen in Malaysia. The raw material is produced by Polyolefins Industries Ltd. (PIL), a partnership company in India. Together with Hoechst Malaysia and domestic partners, PIL founded Polyolefins Pipes Ind., which has recently begun to manufacture high-density polyethylene pipes. These pipes are needed for ore-flotation in tin production. So far as quality and durability are concerned, they are an excellent replacement for the conventional steel pipes. Moreover, their light weight and flexibility make them much easier to handle. Last but not least, they provide some additional export business for India, so badly in need of foreign currency. I particularly enjoy recording these developments because they were not initiated by the parent company but by our people in Malaysia and India.

In 1978, we acquired an interest in Euromedical Industries in Penang. This company produces natural latex catheters for medical use. This investment is in line with the economic policies of the country, which aim at the use of domestic raw materials, increased labor intensity, and higher export volumes.

What will Malaysia's future be like? Its original pro-Western attitude has, in my view, shifted somewhat in recent years. Membership in the Association of Southeast Asian Nations (ASEAN) has caused Malaysia to assume a neutral stance, resulting in a certain cooling of relations with Western Europe. On the other hand, it is apparently being realized that excessive tutelage of the Malaysian people is causing adverse reactions among foreign investors, who are clearly holding back. It is probable that this situation will normalize. Malaysia is a country favored by nature, and it has a peaceful, friendly population. Its Chinese element will continue to ensure dynamism and activity. We can hope, therefore, that in this part of Asia at least, the increase in gross national product will outpace the rapid population growth. There are good prospects here for political stability and economic prosperity, which should also affect the neighboring countries.

Indonesia: Poverty on Tropical Islands

Suharto, the "smiling general" who succeeded Sukarno, had been President of Indonesia for more than four years when he granted us an audience in 1970. To a remarkable degree, he radiated the self-confidence of a man about to lead his country out of the stagnation and early decay of the Sukarno period. At that time, however, the capital, Jakarta, bore the marks of the Sukarno years, and it presented itself as a metropolis with little flair: a conglomerate of villagelike settlements called *kampungs,* of poverty-stricken quarters with picturesque bazaars, with some better neighborhoods and a few modern buildings in the center of the city. There were few indications of town planning.

Since then, Jakarta has developed into a capital with more than 5 million inhabitants. The dreariness of the Hotel Indonesia, which was erected under Sukarno as a giant caravansery, now contrasts with the modern, international style of many new hotels. As a whole, the city has not changed in character, but some streets have been broadened, slums have been cleared, and office buildings, administrative centers, and modern bazaars have been built. What is most impressive, however, are the development zones with their processing industries on the outskirts of town. They are clear signs of the considerable efforts the country is making in the cause of industrialization.

In view of the most recent development, I am reminded of my visit to a batik "factory" some years ago. The workshop was housed in the backyard of a single-story building. Production was still carried out by medieval methods, but that did not detract from the exotic magic of the cloth, used for the standard dress of the Indonesian women. Such workshops can still be found today, but other modern textile processing plants have been established, especially in the vicinity of Jakarta and Bandung. Many of them can readily compete with modern Western plants, and some employ thousands of people.

Bali Has Changed

For the weekend, we traveled to Bali, the "island of the gods and demons." Tourism was just beginning in 1970. There was one acceptable hotel and none of the present conglomeration of modern hotels, many of which do not exactly enhance the magic of the island. The women of Bali are beautiful and naturally graceful. I saw them at work, in their homes, in the fields, in the markets, at temple festivals and village fetes. The dutiful chronicler must record that the days of the "topless" Bali women are largely over.

I have developed a great weakness for Indonesian cuisine and especially that offered in private homes. The famous *rijsttafel* (rice table) contains numerous elements of Indonesian cuisine, but I was told that it has been greatly enlarged by the Dutch. They supplemented the refinements of the Indonesian dishes with the lavish abundance that is so characteristic of the Dutch cuisine. After all, the Dutch were in Indonesia for 400 years. They came as traders, and until World War II, they kept Indonesia under complete control. It is asserted that even the mail carriers were Dutch. That may be an exaggeration, but clearly the Dutch pursued a different colonial policy from the British. Relationships between Holland and its former colony have been normalized by now. Many Dutch people have returned to Indonesia, where they can carry on their business.

The Republic of Indonesia declared its independence in August 1945. It was, however, ill prepared for it, especially in regard to its administration and infrastructure. Because of internal and external circumstances, the young state consolidated slowly. There were a number of serious reverses. Between 1945 and 1949, the Dutch tried twice to regain control of the country by force. Independence was followed by years of disorder and revolt, which former President Sukarno quelled only with difficulty. His foreign policy was a risky balancing act between East and West. Internally, he tried a National Front, in which the Communists participated. Perhaps his greatest service to the country was to unite the various tribes of the widely scattered island states by introducing a uniform state language and compulsory schooling.

Economically, Sukarno led his country to the verge of ruin with his concept of Indonesian socialism. Internally, he soon lost so much support that by 1965, following an unsuccessful left-wing revolt, the military, under the leadership of General Suharto, seized power. After Sukarno's death in 1970, General Suharto was elected president by the People's Congress. The constitution of 1945 provided the president with considerable powers. There are political parties, general elections, a Congress. Nevertheless, the military continues to be the strongest political power in the country. Civilian technocrats, most of them trained in the West, have gained important positions, especially in economic planning. In the economy, the Chinese minority represents an important factor, as everywhere in Southeast Asia.

A Country Looking for an Economic Profile

After Sukarno had been relieved of state control, consolidation of the country and its economy made considerable progress. Indonesia gained a reputation as one of Southeast Asia's most stable countries politically. It looks as though it has averted the communist danger, in spite of continuing social difficulties.

About 90 percent of the population, which is estimated at some 146 million people and growing by 2.1 percent per year, are followers of Islam. With the declaration of independence, attempts were made to industrialize Indonesia, similar to those in other developing countries. In principle, the raw materials situation is favorable. Indonesia has considerable, not even fully explored, reserves of oil, natural gas, coal, tin, other minerals, and timber. There is a lack of capital and of the technical and administrative experience that would be needed in order to exploit these great opportunities. Difficulties range from exploration in inaccessible terrain to the practical implementation of frequently impractical programs.

For a time, an unrealistic euphoria reigned in the oil sector. As oil prices rocketed, the dynamic president of the state company for oil and natural gas, General Ibnu Sutowo, fell victim to the temptation of overexpansion—for instance, in the construction of a tanker fleet of international dimensions. "Big Ibnu," who lacked an adequate and experienced management staff, found himself in serious credit difficulties. The bankruptcy of the Pertamina Oil Company was avoided only because the state intervened through the Central Bank. Additional state debts had to be accepted. The international banking world showed a great deal of understanding, and the crisis could be contained. But enterprising Ibnu Sutowo, who had done much for development in Indonesia, had to be dismissed. The system of sharing production with powerful foreign oil companies, which he had promoted, was modified. The foreigners, who had received 35 percent to the Indonesian companies 65 percent, were now to have a 15 percent share, with 85 percent going to the Indonesians. The companies concerned reacted by curtailing their expensive oil exploration.

Meanwhile, however, compromises have been worked out in the usual Indonesian manner. The production of crude oil had risen to 1.7 million barrels a day at the end of 1977 (1 barrel contains approximately 159 liters). The utilization of the probably very large reserves of natural gas (with the export of liquid gas to Japan) has been initiated, and the prerequisites for a profitable development have thus been created. Oil and natural gas are by far the most important sources of foreign revenue for Indonesia. Another positive aspect is that the Indonesian delegation at OPEC negotiations pursues a middle course and is not one of the hawks. Several chemical projects based on natural gas projects, in which the Japanese are largely involved, are making good progress. But the interesting project of a floating area fertilizer factory had to be postponed. This factory was to be anchored near the natural gas fields in East Kalimantan (Borneo), which are not readily accessible by land. It had been designed to produce more than half a million tons of urea a year.

Production of fertilizers is being promoted. Agriculture has to be rationalized and intensified. This requires fertilizers, insecticides, and herbicides. There is also a plan to limit the export of crude oil and to produce finished products in the country's own refineries. Other energy sources, such as coal and water, are to be better utilized in order to meet Indonesia's own requirements.

Our Debut in Indonesia

Even under the Dutch, Indonesia was an important market for dyestuffs and pharmaceuticals. We therefore tried after the war to reenter this business as soon as possible. At the beginning of the fifties, Indonesian purchasing commissions came to Western Europe, including the Federal Republic of Germany, in order to buy the products the country needed. Naturally, first they turned to the well-known suppliers of the prewar era. We were able to complete fairly extensive pharmaceutical agreements with these commissions. Even the well-known Salvarsan, with which Paul Ehrlich and Hoechst had ushered in the epoch of chemotherapy in 1910, was still popular in Indonesia, although penicillin had in the meantime become the standard remedy for venereal diseases.

Corruption, the Achilles heel of all state-managed economies, had been widespread in Indonesia for a long time and could not be eradicated. A number of major campaigns were waged to eliminate it but did not achieve their goal. The continuing graft did a great deal of harm both to the Indonesian economy and to the country's reputation. In this context, it is of little comfort to know that corruption is by no means confined to Indonesia, or Asia, for that matter.

In 1956, we founded Hoechst Indonesia. Before that, we had cooperated for some time with Bayer and BASF in a joint advisory office. Our main aim was to serve the textile industry. In 1969, we set up Hoechst Pharmaceuticals of Indonesia, our first production plant in the country. We had succeeded in enlisting as partners Mr. and Mrs. Abidin, members of a respected Indonesian family. The Abidins had built up a pharmaceuticals business with the aid of a German pharmacist, Carl Friedrich Schütz. With the help of all three, we started up Hoechst Pharmaceuticals of Indonesia P.T. in 1972. The attractive pharmaceutical plant is located outside the gates of Jakarta.

In 1972, our plans for a synthetic resin dispersion plant also assumed concrete form. The project was started in 1975 with the commercial production of 3000 tons per year. The company involved, P. T. Pulosynthetics, has greatly expanded its facilities since then, especially the installations that manufacture starting materials for the growing domestic textile and paint industries. Hoechst gives employment to more than 400 Indonesians in its plants.

Among Indonesia's many problems is the precarious relationship between the native Indonesian majority and the Chinese minority. The Chinese, who have lived here for generations and often have assumed Indonesian names and nationality, make up about 3.5 percent of the population. But they occupy the important positions in commerce and industry, thereby antagonizing the indigenous population.

Official government policy favors assimilation, but is occasionally forced to take measures that are regarded as discriminatory by those affected. Linked to this is the problem of regulations concerning all foreigners (the so-called *Indonesianisasi*) and foreign investments. Ideally, the Indonesians would like to hold all the economic reins. Some of the more responsible and discerning Indonesian authorities realize, however, that foreign capital and modern technologies are necessary in order to sustain the country's rapid expansion.

The same problems have also confronted our Hoechst branch. Its activities were reduced to those of an advisory bureau. The regulations, which were announced in principle in 1967 and became obligatory in 1977, affect all foreign commercial undertakings in Indonesia. They aim at the Indonesianization of every company, except joint-venture production plants. Nevertheless, both sides are striving for partnership-based cooperation, without which successful long-term development will not be possible.

As the largest and most populous country in Southeast Asia, Indonesia is an important stabilizing factor in this formerly restless region. In August 1967, the ASEAN was founded by Indonesia, Thailand, Malaysia, Singapore, and the Philippines. This association was conceived as an instrument of economic and cultural cooperation. However, real desire for cooperation has become discernible only recently, following the debacle in Vietnam and the overcoming of much mistrust. ASEAN members endeavor to live in peace with their communist neighbors and at the same time to promote their relationships with the West, especially with the European Economic Community.

Australia: Continent of the Future

In the spring of 1970, Fred Millar, the chairman of Australian Hoechst, telephoned me from Melbourne one morning. Such phone calls were a rarity. Indeed, even today I am somewhat loath to make telephone calls over such distances, although they no longer present technical problems. I have agreed with my wife not to call her to announce my safe arrival in some part of the world. Perhaps this attitude is a relic of the times when we were less affluent. Fred Millar was about to acquire a cosmetics company which a Miss Prue Acton had formed a few years earlier with teenagers in mind. Prue Acton was a former beauty queen who had become the leading fashion designer in the country. But her cosmetics venture had taken her rather too far afield. She was therefore looking for someone to take over this business. It so happened that Australian Hoechst had been thinking about diversification anyway. Thus Prue Acton's cosmetics company was acquired by Hoechst.

This undertaking proved to be a doubtful investment. Australia is a huge continent, almost 30 times the size of the Federal Republic of Germany. But its population of only 13.5 million is not increasing at any great rate, and immigration is limited. In addition, there are the huge distances and the fact that more than 80 percent of the population lives in the four large cities of the continent. In short, we did not have much luck with this acquisition and did not hesitate to end our Prue Acton involvement some six years later. Jokingly, I told my friend Fred Millar on the phone, "You fell for the girl." But his possible susceptibility was not the root of the problem. Hoechst simply did not have enough experience in this field to turn the business into a paying proposition.

On the other hand, we were by no means beginners in Australia, where we had resumed our activities shortly after the end of the war. In this, we were helped by Henry H. York. Although advanced in years, he came to Hoechst with his young managing director, Hans-

Werner Luyken, to urge us and the other I.G. successor companies to reestablish the agency relationship that had existed since 1936 and had been interrupted only by the war. Hans Luyken, who had begun as a trainee in Leverkusen in the thirties and had been interned in Australia during the war, provided indispensable support for Mr. York. Hoechst worked for a number of years with the Henry H. York company, in which all three successor companies had a joint interest. Eventually, they took over completely when the old gentleman retired. But in the long term it became apparent that we had more common products than common interests. Hoechst was the first partner to decide upon separation. This was a fairly difficult procedure in which Fred Millar, originally Henry H. York's company lawyer, rendered us a great deal of assistance. A typical Australian and a highly successful self-made man, Fred Millar was and is the chairman of the company to this day. He has also remained on the supervisory board of Australian BASF.

It soon became necessary for Hoechst to establish its own manufacturing plant, since in the long term we could not maintain ourselves in the Australian market through imports alone. Again, our activity started on a fairly modest scale, with Hoechst acquiring an interest in an existing pigment factory. In 1966, we launched the first large-scale manufacture of Hostalen in Altona, an industrial suburb of Melbourne and one of the country's two refinery and petrochemical centers. The other is in Botany Bay, near Sydney, where a fair number of petrochemical industries have settled around one large refinery.

In the pharmaceutical field, we had collaborated since 1952 with Fawns and MacAllan, a company with an established reputation in Australia. Later, we jointly founded Hoechst Pharmaceuticals, in which we acquired the sole interest when our partner, Edward William Lowe, with whom we had enjoyed excellent relations, died in 1968. In line with our policy throughout the world, we brought all our Australian activities together under one roof in Hoechst Australia, founded in 1970. A few years ago, in view of the considerable business volume that Hoechst had achieved, we moved into a modern office block in the center of Melbourne. In 1974, the pharmaceutical sector was reinforced through the acquisition of a modern plant owned by Drug Houses of Australia. This enabled us finally to abandon plans, postponed year after year, of establishing our own production facilities in Australia. Now, we have rather oversized facilities for our pharmaceutical requirement. But in view of the future development of new products from Hoechst research and from our French partner Roussel-Uclaf, which maintains its own center in

Sydney, we shall no doubt grow into those dimensions in the fore-seeable future.

First Visit to Sydney

My wife and I flew to Sydney for the first time at Easter 1956. Al-though the Boeing 707 was already making its first regular flights, it was still a long journey of some 36 hours. Our friends in Sydney had great difficulty in finding accommodations for us because at Eastertime the much-respected Royal Agricultural Show is underway, drawing tens of thousands of farmers to the city. In the end, we had to make do with a small guesthouse at the edge of the town. When we were served lamb cutlets and fried eggs for breakfast each morning, my wife was amazed.

Fortunately, we found time to visit the imposing Agricultural Show. We took particular pleasure in the woodchoppers' contest which was fought with great enthusiasm. Until then, I had believed that such events no longer could be seen anywhere except in Bavaria.

In 1956, Sydney, architecturally, still looked like a British colonial town, although new industrial concerns had sprung up all around it. The petrochemical industry at nearby Botany Bay, in particular, had become firmly established. Today, Sydney presents a completely met-ropolitan silhouette, characterized by the architecturally unique com-plex of the new Opera House, which extends into the harbor area. Fortunately, Sydney has nevertheless remained a city surrounded by green, open spaces.

During the Easter holidays, we drove with Hans-Werner Luyken and his Australian wife to Canberra, the country's capital, and then to Melbourne, 800 kilometers and two and a half days away. This was enough time for us to gain a first impression of the country and its infinite space. En route, we visited one of the large sheep farms, where we were given a demonstration of sheep-shearing. It is difficult to imagine that the Australians succeed in shearing some 145 million sheep a year. The process requires numerous mobile work groups that move from farm to farm. Animal husbandry and grain production are still the backbone of the Australian economy.

Much has changed since 1956. The opening up of mineral deposits in the distant parts of the continent has played an essential role. Coal has always been one of the country's natural riches, but today every kind of ore is mined, including bauxite. In the center and also in the northwest of the country, and off the coast south of Melbourne, rich reserves of natural gas have been found and piped to the industrial centers. Today, three-quarters of the country's oil requirements can be met from its own resources.

Petrochemical capacities are being expanded to a degree that causes some experts to fear overcapacities in the near future. Because of its geographical position, Australia doubtless needs a certain measure of self-sufficiency and is therefore right in pursuing an expansion policy. But domestic labor has always been expensive, and raw materials and energy are no more advantageous here than in other markets. In addition, it must be remembered that Australia has taken over the English trade union system almost in the original. Factories, therefore, have to deal with not just one but many unions representing the various trades. This situation results in many disputes and frequent strikes that lessen the profitability of industry and frequently make Australian industrial products so expensive that they are difficult to sell, even in neighboring markets.

Today, Australia is more outward-looking than ever before. Even the external forms of life have changed as a result of both increased well-being and the influence of immigrants from Eastern and Western Europe. There are modern hotels and restaurants of all kinds. Although restrictions on alcohol exist in certain forms, they are so managed as to have little impact. As a result of the raw materials boom and the policies of the Labor government between 1972 and 1975, the country was pushed into a fairly nationalistic stance. This can be understood to some extent when one remembers that Australia has relatively few inhabitants and must therefore limit immigration.

It is now expected that with a few exceptions, Australian participation in company ownership is ensured whenever domestic raw materials are being processed. Large foreign companies that are exempted from this requirement are expected to have at least an Australian chairman and an Australian majority on their managing boards. The Hoechst companies have no problems in this respect. The number of non-Australians on staff was always limited and we have always endeavored, as is our principle throughout the world, to give able people of the country an opportunity to advance to leading positions.

Hoechst's position in Australia has been greatly strengthened as the result of acquiring the Berger group. Berger, Jensen and Nicholson (BJN) is the leading paint manufacturer in Australia. It has two large companies, Berger Paints and British Paints, which had merged before Hoechst took over BJN. Both these companies have branches at most strategic points in Australia. British Paints has a subsidiary, Selleys Chemicals, with a flourishing do-it-yourself business. It is the custom in Australia for every family to own its own house. And since craft workers are as expensive in Australia as they are in the United States, homeowners often do their own maintenance work. Apart from paint accessories, Selleys also markets accessories for a number of other trades.

In the long run, we shall benefit greatly from the Berger group's membership in our Australian organization. Apart from our headquarters in Melbourne and the large branch office in Sydney, Hoechst also has offices in Brisbane, Adelaide, and Perth. The sales organization of the Berger paint factories is even more extensive and may be of help to us when opportunities arise to open up markets that have so far been exploited inadequately for reasons of profitability.

The Australian continent has remained firmly linked to the West. It is a country for young people looking for new horizons and ready to test their strength in a land with apparently unlimited potentials. At any rate, our experience with our own staff is that most of them want to stay in Australia, where the young among them have started families. They have no intention of returning to Europe voluntarily or without overriding reasons.

New Zealand: The Green Isles

A trip to Australia should also include a stop in New Zealand. Therefore, we flew from Melbourne to Auckland in 1956. The flight took about five hours in an Electra. New Zealand offers everything that a stranger would not expect to find in the Pacific Ocean region—a moderate climate, green meadows, dark pine forests, high mountains covered with snow during the colder months, beautiful beaches, and sheer rock cliffs. Most of the people are European, largely English. The indigenous population, the Polynesian Maoris, account for about one-tenth of the 3 million people on the two main islands and are a characteristic feature of the cities. We also met them during our overland trip from Auckland to Wellington, especially in the neighborhood of Rotorua.

Rotorua, halfway between Wellington and Auckland, is a natural stopover for the traveler. A particularly attractive spot, it boasts some hot springs and extensive pine forests. The hot springs are used, though only to a moderate extent, for heating and cooking purposes and for hot-water supplies. From an energy point of view, the hydro-power stations of the southern island are of far greater significance. In Taupo, for example, one of the largest geothermal power stations in the world, the Wairakei Power Station, has a capacity of more than 200,000 megawatts.

We had the impression during our visit that the people of New Zealand lived a rather tranquil existence. Apart from consumer industries and processing plants for agricultural products, there was hardly any other industry. The country lived off its agrarian economy—off its 56 million sheep, 10 million cattle, and the products of dairy farming. Agriculture still accounts for some 80 percent of exports, going largely to England.

Since our 1956 visit, however, the country has developed substantially. But progress there has been more organic and less hectic than in Australia. In New Zealand, too, a Labor government has greatly strained the resources of the welfare state, with resultant inflation and unemployment. The present premier, Robert Muldoon, succeeded in leading the Liberal National Party back to power in 1975. Since then, considerable efforts have been made to repair the damage that was done by the labor "reformers" and to put the country on a sensible economic footing once more.

The processing industry, as well as the basic industries, have developed considerably. A refinery in the north supplies New Zealand with the required oil products, and there are concrete plans for utilizing the natural gas reserves on North Island near New Plymouth, to establish a petrochemicals industry. Whether this is worthwhile in view of the limited consumer market in New Zealand will have to be studied very closely. The country's oldest industry is the textile industry, mainly processing wool. New Zealand is the world's third largest producer of wool. The cheap energy available in the south has led to the establishment of an aluminum smelting plant that operates with bauxite imported from Australia. Because the country is rich in timber, there is paper and pulp production as well.

Fresh Start with Old Friends

When we arrived in Wellington, we met Herbert Melchior. He had also come from the former I.G., where he had completed his training. After internment in Australia, he had gone to New Zealand at the end of the war to attempt to reestablish the chemical business in the Wellington branch of Henry H. York. This was very much in line with old Henry York's successful efforts to reestablish connections with the I.G. successors.

Initially, this approach was quite successful in New Zealand. Melchior, whose I.G. career had been similar to that of Hans-Werner Luyken, performed sterling work. But in 1963, Hoechst was obliged to set up its own pharmaceutical company. At first, this company worked together with partners from New Zealand. Then, parallel with the development in Australia although somewhat later, we conducted all our operations in the relatively small New Zealand market through Hoechst New Zealand. In addition to its headquarters in Auckland, this company has branches in the two main centers, in Wellington on North Island and in Christchurch on South Island. Because of the structure of the country, there is relatively little room for industrial activities. Auckland, on North Island, has increasingly become the industrial center of New Zealand, and it accounts for more than 60 percent of the turnover of Hoechst New Zealand. Agrochemicals, ozasol chemicals, pharmaceuticals, Panacur, and textile auxiliaries are manufactured there. We have also acquired a cosmetics plant through Schwarzkopf, our Hamburg affiliate. In total, some 120 people work for Hoechst in New Zealand.

It was feared in some quarters that Britain's entry into the Common Market would have repercussions on New Zealand's agriculture. After all, England, certain European countries, and Japan were the chief importers of New Zealand's meat and dairy products, the country's main commodities. Indeed, New Zealand butter and Commonwealth preference were two of Britain's main arguments against joining the Common Market. As it turned out, the impact was not so dramatic as had been feared. Transitional solutions were found that prevented any long-term damage to New Zealand's agriculture.

There is much talk today about the recreational value of a town or a country. In my view, few countries offer better recreational value than New Zealand. The two islands together are about the size of the German Federal Republic, but they are sparsely populated. Their three large cities are far apart. Except in the centers of the cities, there are no large blocks of apartments. The typical New Zealander owns his own home, and often a boat as well. The beautiful coasts and beaches offer infinite leisure activities for people who love nature and want to live a quiet life.

New Zealand is far more British than Australia. But for its occasional political problems, it would be a little paradise in the Pacific Ocean.

17
Emancipation of the Middle East

In 1961, some of my colleagues at Hoechst and I went to the Middle East on a chartered Fokker-Friendship. Oil had already become a powerful economic factor in many countries, but it was not so much an instrument of politics as it is now. Utilization of this most coveted of raw materials proceeded in peaceful cooperation between the owners and the international companies which ensured that the oil was sold in the marketplaces of the world.

At that time it still looked as though Syria and Lebanon formed a link between Europe and the Oriental world. Everywhere there were still traces of centuries of Turkish rule, becoming more lenient toward its end. Except for sizable Christian populations in Lebanon and Egypt, this is essentially an Islamic world. This, of course, does not mean that there are no conflicts among Islamic countries. In fact, they were frequent and extremely violent at times. The world of Islam is not a homogeneous world, as demonstrated by the religious rivalries between the Sunnite and Shiite sects. The quarrels date from the beginning of Islam and have often had political consequences. Nevertheless, both groups continue to represent Islam, even though the Sunnites account for 92 percent of all Moslems, while the Shiites have to divide the remaining 8 percent with various other small groups.

With the end of Turkish rule, England and France assumed control over the Middle East, exerting strong military, economic, and cultural influences. These influences remained even after the the two powers had abandoned their role, more or less voluntarily, after World War II.

Since the oil trauma of 1973, it has become clear that the Islamic world is divided into at least two groups, the haves and the have nots. On the one hand, there are the rich oil countries, which are steadily getting richer; on the other hand, the countries without oil are therefore becoming more and more dependent economically on the oil-producing nations. This contrast has not only economic but also political consequences, which will no doubt emerge in time.

Does Turkey Still Look to Europe?

I remember an early summer morning in 1959. From my room in the Hilton at Istanbul I had a wonderful view over the Bosporus. It had been a short night, and I witnessed a sunrise over Anatolia of a beauty that is possible only in these latitudes. Before the business of the day was due to start, I had some time to reflect on what had come to my mind during my visit. As a German, one tends to assume the friendship and sympathy between Turkey and Germany as a matter of course. This assumption arises from the history before and during World War I, the bonds forged while we were brothers in arms, and the temporary identity of interests. What is the position today? In Turkey's economy, politics, and commerce, one encounters many educated people who have studied in Germany and have a command of German. We naturally assume that these people are also Germany's friends, and usually this is true.

Education in a European country is the surest way of guaranteeing that people from other areas will get to know the country and also become its friends. Perhaps we do not pay enough attention to this fact. At any rate, the United States, the Soviet Union, the German Democratic Republic, France, and England have developed substantial programs to attract foreign students.

Turkey caught up with the West, especially with its civilization, at a relatively late date. Kemal Ataturk engineered the orientation toward Europe between 1923 and 1938. He introduced the European

alphabet into Turkey and shifted education away from the Koran
schools to a secular school system. This emphasis was of decisive
consequence for the cities, greatly changing the urban face of Turkey
usually seen by Western visitors. In the countryside, on the other
hand, the muezzins, the heralds of Islamic belief, have retained their
influence. Today, it is regarded as good taste in Turkish politics to
advocate, more or less radically, a return to Islam. At the same time,
the question is raised whether Turkey is right in linking itself to the
West politically and to NATO militarily.

Oil is no doubt an important element in these considerations. There
is a substantial body of political opinion advocating that Turkey
should look to the Arab world for greater and more timely aid in rising
out of its economic misery.

I do not regard such considerations as realistic. Even the occasional
flirtations with the Soviet Union are of no benefit to Turkey either
politically or economically. I hope that Turkish politicians will realize
in the end that the future of their country is indissoluably linked with
the Western world, for better or for worse. Of course, there are huge
economic problems. The country's wealth lies not in its natural resources but in its agriculture. Anatolia is the large part of Turkey that
is in Asia, and it contributes the greatest share to the country's economy. But the Turkish population is increasing rapidly. It is estimated
that some 60 million people will live in Turkey by the year 2000.
Turkey's agricultural exports consist of only a few products, such as
raisins and almonds.

Dyestuffs and Pharmaceuticals in Turkey

Of course, the country also has industry. Under Kemal Ataturk, the
textile industry was systematically established with state aid. It forms
the backbone of the Turkish economy, even today. As a result, Turkey
has for decades been an attractive market for Western manufacturers
of coal-tar dyestuffs. At the same time, the country has set up its own
man-made fibers and dyestuff production, carefully protected against
competition from imports. Not only are the tariffs prohibitive, but
also, no foreign exchange is allocated for products that can be manufactured in the country itself.

Another highly developed branch of industry is the pharmaceutical
sector. All the major pharmaceutical companies of the Western world
produce in Turkey, but the country also has its own pharmaceutical

companies, which are serious contenders. Life is not easy for the pharmaceutical industry. The shortage of foreign exchange makes it difficult to import the necessary base materials. In addition, administrative obstacles have increased over the years. Approval of new products is a protracted process, subject to bureaucratic delay and often capricious decisions. Moreover, rigorous price controls permit price levels to reflect increased costs only after long delays. This means that price rises lag far behind the double-digit inflation. The production of pharmaceuticals is a red-ink proposition for almost every company. A long list could be made of companies that have thrown in the sponge because they lost all hope of ever achieving reasonable profitability.

Our Interests on the Bosporus

Turk Hoechst had a difficult birth. Its precursor was the so-called Near East Office in Cairo, which had cooperated with Turkish import companies separately for the dyestuffs, industrial, and pharmaceutical lines. But it soon became evident that we would not be able to establish any sizable pharmaceutical business unless we produced in the country itself. For this purpose, Turk Hoechst was founded in partnership with a knowledgeable Turkish businessman. It started work in 1963 in modest rented offices. Eventually, after difficult negotiations with the partners concerned, we integrated our remaining industrial business with this pharmaceutical concern and established a central company for Hoechst.

This new Turk Hoechst, which soon moved into a more suitable office building in the city center, erected a small industrial complex on an industrial site twenty minutes by car from the city gates. In 1967, we inaugurated a modern pharmaceutical factory on this site. Turk Hoechst also produces plastics dispersions, surfactants, and other products for industry. These manufacturing activities—aside from the relatively unprofitable pharmaceutical operation—have become the cornerstone of our Turkish business.

Importing into Turkey has always been troublesome. The shortage of foreign exchange continues. The Turkish currency is relatively unstable, and every devaluation of the Turkish pound results in a new profit squeeze. But, we are fortunate in having a staff of very efficient Turkish people, many of them trained in Frankfurt. By having a flexible commercial policy and taking great care to keep within the Turkish laws and regulations, we shall no doubt maintain our position in the

country. Turkey is an enticing market that will doubtless continue to
develop in spite of its various problems.

Lively Orient

From a business standpoint, visits to Istanbul were often fairly strenu-
ous, but culturally they were always extremely interesting. The city,
whose Oriental "inner life" is best ignored, is beautifully situated on
the Bosporus. The journey to the Black Sea is filled with history and
culture, and Istanbul has many buildings that show both. The Hagia
Sophia, the Blue Mosque, the Topkapi Museum, the medieval fortifi-
cations of the town—they are all worth a trip.

A journey to Smyrna and visits to Pergamon and Ephesus are unforgettable experiences for every visitor to Turkey. Pergamon bears eternal witness to Greco-Roman culture, being one of the sites where
Magna Graecia is literally tangible. I had seen the famous altar of
Pergamon in Berlin before the war. It was later dismantled and stored
safe from air raids. After the end of the war, it was reconstructed in
its former glory on the museum-island in East Berlin.

The Hittite Museum is also part of Turkey's cultural wealth, and I
always visit it when I go to Ankara. It displays the products of a
civilization that contributed eternal works of art to Turkey's heritage.
They appeal to us immediately, even today.

Earlier, I asked whether the Turks are still friendly toward Germany.
Probably, the question was not properly phrased. Their past history,
religion, and civilization have given the Turkish people characteristics
that are different from those of the central Europeans. In addition,
there is a deep gulf caused by the feudal and patriarchal rule that
survives in Turkey in spite of its parliamentary democracy. I do not
think that the Turks are by nature very friendly to foreigners. Their
concern is to maintain national independence and national dignity, as
they showed during the Cyprus conflict. It can only be hoped that the
hundreds of thousands of Turkish guest workers in the Federal Republic and the growing wave of German tourists in Turkey will help
to portray Germany, its economy, and its people in a friendly light.

Iran: A Transition Full of Problems

When I visited Teheran for the first time in 1953, I had only a vague
idea of this Oriental metropolis. Sand-colored Teheran lay in the

middle of what appeared to be an endless desert. The city is a young capital of an old empire that, as the result of its income from oil, has recently been enjoying considerable well-being. But at the time of my visit, the streets were lined by relatively modest but partially modern office and commercial buildings. The hotel situation was similarly modest. The best available was the old Park Hotel, a five-story hostelry that was not exactly fitted out with the latest comforts.

All this has changed since my first visit. Today, one arrives at Teheran's modern airport and drives on broad roads to one of the hotels located on the hills to the north of the city. Business discussions are conducted in offices and commercial establishments no different from those in Western countries. The silhouette of the city is dominated by modern skyscrapers. Fortunately, Teheran has surrounded itself with a green belt. The slopes around the city have been planted with trees, favorably influencing the climate. Except for the short, cold winter months, Teheran is dry and hot. Now there are some rainy days, even rainy weeks, that were quite unknown previously.

In 1953 our agency had its offices in the bazaar quarter. When I arrived, I had to find my way through a herd of goats that were being offered for sale. Eventually, after ascending a lot of narrow stairs, I found our modest offices and faced Khamney, the chief of our company. He was an Iranian in his middle fifties and spoke only a little French. Conversation was therefore rather difficult, and we would not have advanced very far without the Hoechst delegate, Günter Rexroth, who had acquired a fair knowledge of Persian by then. As a result, we were soon engaged in a lively exchange of ideas about ways to develop the Hoechst business in Iran. It was not surprising that the views of the trader and moneylender Khamney and of his Iranian associates did not coincide with our own. Our first purpose was to invest money in the business, to attract staff, to create a sales organization, and in this way to achieve a volume of business that would provide all concerned with a reasonable profit.

A Hard Fight for Every Mark

We had to fight hard for every Deutsche mark in our discussions with our Iranian partners, but their hospitality was extremely generous. Hard and tough bargaining is simply a part of business for the Iranians, who believe that this is the only way to get reasonable terms. Nevertheless, pleasant human relationships, marked by outstanding courtesy, were established during this bargaining process. Most Iranian business partners are highly educated. This education of the older generation and also of the younger people has often been acquired or perfected at foreign universities. Knowledge of foreign languages, especially of English and German, has increased greatly. As a rule, French is spoken only by the older generation—a phenomenon also observed in the countries of the Near East.

At the time of my visit in the 1950s, Iran was by no means poor. Oil was flowing plentifully, and the wells were owned by a company most of whose shares were in Iranian hands. The unrest of the Mossadegh era had passed, and the country had a stable government.

Shah Reza Pahlevi had set himself two goals. One was land reform. Large property holdings were to be divided, agricultural cooperatives were to be set up, and independent peasants were to be trained. It took a long time until part of these plans could be realized in the face of sustained opposition from the feudal landlords in the large cities. Apparently, the plans seem to have been carried out in the end.

The second goal was industrialization. The money was available, and a large planning authority drew up a five-year plan that aimed at an ambitious project for broad-scale industrialization. This five-year plan went far beyond what was feasible for a country that lacked the infrastructure, the skills, and the structural prerequisites for industrialization. Nevertheless, many projects were carried out, though after much delay.

Conquest of the Alphabet

This short description of Iran's development would be incomplete without mentioning the "white revolution." After the Parliament had been dissolved on May 9, 1962, the government tried to reduce social tensions through a comprehensive reform program. On January 9, 1963, the first points of the "white revolution" were announced, which, on its tenth anniversary, was renamed the "revolution of emperor and people." The most important measures, in the view of the Iranian head of state, were land reform, the nationalization of the forests which were almost entirely privately owned at that time, the return of state-owned companies to private enterprise, and the introduction of profit sharing for the workers. Also, the election laws were to be changed and an "army of knowledge" was to be set up. This unit of the Iranian armed forces was to conquer illiteracy in the countryside. Subsequently, the reform program was enlarged by almost a dozen further points.

Frost on the Blooms

The oil boom of 1973 greatly assisted the Shah's plans. Unfortunately, the possibilities of processing oil in Iran itself were greatly over-estimated. Talks with several countries, including the Federal Republic of Germany, resulted in agreements in principle for establishing a petrochemicals industry in Iran. But the sums did not add up. To use oil in the country itself at world prices simply could not be economic. Some of the dreams of rapid industrialization on the basis of petro-chemistry were therefore rather rudely shattered.

Many of the plans that are still beyond reality will no doubt be put into practice one day. And that day will come when Iran (and many other oil-producing countries as well) recognizes that it is totally im-possible to have both unlimited profits from oil and, at the same time, benefits from the manufacture of products for which the domestic market offers only limited sales opportunities. Exporting to third countries is expensive, partially because of transport and partially because of the organization needed for selling raw materials destined for further processing in the importing countries.

These comments do not mean that I regard plans for processing part of the Iranian oil in the country itself, and for exporting petrochemi-cals instead of oil, as unrealistic in the long term. But it will take careful preparations to create the necessary conditions, especially with regard to the infrastructure. An interministerial office is responsible for the implementation of the five-year industrialization plan. We con-ducted many negotiations with this authority, but never succeeded in realizing any large projects, because we could not agree on conditions.

The Oil Boom Clouds All Reason

Iran's ambition to industrialize gained new inspiration after the oil crisis of 1973. The country's income rose suddenly to 3 or 4 times its former figure. The Iranians believed that industrialization would now proceed at top speed. Conversely, many foreign countries and compa-nies thought that decisions to invest in Iran had to be taken as quickly as possible in order to get a secure raw material supply and an infi-nitely rich partner.

Such views were also put forward in West Germany. As members of a large delegation from the Economics Ministry, German industrialists negotiated in Teheran. The objective was the establishment of a petrochemical company, or more precisely, the production of aromatic intermediates for which a plant was being planned as a corollary to the so-called export refinery in Bushir. Bayer and Hoechst were to take on the intermediates project together with the state-owned Iranian petrochemical company. The German federal government would have to recruit a consortium of oil companies that would operate the export refinery together with the Iranian oil company.

The reason why this project, like so many others, came to naught was that the various partners had divergent views. The Iranians believed, on the one hand, that they could make their oil available for further processing at world market prices, and on the other, that they could participate in the profits that they hoped would result from further processing. Also, they overlooked the fact that the infrastructure (communications, accommodation, trained personnel) and other prerequisites for industrial activity did not exist in the country. This condition was true in particular at the very locations designated for the new investments that were to contribute to the economic decentralization of the country. The Iranians did not anticipate that the costs of the infrastructure had to be met beforehand. They could not be debited to the initial phase of a project that was to be accomplished with a 50 percent contribution from private industry.

Nevertheless, I believe that several projects which have now been shelved will be realized at some future date. In contrast with most oil countries, Iran can reckon with an increasing domestic market. It is estimated that its population of 40 million in the 1980s will have increased to 60 million by the year 2000. In many respects, things will then look different from today. The important thing is to find the right moment for catching up with the changes.

Our Start in Iran

There was a long road to be traveled before we established Iran Hoechst in 1956. We had started in the field of dyestuffs and industrial products, where the processing industry offered favorable conditions. As was also essential, we had at a fairly early state found a group of young Iranians who had completed their studies in Germany and therefore knew the German language and mentality.

We soon abandoned our purely merchandising activities. The export of pharmaceuticals to Iran did not make much sense in the long term because of the geographical conditions as well as the commercial situation. It was logical that Hoechst should build a pharmaceutical factory in Iran as a first industrial step in that country. This plant was inaugurated in 1969 in Teheran-Pars, an industrial suburb. We had also erected a central warehouse there and provided manufacturing facilities for plastics dispersions and auxiliary products. In this way, we provided a sound complementary manufacturing basis for Iran Hoechst, which became a Hoechst subsidiary after a complex set of negotiations.

New Factories outside Teheran

Further possible expansion will raise the problem of location, since no further industrial projects will be permitted in Teheran. We have therefore acquired land about 120 kilometers west of the capital in Ghazvin, where new factories are to be built in the next few years.

Western entrepreneurs are not exactly encouraged to invest in Iran. A majority interest for the Iranian partner is a prime condition. In most cases this is feasible if the partner is reasonable and has some economic experience. Matters become more difficult, however, with the so-called People's Share Law. Passed in 1975, this law requires that the employees of the plants and the public should acquire a 49 percent interest in industrial companies in the form of shares.

At first sight, this seems unrealistic because there are not likely to be many employees with either the necessary funds or experience. However, there was much talk about auxiliary measures and modifications for allowing that law to be applied in practice. But some time is bound to pass before a practical solution is found. Investing in Iran has become problematic for many reasons, indeed. One aggravating factor is that the foreign share in new investments has been generally restricted to 25 percent, and in a few cases that involve high technology, to 35 percent.

From Isfahan to Persepolis

I have always looked forward to visits to Iran. They gave me an opportunity of getting to know places outside the metropolis where one still encounters the great architectural testaments of the past. I like to recall Isfahan and the beauty of its mosques, with colored domes richly ornamented. Isfahan is a city whose external aspects reflect the embodiment of Islamic art in a unique way.

The early history of the country comes to life in the rose-colored city of Shiraz, the home of Persia's famous poets Saadi and Hafiz. It was Hafiz's poems that moved Goethe to write his "West-East Divan." From Shiraz, it is only a short distance to the ruins of Persepolis. The name recalls the political and cultural importance of this city for more than 2500 years, from the time when Darius the Great established Persia's hegemony far beyond its own boundaries. Those historic events were recorded in the beautiful half-reliefs of the palace in Persepolis.

Among the oil-rich countries of the Middle East, Iran occupies a special position because, beside its efforts to industrialize its own land, it has impressive, indeed sensational, investments in the Western world. The Federal Republic of Germany welcomed Iran's investment policy when it found suitable German targets, such as Krupp and Babcock, for Iranian investment. The policy is not only an expression of political and economic self-confidence but also a sign that Iran believes in a permanent economic partnership with Western Europe. Iran will, however, have to adapt its policy to this situation in respect to oil. Price increases must be kept within bounds if the economic equilibrium is not to be disturbed at a cost to both sides.

West European countries are realizing increasingly that Iran should be a power with a great future, though recent events may have greatly shaken this belief. Once more, we have had to recognize that we have to live with a rather unsettled Middle East. Still, taking a realistic point of view, the Western world will have to maintain and develop economic and political links with this ambitious country.

Iraq: Socialism, Oil, and Desert Sands

What Rome has been to Christendom for 2000 years, Baghdad was for 500 years to Islam: seat of the caliphs, treasurehouse of Oriental splendor, and center of a culture far ahead of that of the Occident. But in the thirteenth century, all its power and glory were turned to rubble by the Mongolians.

To our ears, Baghdad represents the picturesque Orient, and for the people of my generation, it has literary associations with such accounts as the travelogues of the French writer Pierre Loti.

Baghdad—how many memories the name of this city and this country, evoke: Mesopotamia, the country bounded by the Euphrates and the Tigris; the Old Testament theater of the Babylonians and the Assyrians; the Tower of Babel; Ur, the home of Abraham; the decline of depraved Nineveh. Travelers must shut their eyes and evoke the fantasies of their early Bible lessons because little of that historic scenario has remained under this hot sky. Baghdad once helped to mold world history, but the surviving sites are of interest only to the archeologist.

The Baghdad that I came to know for the first time about twenty years ago had very little to do with either the Bible or the romanticism of literature. Today the character of the city is determined by uniform, sand-colored structures which, if they have any charms at all, hide them in shady courtyards behind high clay walls. In spite of its dusty monotony, Baghdad has maintained a certain dignity wherever it presents a view of the imposingly broad Tigris River.

From my first visit to Baghdad, I particularly remember the great Shiite Mosque on the edge of the town in Khadmyia. Its gilded cupolas and minarets, surrounded by a loose circle of date palms, send glistening, golden greetings to the traveler from Mosul or Samarra. I also like to recall the labyrinthine, gaily colored bazaars—the souks—where coppersmiths and spice traders, tailors, shoemakers, cloth and carpet merchants, money changers and gold- and silversmiths still offer their goods and services in narrow, covered alleys, haggling patiently, joking, cursing. They are friendly people who regard the stranger openly and without enmity. Confidence is rewarded with even greater confidence. Traders will readily let shoppers they know have a valuable item for inspection, allowing them to keep it for days or even weeks without any kind of security.

A visitor to Iraq should forget any dietary resolutions. To try to escape the overwhelming hospitality of the Iraqis would be both a marvel and an insult. And, indeed, the Iraqi cuisine offers many delicacies, such as *kusi,* for example, a crisp-fried half-side of mutton served on a mountain of spicy rice with almonds, raisins, and saffron. Another dish is *masquuf,* a fish speciality in this city in the middle of the desert. The *masquuf* is a large carplike fish from the Tigris. It is slit open at the back and then, flat like a sole filet, fried on sticks over roaring fires. Sitting in a circle of friends by the river during one of Baghdad's beautiful evenings, one eats *masquuf* with the fingers, together with brown pittah, the local bread. This is but one of the scenes that make this country, which often seems so unfriendly to outsiders, particularly alluring.

On the Road to a State Economy

Baghdad hit the world's headlines when the monarchist regime was toppled in 1958 and its pro-Western prime minister, Nuri al-Said, who had held office for many years and had enjoyed almost dictatorial powers, was removed from office. The military regime established under General Kassim turned toward socialism and a corresponding foreign policy. In addition, there were nationalization trends that did not leave much room for the business that Hoechst, like many others, had started in the customary manner, through cooperation with local merchants.

Traditionally, Iraq had been a good market for dyestuffs and pharmaceuticals. Those were, therefore, the fields in which we started again after the war; we were able to build on the good name these Hoechst products had won in the country before the war. Because of the political conditions, we were never able to establish a Hoechst sales and production company, as we had done in the neighboring countries. We therefore collaborated with local merchant houses, all of which have since been nationalized.

Today we operate an advisory office, whose young, well-trained staff is endeavoring to generate the appropriate sales of Hoechst products to state organizations and private industrial companies. In this, we have so far succeeded, although it should be understood that the market cannot be fully penetrated through this system.

Industrialization in Iraq has concentrated on the textile and the plastics processing industries, which did not even exist two decades ago, and on the exploitation of the oil reserves, which were nationalized in 1972.

Uhde made a great effort to obtain engineering orders in Iraq, and was about to sign its largest contract in its most traditional field, the construction of ammonia and nitrogenous fertilizer factories. Then a period of internal political upheavals, accompanied by financial self-examination and retrenchment, relegated this project to the distant future.

With a current population of about 12 million, probably increasing to 25 or 30 million by the year 2000, Iraq is clearly a promising market. This is true more than ever today, since it may be assumed that the civil war with the Kurdish tribes has ended. It is my hope that the oil wealth will promote a more liberal import policy and that the country will continue to be concerned with expanding its industry.

Such progress should provide the chemical industry with an expanding field of activity in the future. Although, from a Western point of view, Iraq is not playing so spectacular a role as some other oil-rich countries, its tremendous potential should not be underestimated. Its wealth is currently used to effect reforms in the country itself and to exert political and economic influences on the world outside. I believe there is every reason for fostering our trade relations with Iraq. Its isolation from the Western world would be detrimental to the balance of power in the Middle East.

Egypt: From the Pharaohs to the Aswan Dam

Like millions of other people, I too have followed President Sadat's work of recent years with sympathy and concern. In contrast to his predecessor, Sadat is following a clearly pro-Western political course. In the Middle East conflict, he boldly assumed the initiative and attempted to negotiate a peaceful solution by means of a compromise that would do justice to all parties. In this mediating role, he depends particularly on the support and backing of Saudi Arabia, whose help he will be able to rely on only if he succeeds in improving the economic situation and in creating stable conditions in his own country.

In economic respects, too, President Sadat has opened up Egypt once more to the West. With the aid of Western technology and Arab capital (probably mainly from Saudi Arabia), the industrialization of the country is to be promoted; existing raw materials, such as crude phosphate, are to be better exploited; oil exploration and production are to be intensified; tourism is to be expanded; and the Suez Canal is to be enlarged.

With these plans and projects, which also include a large-scale Egyptian-Iranian joint venture in the fertilizer sector, Egypt's economic planners hope, by the beginning of the 1980s, to restore the Egyptian economy, which suffered greatly from the war and which carries an enormous social burden. Egypt is facing a precarious internal situation. The poverty of the population, the low standard of living, 40 million people living on only 3 percent of the land area, and the population increase of 1 million every twelve months—all these weaken Egypt's position, both absolutely and in relation to its Arab neighbors. The deepening poverty of the masses and an infrastructure that cannot cope with the population explosion swallow up millions in subsidies—at the expense of the economic expansion plans—to prevent unrest.

Egypt's new economic concept does not have a firm basis because Western industry did not prove nearly so investment-happy as the planners had hoped. The West feared the political risks and, in my view, the investment inducements themselves were not inviting enough. However, now that investment laws have been reformed, a more favorable climate seems to have been established. The number of projects that are being discussed and even carried out by Western companies with the aid of Arab capital has greatly increased in recent times. On the economic credit side, there are the positive yields from the foreign currency sources on which Egypt's new economic approach is based: proceeds from tourism and the Suez Canal, income from the newly erected Sumed oil pipeline (the first real Egyptian-Arab joint venture), and above all, oil. Egyptian oil production has now reached a point at which the country no longer needs to import oil except for changing different grades. There may already be some modest export of oil. But that does not mean that Egypt will become one of the wealthy oil-producing countries in the near future. Nevertheless, the oil that has been found should contribute significantly to stabilization of the Egyptian economy in the 1980s.

Once an improvement has taken place in the economic area, and once the internal political situation has been consolidated, Egypt will be able to pursue its peaceful mediator role in the Mideast conflict with far greater authority. In the longer run, it will no doubt be able to rely on the support of other Arab nations, especially Saudi Arabia. The oil-rich Arab countries will be more ready to invest their vast sums in a country that speaks their own language and shares their common culture if Western technology ensures adequate returns. For the West, such a development would mean a new Afro-Arab market.

A Pharmaceutical Factory on the Nile

It was in the spring of 1963 when a group of Hoechst executives accompanied Professor Winnacker to Cairo to launch Hoechst's new pharmaceutical factory in the industrial suburb of Zeitoun. It was a proud moment for Hoechst because there were only two other companies—one American and one Swiss—that were being allowed to establish a pharmaceutical company with a foreign majority holding. The Egyptian Minister of Health, Dr. Nabawi El Mohandes, took part in the inauguration ceremony and made a thoughtful and encouraging speech in which he expressly confirmed that research-orientated foreign companies in Egypt were highly desirable and would have their efforts supported by the state.

Almost two decades later, we know that this encouragement was sincerely given. In spite of many difficulties over raw material supplies, the transfer of dividends, the exchange of private Egyptian minority shares against a state bank, and the development of an Egyptian management, Hoechst Orient has attained a position that is appropriate for a large, Western European undertaking.

An essential feature of this progress was Hoechst's establishment, right at the beginning, of a pharmaceutical research and development center alongside its production plant. This center is concerned with the specific diseases of the country, primarily bilharziosis, a parasitical disease named after Theodor Bilharz, who described it for the first time in 1852. Its pathogens are parasitic worms whose larvae live in Egypt's waters. In the water, certain snails serve as intermediate hosts. When infected water is drunk, or when the larvae encounter people bathing, they will bore through the skin into the venous system, where they develop into sexually mature worms and multiply rapidly. The parasites and their eggs produce fever, inflammation, abscesses, and damage to intestine, liver, spleen, lungs, and other organs. Many infected people die or are condemned to long years of decrepitude.

Specialists in tropical medicine have estimated that some 300 million people in Africa, Asia, and South America suffer from this parasitic disease. There are some promising chemotherapeutical beginnings, and Egypt is making great efforts to educate the public to the dangers of infection in water. Snails, the intermediate hosts, are being energetically attacked with chemicals. But it seems probable that the disease of the Pharaohs will be conquered only by a preventive vaccine.

To realize such possibilities of active immunization is the function of our Behring research laboratory, housed in the pleasant Hoechst Orient complex, which is surrounded by green meadows. The Behring research workers are confident that a successful vaccine can be developed. They are cooperating with Egyptian hospitals and the Theodor Bilharz Institute in Cairo, which is supported by the Federal Government of Germany.

Working round the Clock in Aswan

I had discovered Egypt quite some time before the inauguration of the pharmaceutical plant and the research center. In 1966, in the course of an extended Middle East trip lasting several weeks, we had visited Cairo and Aswan, where work to complete the high dam was in full swing. Work was going on even during the night, not because it is less hot then, but because working in shifts made it more likely the deadline would be met.

It was most impressive to view this vast site with its thousands of Egyptian workers and hundreds of Soviet experts. We were shown the plans of what was to be completed a few years later: the enormous dam and the huge Lake Nasser, as it is still called today, the power station and the industrial installations planned in conjunction with it. At the time, the project was not merely armchair planning. The industrial installations were already under construction.

Uhde, Hoechst's engineering subsidiary, played an active role in Egypt's industrialization, particularly in its special field of plants for ammonia and nitrogenous fertilizers. The Aswan Dam is to serve two vital purposes: to facilitate improved irrigation, and thus expansion of the agriculturally usable areas of the Nile valley and delta; and to provide power stations with sufficient energy for the establishment of a variety of new industries.

It is not easy to assess the extent to which these ambitious goals have been achieved. In agriculture and industry they have been realized to a large degree. It is certain, however, that the construction and successful completion of the Aswan Dam have provided the country with a tremendous moral impetus. Above all, it has been an inspiration to Egypt's younger generation and has stimulated an élan that will be influential for a long time to come.

The Formation of Hoechst Orient

We had started our Egyptian business with large nitrogen exports. These were transacted through the Sabet Frères company in Cairo, owned by Egyptians of Syrian origin. During the reign of King Farouk, the Egyptian import business was run largely by citizens of neighboring countries, especially the Syrians and the Lebanese.

It was good business for Hoechst. The nitrogen export to Egypt was probably one of the first large export transactions that became possible after the war. Because at that time I was temporarily in charge of the agriculture sector and thus of nitrogen exports, I had frequent opportunities to make on-the-spot observations of Egypt's economic, political, and social situation. On one memorable trip, my colleagues and I were accompanied by the two brothers Sabet. Using a single-engined charter plane of Misr-Air, the internal Egyptian airline, we flew to the Valley of the Kings, which made a tremendous impression on me and which, since then, I have seen several times. When one considers that the tombs there have been systematically plundered for hundreds of years, one marvels that so many treasures still remain. Much has been either left or reconstructed, giving a true picture of a City of the Dead that dates back more than 3000 years.

Saudi Arabia: Pivot of the Oil Empire

About a quarter of all known oil reserves in the world are found in this Arab desert land with about 8 million inhabitants. In past decades, when Saudi Arabia's oil wells had already become a significant economic factor and were being exploited by ARAMCO (Arabian American Oil Company) in cooperation with the Saudi government, the country did not play any discernible part in world politics. However, this isolation changed fundamentally after the oil crisis of 1973. After Iran, Saudi Arabia today is the decisive power in the Persian (or Arabian) Gulf and, together with Egypt and Syria, the leading force in the Arab camp.

Saudi Arabia is a country of contradictions: Its problems and internal stresses cannot be fully understood unless one realizes that this country grew out of the religious revival movement of the Wahhabites. It only gradually assumed its political contours in the years between the two World Wars, under the leadership of the Saud family and after tribal feuds lasting decades. Even today, the missionary zeal and the ideals of spartan Wahhabites are outstanding factors behind Saudi Arabia's difficulties.

Immeasurable oil wealth has understandably presented this highly religious, puritanical state with severe problems. Its reaction has been to tighten rather than loosen its ties with the laws of the Koran, thus creating a widening schism between official morals and the aspirations of a rising middle class of officers, officials, and technicians who have encountered the Western way of life.

Foreign Advisers

Recreational facilities for foreigners are absent in Saudi Arabia. A good friend who had just returned from Riyadh told me that at a reception at the German Embassy, to which many young German couples had been invited, he had never seen so many expectant mothers together in one place. Saudi Arabian politics, under Ibn Saud and his successor, King Khaled, were pro-Western, and were particularly aimed at a close association with the United States. It is estimated that Saudi Arabia has currency reserves in the United States in the order of U.S.$50 billion.

As in all Arab countries, industrialization is a prime goal. But, as in the other countries of the Persian Gulf, the appropriate infrastructure is largely nonexistent. It will have to be created with the aid of the government before plans to industrialize in conjunction with Western industrial companies can be envisaged on an economic basis.

Heading these efforts is the exploitation of natural gas. Found both alone and together with oil, it now is simply being burned. To use the energy of this gas for heating would make no sense, given Saudi Arabia's climate. But to use it as an energy source and as raw material for the chemical industry requires conditions that do not as yet exist here. Negotiations in this direction have gone on for many years with various consortiums and large chemical companies. It is to be hoped that a solution will be found in the foreseeable future so that at least some of the gas will be used productively.

It may be of interest to add a word about the role of chemistry in oil production. Chemical auxiliaries are used to an increasing extent in oil production, especially when the wells must be sunk deeper and deeper or when other geological problems arise. Moreover, the addition of chemicals increases the utilization of oil deposits.

Up to now, only 30 to 45 percent of the deposits have been used. Through testing production measures, this figure can be raised by 10 percent. Like many other companies in the chemical industry, Hoechst has so far not exploited this development directly. Business with the oil companies is conducted through service companies concerned only with the oil industry to which they supply everything that may be required in technical or other respects. They also provide the experts who can offer the necessary advice. Chemical companies like Hoechst can probably realize the vast potential of the oil business only in collaboration with such service companies.

One pressing problem in this desert country is the provision of water for drinking and for irrigation. If this provision could be achieved, the country could be transformed into a flowering garden. Up to now, only expensive desalination plants for converting seawater into drinking water are available. They can provide a sufficient amount of drinking water, but they cannot supply the vast quantities of water that would be necessary for agricultural purposes. No doubt this will change in the coming years.

Icebergs for the Desert

One project sounded like a fairy tale but turned out to be possibly quite realistic on closer inspection. Large icebergs in the Arctic and Antarctic, which present hazards to shipping, are to be towed to Saudi Arabia and other Persian Gulf states with similar water problems. Experts have calculated that such an operation could be carried out without difficulty and would take about six months. Since nine-tenths of an iceberg is under water, it would lose only about one-third of its volume during the journey. Moreover, icebergs consist of sweet water, which could easily be purified to render it fit for human consumption. The realization of such a project would fundamentally alter the settlement possibilities and agricultural structures of these countries.

Hoechst, like the chemical industry as a whole, is closely following the development of petrochemistry in Saudi Arabia. Through consortiums, we have also been involved in discussions on a number of projects. However, these projects have usually been abandoned because of the lack of infrastructure and the inflexibility of the national partners. But they will be revived again in one form or another, and nitrogenous fertilizers, plastics, and fibers will probably be in the forefront of planning.

In the meantime, our business is following traditional lines. Pharmaceuticals, dyestuffs, and chemicals for Saudi Arabia's small domestic industry are supplied through a trading company. This company is supported by experts from Hoechst. If one takes the neighboring Gulf countries into account as well, these activities should one day result in a common trading policy.

Regarding the political situation, Saudi Arabia, with its balanced, realistic, and friendly attitude to the West, represents a haven of peace in the middle of a turbulent world. It is also important that Saudi Arabia can exert political influence on its neighbors and can further cement this policy with loans and other monetary contributions. The

nation's influence is therefore growing continually. Egypt would pre-
sumably have landed in many political difficulties if the Saudis had not
helped it with monies and credits. Its release from the Russian embrace
would hardly have been possible otherwise.

There is no shortage of political problems in the Middle East. Therefore it is to be hoped that Saudi Arabia will continue to be a stabilizing
factor in the future. Its moderate stance within OPEC has greatly
contributed to keeping oil prices within some limits so that the economy of Western consumers has not become entirely inflationary. If
not, the consequence would have been extremely harmful to both the
Western consumers and the oil countries that depend on this oil.

Kuwait: Largest Gross National Product in the World

As we approached Kuwait in the small Fokker-Friendship, we saw a
city set among only a few green patches, beyond which stretched an
apparently endless desert. We also saw huge installations near the
port, which were explained to us as we drove into the city. These
plants desalinated seawater, supplying the city with drinking and
industrial water. They satisfied water requirements at that time, albeit
in a very expensive manner.

However, Kuwait can afford to spend money on such installations.
With the country's ample flow of oil, the per capita gross national
product is the highest in the world—U.S.$12,000 in 1976. The origi-
nally small population of half a million has doubled in the intervening
years. The increase consists mainly of foreigners, some from the West,
but most from surrounding countries. In Kuwait, these neighbors can
earn many times the wages that they would receive at home.

From the beginning of this century until 1961, Kuwait was a British
protectorate. The English left their mark on the city in many ways. It
has an English touch, as have its hotels. The commercial people speak
English. The politics that the country pursues in the Arab League,
whose members are very much at odds with one another, is both
conservative and outward-looking. Rich Kuwait provides financial aid
to its poorer Arab neighbors.

When the Oil Dries Up

The rulers of Kuwait, a sheikhdom or monarchy, consist of a number
of family groups. They exercise their power gently, however. Their

aim is the industrialization of the country. Education, schools, medi-cine, hospital care, and medical services are free. Like its bigger neigh-bors, Kuwait wants to divert part of the available oil to industrial projects, so that the small state will have a basis for existence even when the oil begins to dry up. While the exhaustion in oil reserves may take longer than the pessimists forecast, eventually the day must come when no further deposits can be found. It is therefore valid for the government to concern itself not only with today's well-being, but also with the living standards of tomorrow.

In the development of industry, low-population countries are beset by the classic dilemma that native labor is either unavailable or unwill-ing to perform the less skilled, less well-paid jobs. For this reason, Kuwait must rely on foreign labor. Much attention is paid to the fact that half the present population is foreign. Such large numbers of foreign workers form a constant risk of political subversion. Hence, most of the Gulf states, including Saudi Arabia, prefer the employ-ment of politically abstinent Pakistanis and Indians instead of Pales-tinians, Egyptians, and Lebanese. The latter might easily transmit the Pan-Arab germ, together with a number of left-wing variations, to these small—and particularly susceptible—feudal states.

Hoechst is following the economic development of Kuwait very attentively—especially the industrial plans that are being discussed with numerous foreign groups. I do not believe that these plans will be realized quickly. The raw material bounty has its own weight which will continue to increase but, it is fervently hoped, not beyond a measure that might do harm to the world economy.

Hoechst is also established in Kuwait. We are represented by a well-respected local trading concern, some of whose staff were trained at Hoechst headquarters. With the help of our experts, they try to take care of all realistic economic opportunities. Selection of the right local partner is decisive in all these countries unless, as in Iraq and Syria, they have a centralized economic system.

Whether cooperation proves fruitful depends on a variety of fac-tors. If the indigenous partner has experience in dealing with West-ern industrial companies and has been educated partially in Europe or the United States, fruitful cooperation is quite easy. Our Kuwait partner, Yussuf B. Alghanim, is a good example of this. His elegant king's English is the envy of many a Hoechst employee. It is also of some importance that the partner or his or her family occupy a prominent role in economic life and in politics. Both these factors usually apply in these countries where the social structures are only gradually beginning to change. On the other hand, too close an interest in, let alone association with, political leaders, may become dangerous, since the political constellations and alliances tend to shift rather frequently.

In conclusion, it is very important to find a partner whose mentality complies with the Western way of acting. Understanding of the well-organized life in the camps of rich oil companies, as well as of the activities of temporary delegates from Western industrial groups, is a prerequisite, even if different private and social attitudes sometimes present a problem.

Bahrain, Qatar, and the United Arab Emirate

Anyone traveling in the Middle East should find time to pay a visit to the small, formerly independent, emirates that, in 1971, joined to form the United Arab Emirates. Visits to Bahrain, the island with the largest refinery in this area, and to Qatar are worth the effort from both a business and a tourist point of view. I was there in 1975, and I was much impressed by the oil wealth which has completely transformed this previously rather insignificant area both politically and economically.

Abu Dhabi and Dubai are the most important of the seven federated emirates. Abu Dhabi has the most oil; Dubai is the largest commercial center of the Persian Gulf as well as a historical center for smuggling. The Sheikh of Abu Dhabi is president of the federation, and it is difficult to avoid parallels with the European Economic Community members, which also jealously guard their independence. This is evident even in passport and custom formalities.

All these Persian Gulf nations, blessed by oil wealth, have one aim: to meet the pent-up demand for industrial and consumer goods as quickly as possible. They also are careful to expand schools, universities, hospitals, and infirmaries as fast as they can. Little or no income tax is levied in the emirates and import duties are very low. Foreign companies can conduct their business practically everywhere. They are, however, required to cooperate with local partners who must hold at least 51 percent of the shares. In view of the booming commercial activity, it is not surprising that the inflation screw has begun to turn, raising prices by about 25 to 30 percent in the last few years. But as far as the purchasing power of the population is concerned, rising costs are compensated for by the rapidly increasing national income.

Steel, a National Status Symbol

There is no shortage of industrial projects in the countries around the Gulf. Of course, the first priority is petrochemistry, although it suffers from the absence of an infrastructure, organization, and markets. The next priority is enjoyed by the steel producers. It seems to me that having a steel industry is almost a national status symbol. But here the energy conditions are good. The iron ore is found either in the country itself or is imported. Natural gas is replacing coal and coke in the direct reduction process.

Several Gulf states have launched large-scale projects that are perfectly logical in economic respects. For example, in Bahrain a large aluminum foundry has been established. It presses crude aluminum from Australian bauxite, thus cleverly exploiting the local energy sources. The project also makes sense because the large investment and capital requirement can be met from local resources; furthermore, little local labor, which is in very short supply, is needed.

From the point of view of the labor market, two further projects —shipyards in Bahrain and Dubai for the construction of supertankers—must be judged with rather more reserve. Yet the idea of shipping their oil in their own tankers, constructed in their own country, is a tempting idea for the financially powerful oil sheikhdoms.

In this part of the world, molded for centuries by the presence of the British, the main preoccupation today is the development of a way to overcome the inferior infrastructure as quickly as possible. Everywhere, foreign construction companies, including many German ones, are at work. They are building new roads, constructing or expanding harbors, and improving telecommunications and electrical supplies. An essential role is also played by the construction of water-desalination plants and houses and apartments, and the expansion of health and education services. A continuing important goal is the consolidation of the functions of trading and transshipment centers, which for many centuries secured a certain degree of well-being for these countries.

As indicated, labor also is a central problem in the Gulf countries. The population of the individual emirates varies from 4,000 to 50,000, and totals some 250,000 people. In many places, the native population is matched by an equal number of foreign inhabitants, who have come mostly from the poorer neighboring Islamic countries.

Stresses under the Surface

This mixture of older and younger, native and foreign people results in political risks. The inherited feudal system of the emirates is increasingly being questioned by the younger generation, and even more often by the immigrants. The contrast between the conservative thinking of Islam and its traditions on the one hand, and the necessity of creating entirely new conditions for improved standards of living in these countries on the other, is obvious. Nevertheless, the per capita income no longer lags behind that of a Western industrial country. I only hope that the changes in political structure, which eventually will take place in this part of the Middle East, will not involve too much political interest.

It is undoubtedly wise and proper that the Gulf states, and particularly the United Arab Emirates, use part of their enormous oil incomes to help the less affluent Arab nations. Whether this aid will be sufficient to satisfy the needs of the more populated but less oil-rich countries in the long term is another question. I always remember a conversation that I once had with an old acquaintance in Bahrain. He believed that the real problem of both his country and himself was that it had become too easy to earn a lot of money. He did not know what to do with it all. Culturally, his home was England. There, he said, inflation was so rampant that it was difficult to find financial investments that would be really profitable. The question of how to find investment opportunities with a secure income in Western Europe was his real problem for the future, especially for the future of his children.

I hope that this friend will never have worse worries.

At the Portals to the Near East

During the first twenty years after the war, Lebanon was the Switzerland of the Near East. The beauty of the country attracted tourists from all over the world, especially from Western Europe. Lebanon touted the powerful advertising slogan that it was the only country in the world where one could bathe in the sea and ski on the snow slopes on the same day. Although this claim corresponded to the facts at a certain time of the year, it has probably not been tried too often.

Lebanon is a very attractive country; its soil is rich in history and culture; its mountains are beautifully structured and its valleys fertile. But it was not these external characteristics that suggested its charac-

terization as the Switzerland of the Near East, but the much reported
fact that here Christian and Mohammedan citizens were living in
equality and had formed a joint government. This was based on a
clever and flexible proportional system which, in spite of occasional
friction, functioned well for many decades.

Thus, even during my first visits, I had the impression that Lebanon
was a small and peaceful country with an industrious population and
that it might well be a stabilizing factor in the area. My numerous
visits to Beirut and to the Lebanese countryside were a great enjoyment not only in business terms. The diligent population had created
a flourishing industry and, at the same time, the country was fulfilling
a transit function for numerous neighboring nations. The business
world was essentially European in character. In the early postwar
years, Hoechst had cooperated with a group of Lebanese who had been
closely linked to Germany and who, in some cases, had formed family
relationships. Dr. Ara Hrechdakian was our first representative and,
later, a partner in Hoechst Middle East. With considerable hospitality,
he helped many visitors to Lebanon to get an idea of the beauties of
the country.

These beauties were not only the famed cedars of Lebanon, not only
the old Phoenician ports like Byblos, Sidon, and Tyre, or the numerous
relics from the distant era of the Crusaders. They were also the spectacular sights in the interior, the fertile Bekaa Valley and the imposing
ruins of Baalbek, a sight to which I enjoy returning again and again.
Outside of Rome, there are few such well-preserved and impressive
remains from the flowering period of the Roman Empire. But en-
tranced as I was by these treasures, I was captivated even more by the
charm and geniality of the people, their readiness to help, and their
friendly cooperation.

Lebanon: Crisis Center

These idyllic conditions were abruptly upset when the unrest insti-
gated by the Palestinians broke out in 1975. Potentially, the threat
from this quarter was always present. The conflict was inevitable from
the moment when the precarious equilibrium between Christians and
Moslems was disturbed by the massive immigration of Palestinians,
first after their expulsion from their Israeli homes in 1948 and then
from Jordan in 1970. In addition, the difference in birthrates between
the well-to-do Christians and Sunnites and the underprivileged

classes gradually led to a further shift in the equilibrium. Finally, as in most developing countries, there was considerable rural exodus, leading to slum development around the metropolis and consequent impoverishment. Taken together, these factors created dangerous social dynamite.

At first, life in Lebanon proceeded peacefully and quietly; visitors were inclined to forget the continuous latent threat. It was not until 1975 that the tragic conflict, marked by so much violence, between Christians—mainly Maronites—and Moslems, Palestinians, and left-wing activists broke out. It will take a long time before the wounds of that conflict are healed.

The situation in Lebanon continues to be precarious. Beirut, the once beautiful city whose center was so brutally destroyed, is being rebuilt. I have no doubt that it may regain its former splendor. Yet, it is questionable whether it will ever recover its function as a tourist center, as a crossroads of the Middle Eastern world, and as a junction between the West and the Near and Middle East. Indeed, none of the causes of the catastrophe appears to have been overcome. The country will gain permanent peace only within the framework of a solution to the Middle East problem in which the question of the Palestinians is resolved.

The small group of German people at Hoechst Middle East, together with their families, had to leave Beirut during the period of the worst fighting. Many of their Lebanese colleagues, too, sought safety in nearby Damascus or in Europe. Most have since returned and our company is in business again.

During our first development phase, the assistance of colleagues ·from the parent company was necessary. However, their tasks are gradually being taken over by indigenous experts. We have had particularly good experience with Arab people who have become familiar with the German language as a result of their studies in Germany, their work for Hoechst in Europe, or their marriage to a German national. They are most valuable as liaison with the local market.

The Pioneering Spirit Wins the Day

Particularly during the initial phase after the war when there were no Hilton or Intercontinental hotels, our German representatives often displayed a remarkable vitality. This enabled them to succeed even under rather primitive living conditions and in an entirely new environment. They were helped by a spirit of adventure and a desire to see the world, attributes that have now become somewhat rare.

Hoechst Middle East, with its headquarters in Beirut, was conceived as a regional center for our commercial activities in the Middle East. It is not now possible to say whether it will ever again assume this role. For the time being, we are actively pursuing our business in Lebanon. The functions, both commercial and advisory, that our headquarters in Beirut fulfilled for other countries have largely been transferred to those countries. In cases where there is not sufficient opportunity for direct activities, we have developed a kind of neighborhood help.

The courageous people of Lebanon have our sympathies. I believe that the Western world should do all in its power to renew the sensible coexistence between Christians and Moslems which prevailed in Lebanon over many years.

While there is no oil in Lebanon, the country has profited from the oil riches of its neighbors because of its transit function. There is a fertilizer industry that concentrates primarily on phosphates which has replaced imports. It was of great help to the domestic markets. I do not believe that petrochemical products will be manufactured in Lebanon in the foreseeable future. I can imagine, however, that Lebanon will play a mediating role in the development of raw material deposits and the consumer markets.

In the last few years, a promising processing industry has been established, frequently as a result of private initiative which exists in nationalization-oriented neighboring states like Iraq, Syria, and Libya. This industrialization was made easy by the availability of experts— Beirut alone has four universities—and access to low-cost labor, essentially from Syria, the Palestinian camps, and Egypt.

Lebanon was not only the Switzerland of the Middle East but also the Paris of those who earned much money in neighboring countries, often under rugged living conditions. The casino, which enjoyed a noted reputation in this region, was never short of visitors. A cosmopolitan center of many minorities, languages and religions, catalyst for Eastern and Western intellectual thoughts, an important financial center, Beirut had become all this only after the elimination of Alexandria and Cairo. Beirut played an outstanding role in the fields of education and the press even when under French mandate between the World Wars. Before its civil war, Lebanon was more or less the only Arab country with a free press and an education system administered by qualified people. It was thus a valuable environment for intellectual and political ideas in the area and the "high school" of the Arab intelligentsia.

Birthplace of Near Eastern Socialism

The route from Beirut to Damascus, on a comfortable, well-constructed motorway through the Lebanese mountains, can be covered effortlessly in less than an hour and a half. In the "good old days" when Damascus was then still part of the "European" Middle East, it was ruled by a group of politicians and merchants, most of whom had been educated in the West. In those years, Syria was a particularly enjoyable country where we could find partners who thought like ourselves. With their help, we were able to rebuild our traditional business rather quickly. The textile industry formed the foundation and was followed by a very active building industry.

The ambience changed when the feudal system was abolished in 1955. The socialist Baath party gained power, and military governments with varying policies followed one another in rapid succession. What finally remained was an Arab socialist system similar to the ideology that prevails in Iraq. The Syrian state, originally more or less parallel to Lebanon, was fundamentally changed. The conservative element, which had occupied the leading political and economic positions, left Syria. It was replaced by the military. No doubt, the bourgeois rule also had its negative aspects, and it certainly clung shortsightedly to its privileges.

Hospitals and Schools

The Baath party, in contrast, advocated improvement in the standard of living of the general populace, and it can record some achievements. Wherever one travels, whether in the cities or in the remotest provinces, true poverty is rarely encountered, compared with neighboring countries. There are no begging children in the streets. The numerous, if modest, schools and hospitals are also remarkable signs of progress.

Until recently, Syria had no efficient ports and had to rely largely on Beirut. The government has made great efforts to expand the Syrian harbors of Tartous and Latakia. The railway line from Latakia to Homs will soon be completed. Syria's oil deposits are not very important; yet, it has been the transit country for oil, especially from Iraq.

Presently the Oil Bypasses Syria

Syria's transit role also has changed provisionally. Several Arab countries have taken it amiss that the country assumed the role of peacemaker in Lebanon, putting a limit to the excessive power ambitions of the Palestinians. While this altruistic attitude should have gained Syria a great deal of prestige in the West, the Arab world viewed it rather differently. That the Alawites, a minority community whose members control the government and army, rushed to help a Christian neighbor, and not the primarily Moslem-Palestinian coreligionists, was not easily explained to an Islamic world where religion still plays the dominating role. The Saudis, too, seem to have turned off their financial support of the Syrians. This could be a bitter blow because many Syrian development projects, grouped into five-year plans in the socialist manner, depend on the money which they obtain from the Saudis and oil-rich Iraq.

It remains to be seen whether the diversion of Iraqi oil through Turkey will be maintained for any length of time. It would be desirable for the acclaim that President Assad gained for his successful peace efforts in Lebanon to continue. A lengthy Syrian occupation of Lebanon, and the consequent high military expenditures, will place a heavy burden on the economic development of the country.

The Soviets Help with Irrigation

The backbone of Syria is its agrarian economy, which has been promoted under the new economic system. It will experience a considerable upsurge once its full potential can be realized. This goal presupposes that the irrigation project of the Euphrates dam, which is being designed and constructed by the Soviet Union, will be completed in the foreseeable future. Characteristically, the cotton harvests, according to unofficial information, have decreased for many years in spite of massive government investment. The earlier private initiative cannot be adequately compensated for by such measures. A similar decline has also occurred in grain crops, where Syria barely meets its present-day requirements. Not long ago, it was still an export country, to say nothing of the Hellenic-Roman eras when Syria was regarded as one of the granaries of the eastern Mediterranean.

A not inconsiderable advantage for Syria is its light population density, so unlike Egypt with its insoluble problems of overpopulation.

In contrast with many other countries of this region, Syria has never accepted full dependence on the Soviet Union. Contacts with the Western world, especially the United States, have been kept open. Syria's attitude will therefore play a considerable role if it should prove possible to find a solution of the Middle East problem that is generally acceptable for the long term.

The Wise Brave King

A visit to the Middle East is incomplete without including the Kingdom of Jordan. Driving from Damascus on a bumpy, narrow road full of potholes along the Israeli border and behind columns of trucks in the direction of Jordan, one immediately notes the cleanliness and order that prevail once the border into Jordan is crossed. The roads are in better repair, the uniforms of the Bedouin soldiers with their kefiya look freshly pressed, and there are extensive reforestation projects on either side of the road. What is the reason? Could it be that, apart from the socialist, or rather more capitalist, present, the differences in the colonial pasts of Syria and Jordan also play a role?

Jordan has had no shortage of political crises. But, thanks to its courageous and intellectually versatile king, it has survived these crises without significant harm, although it has suffered the loss of the West Bank of the Jordan River. This is all the more remarkable in view of the fact that, among all the Arab states, this small kingdom of only 3 million inhabitants houses probably the largest percentage—about half—of all the Palestinians.

The Palestinian Question

These Palestinians have never become a real part of the Jordanian state; they remain a state within a state. The older generation passed on the trauma of separation and the demand for its own homeland to the younger generation, which largely lives in camps supported financially by the United Nations. In many cases, it is already the third generation. The Palestinian question has become a problem of existence for Jordan. The way in which Hussein and his followers have coped with the question is almost a miracle. Many compromises have been made. Solutions have always been found by which the Palestinians were excluded when they could no longer be tolerated within the state system. On the other hand, care has been taken not to turn them into such bitter opponents that they might endanger the existence of the Hashemite kingdom. This was noticeable when I traveled through Jordan and suddenly came across extraterritorial enclaves the extent of whose control by the Jordanian state was impossible to measure exactly.

Since Black September 1970, major confrontations have been avoided, and Jordan has made a serious attempt to integrate the Palestinians living in its territory. To what extent this effort will succeed under the emotionally charged conditions is a fundamental question. I have witnessed with admiration how the Jordanians are striving to face, and to master, the strains to which they are exposed on all sides. Jordan's ability to reach a compromise with Israel that also adequately respects the existence of the Palestinians is essential for the maintenance, or better, the restoration, of peace in this part of the globe.

The small Kingdom of Jordan is not one of the nations of the Middle East that will join the ranks of the industrial nations in the immediate future. Nevertheless, it is rich in trade and commerce. Its success in the rapid expansion of Aqaba on the Red Sea, the only harbor in the country where industrial development is beginning, deserves recognition. A German consortium is extending the railway line beyond Amman to Aqaba. The industrial development potential is based on the intelligence and craft skills of the inhabitants of Jordan. Here, too, it is essential to maintain economic links and to design relationships with the West so that the people who know the Western world and share its intellectual life—and there are now a few of them—do not lose their belief in a joint future.

Israel—Hopes of Agreement

People from many lands have come together in Israel. With unequaled diligence they have created from an arid fragment of land a country that, even without oil and subsidies, would be viable economically if it did not have to bear the giant burden of defense. Economically, Israel has become a factor whose influence extends far beyond the Middle East. For the Federal Republic of Germany, Israel is a reliable export partner whose importance easily matches that of other countries with larger populations and richer mineral wealth.

No one is in a position to say how peaceful coexistence between

Arabs and Israelis in the Middle East may one day be achieved. It is my belief, based on business and personal observations over many years, that nothing should be done to weaken Israel. Any such approach could result in a conflict which would have catastrophic consequences for a large part of our globe. The Anglo-Australian writer Nevil Shute has written an apocalyptic vision of this danger in his novel *On the Beach*. Recent developments have not produced any reasonable answers to the question of how Israel could live in peaceful coexistence with an Arab world. Continued political dialogue and the courageous initiative taken by Egypt's President Sadat—including his memorable trip to Israel in 1978—have at long last led to the conclusion of a peace treaty between Egypt and Israel, even if there is still disagreement on many important points. In spite of the disapproval of the Arab Union, this treaty is a historical fact and, it is hoped, the starting point for a peaceful development so badly needed in this part of our world. Perhaps the oil riches, which have enabled the Arab nations to take such great leaps forward, also can contribute to overcoming the contrasts that now beset the area.

For us Germans, it is time to rid our relationship of the guilt of a terrible past. The generation responsible for the fate of the Jews under the Nazis no longer holds political or economic power in West Germany. Both sides should find the courage to overcome well-justified resentments and to take a positive look into the future.

18
Africa on the Move

For a long time, Africa was *terra incognita* for me. Apart from flying visits, my more extensive discovery of the black–white–brown continent began only during a wide-ranging African tour with Karl Winnacker in 1962. Our first stop was Abidjan, capital of the Republic of the Ivory Coast. Our business had developed to an extent which required a small office with an adjoining warehouse at Polychimie, the predecessor of Hoechst Afrique de l'Ouest in Abidjan. There were some Germans and French, but the majority of the staff were nationals who had been partially trained in France.

Our friends had arranged an audience with the President Félix Houphouet-Boigny. The discussion was most stimulating and of a quite different kind from what I had imagined such events to be in Africa. Houphouet-Boigny, who had studied in France, is not only well-educated but also thoroughly acquainted with the economic and political problems of his country. In addition to the Senegalese President Senghor, who has also acquired a reputation as a poet, Houphouet-Boigny is regarded by many in Africa as an exponent of France and the West. He is a physician and *planteur,* and in prewar times he was also a minister in the French government.

From an economic point of view, the priority of the Ivory Coast was to integrate the plantation economy, founded by foreigners, into a partnership with the Africans and to provide the population with an opportunity to learn modern methods of cultivation. We visited a newly established pineapple plant, which proved that such projects can be readily realized. The country also needed to expand its existing processing industry and to improve its foreign exchange situation by attracting more tourists. In this respect also much success was clearly in evidence. There are modern, fully air-conditioned hotels equipped with up-to-date facilities in Abidjan and other locations along a coast blessed by tropical vegetation.

Starting with Abidjan, Hoechst soon opened branches in Dakar and Douala. Today in this part of the continent, we have become equal partners of the old, established branches of French companies that have retained a good deal of their influence.

Black Africa's Rich Countries

Hoechst had reestablished itself in Nigeria and Ghana, the most important English-speaking countries of Western Africa, well before our 1962 visit. In these countries, the combination of a comparatively high per capita income and an above-average population density provided the conditions for a promising engagement. In addition, the relatively recent oil finds in Nigeria were an unexpectedly strong attraction for the chemical industry. A prerequisite was the development of crackers and refineries which might be completed by the end of the 1970s or the beginning of the 1980s. It is hoped that Nigeria, apart from its oil which is predicted to last until the year 2000, will continue to concentrate on its original agrarian products, and thus to stabilize the economic basis required for its positive development.

We have made special efforts in Nigeria to continue our commercial operations and, increasingly, our industrial activities in spite of political upheavals, such as the Biafra conflict. The production in Lagos of polyvinyl acetate dispersions, pigment preparations, tensides, and auxiliaries should provide a suitable means of participating in the industrial development of the country.

The groundwork for our pharmaceutical activities was provided by years of collaboration with Major & Company, headed by John Fielding, who drew upon tremendous experience in Africa. Eventually, both Major and Hoechst came to the conclusion that the time had come to go our separate ways. This meant even more direct and systematic introduction of Hoechst products into comparatively young markets. It was done with such success that the conditions for pharmaceutical production in Nigeria were established.

The sales of Roussel products, taken over by Hoechst in Nigeria, developed equally successfully. Hoechst has followed the booming economic development of the country in recent years. The paint business of Berger Paints, a subsidiary of Berger, Jensen & Nicholson in London, has enjoyed similar success. With the largest market share in the paint sector, Berger operates plants in Lagos and Port Harcourt. On the other hand, the increase of confidence in Nigeria and the country's desire for political independence have meant that, since 1973, most foreign companies have one or more local partners.

It was therefore logical for Hoechst to invite Chief Ashamu to be a local partner of Nigerian Hoechst Ltd. We had to abandon our majority position, but we have retained management and have secured all the rights accrued to us from earlier agreements. Future quotations of the shares of Nigerian Hoechst Ltd. on the Stock Exchange will give Hoechst, as the largest single shareholder, a degree of flexibility that is desirable for various new projects and for a sound future of our business.

Nigeria, an agrarian country renowned for its cocoa and palm-kernel production, was able to boast of considerable wealth even in the past. The need now was to create an infrastructure that could cope with the sudden benefits of the recently discovered oil. A vast building program was being executed everywhere. Municipal roads have already provided noticeable relief to traffic. The harbor of Apapa is now much too small. Ships wait for weeks and even months before being unloaded. The port on Tincan Island also needs to be modernized. Electric power failures and the usually inoperative telephone and telex systems frequently cause additional problems. At the time of our trip, the city center was paralyzed by traffic chaos, and the hotels had in no way kept pace with business requirements. Our entire trip was almost ruined when more than half its members contracted food poisoning in Lagos.

The day before, we had made a day trip to the country's interior to get an impression of the living conditions. We visited one of the hospitals erected in the villages, some of whose personnel dated back to the missionary era. The doctors told us that only those children whom nature had endowed with a great deal of resistance had a chance to survive the health conditions that existed in the countryside. It can be readily imagined how difficult it will be to solve these problems.

With an extensive textile industry as a base, Nigeria had developed numerous related industries. We soon engaged many Nigerians and developed a training center in which they could acquire the necessary skills.

Paradise for Tourists

When the British talk about East Africa, they are not referring to the whole of eastern Africa, but only to Kenya, Tanzania, and Uganda. From the point of view of climate and landscape, this region is one of Africa's preferred areas. British colonial history is witness to this, as are today's tourists who trek to the Indian Ocean for photo safaris and bathing holidays. Nairobi, the capital of Kenya, was the center of the business that we built in this region. In our plans to exploit the favorable conditions and to participate in the industrial development of the country, we found once more that the future of the chemical industry begins more slowly in black Africa than in other parts of the world.

During the colonial era, Kenya, Tanzania, and Uganda were in economic association. Currency, jurisdiction and administration, postal services, railways, aviation, and customs were centrally organized. This union alleviated the disadvantages of a relatively small economic area. Since the countries became autonomous, however, their political leaders have drifted ever further apart. What might still have been possible with political understanding seems to fail in the face of the pronounced tribal egoism of the individual countries. A common economic area as large as central Europe, with almost 45 million inhabitants as well as excellent agricultural conditions, would have permitted a far greater degree of industrialization than has taken place so far, even in the forward-looking and most highly developed state of Kenya. Kenya has production capacities twice as large as those of Tanzania and Uganda together.

A certain parallel to this disintegration of former British East Africa can be observed in the neighboring southwestern area. While the Federation of Rhodesia and Nyasaland, when British colonial territory, enjoyed a degree of prosperity, Zambia, Southern Rhodesia, and Malawi have developed as three entirely separate and partially estranged countries. In spite of several attempts by our agency, operating in the Zambian mining town of Kitwe, we have been unable to achieve an industrial foothold. As before in other countries, local difficulties in financing Zambia's share and other investment conditions did nothing to promote the project. Only in the Zambesian part of the copper belt were we able to supply chemicals for mining companies. The country was then overtaken by serious economic problems which arose in the wake of declining copper prices, the inadequate diversification of Zambia, and—not least—the nation's confrontation with Rhodesia.

President Kuanda, who once visited us in Hoechst many years ago, will have to apply his entire statecraft and authority, much effort, and some severity in order to return the country to normalcy. An early solution of the Rhodesian problem would provide a good beginning.

King Léopold's Dream

To the west of Zambia lies the Shaba province of Zaire, the former Belgian Congo. With its rich copper mines, Shaba (formerly known as Katanga) is the industrial core of Zaire, although the province is situated at the the extreme southeast corner of this troubled country. Zaire, so rich in natural wealth and economic problems, is geographically the heart of Africa.

Two years before our excursion via Bal Air, I had been in Léopoldville for the first time. I like the city for its picturesque location on the mighty Congo river and for what seemed to be a good understanding between the black people and the numerous Belgians. The Belgians proudly showed us a project to build a huge Catholic university for the Africans in the vicinity of Léopoldville. The European population was in no doubt that the wave of self-determination flowing across the continent would also reach the rich and peaceful Congo. But no one expected it to happen so soon. Only one year later, Belgian rule was suddenly at an end. The former colonial rulers and their families had to flee the country, at times under very disagreeable circumstances. When the Belgians had been driven out, there was no one as yet to take their place in administration or industry. The number of Congolese educated in the West was very small. Bloody riots, smoldering civil war, and attempts at secession from the newly formed Republic of the Congo were the concomitants of an independence for which the area was not prepared.

In spite of all the continuity in the leadership of the country, there was no strict policy which would have offered reliable framework conditions for economic activities by foreign investors. In most cases, including Hoechst, activities were therefore confined to commercial ventures which achieved a considerable volume, due not least to local support and advice. Our colleagues in Brussels also gave us considerable help.

It is remarkable that in spite of sustained quarrels and ill will, the relationship between Zaire and the former Belgian mother country of the Congo have remained fairly close both culturally and economically. Even today, 50 percent of Zaire's exports go to Belgium, and the Belgian universities are preferred by students from Zaire. Happily, many things have become normalized and the Belgians are again the foreigners most readily tolerated in the country and whose return is even promoted, whether for linguistic or historical reasons.

The importance of Zaire as a raw materials source is likely to be significant for its future. Apart from its copper wealth and its position as the largest world supplier of industrial diamonds, the available agrarian area in the various climatic belts has yet to be put to beneficial use. On the lower reaches of the Congo, a huge dam is being completed which will secure for Zaire considerable independence in energy. The industrial center is in Shaba province, where copper and cobalt mines are concentrated. The mines are operated with substantial support from Belgian and French engineers, technicians, and merchants.

Although the invasion of Shaba in 1977 was largely a storm warning from the outside, the country is threatened also by internal dangers, as recent unrest has shown. It is not yet clear whether the 1978 assault on Kolwezi, which was apparently conducted with Cuban and Soviet support, was aimed at expelling the essential Europeans. Their departure would have ruined Zaire economically and weakened the government considerably. Western intervention, which aided a government weak in military power, put a quick end to the unrest. The political importance of Zaire as a possible stabilizer and dam against communist infiltration of the eastern and western African coasts has once again been clearly shown during these political events. What will be the future life of this country if the big political power blocs cannot agree to respect its territorial integrity?

Portugal's Former Overseas Provinces

In neighboring Angola, as in Mozambique, the other former Portuguese possession in southern Africa, separation from the mother country took place in a different, but no less dramatic, manner. Portugal, itself still unsteadily seeking a new political identity, released the overseas provinces after many years of guerrilla war. The colonies presented a considerable burden on the budget of the small country.

Hoechst operates small production plants in the two countries, both of which were at first unaffected by the turbulence. Their ability ever to be economically viable has to be doubted in view of the lack of raw materials and foreign exchange. The two paint factories founded by Berger, Jensen & Nicholson years ago have so far proved the most resistant to these adverse conditions.

In Mozambique, many readers may recall, a hate-filled campaign had been mounted against the Cabora Bassa Dam. Now that the country has gained its independence, the critics—from animal lovers to environment protectors and ideologists—have fallen silent. The project is now being praised as providing a guarantee of future industrialization and a main source of foreign exchange. After all, the major part of the current made available by the dam goes to South Africa in return for payment in hard currencies.

Madagascar, the Non-African Island

During our 1964 trip, we left the African continent for two days and spent some interesting and stimulating hours in Madagascar. We encountered a world that, although geographically close to Africa, is nevertheless vastly different from black Africa in ethnology and climate, in vegetation, and in the customs of its inhabitants. The capital, Tananarive, is a modern city whose design and construction reveal the long period of French presence. The majority of foreigners living there are French; they are reluctantly accepting the fact that they must slowly pass to the Malayan-Indonesian population the economic benefits for which they have worked. The visitors from Hoechst were interested in the rich tropical vegetation because, in addition to our small trading organization, we had an expert in the country who systematically collected herbs of possible interest as base materials for the pharmaceutical industry. However, these experiments were abandoned after a number of years because there was no indication of any success. Similar work is being continued by our pharmaceutical research in India.

New Problems: Ethiopia, Somalia, Sudan

In the northeastern part of the broad black belt of Africa, there is another triad of countries that exhibit peculiarities in racial respects: Ethiopia, Somalia, and the Sudan.

Until modern times an empire and, according to legend, the oldest dynasty in the world, Ethiopia has disproved the theory that economic and social stagnation is the result only of colonial rule. Apart from urban areas, this empire of the "Lion of Judah" is still in an unbelievable state of underdevelopment. The feudal system of this multipeople state is without equal on this continent. It is all the more surprising, then, that Addis Ababa was to have become the seat of the Organization of African Unity, somewhat like a capital of Africa. In contrast with many other African capitals, Addis Ababa in 1964 had the look of an imperial metropolis whose architectural profile had been determined mainly by the new buildings erected for the meetings of the Union of African States.

Hoechst started to operate in Ethiopia many years ago. It faced no great problems in the industrial field because the processing industry is located around the capital. The pharmaceutical and agricultural sectors, however, proved much more difficult. Under present-day conditions, this problem is hardly capable of solution. Ethiopia has been wasting its limited resources for many years now in the frontier dispute with Somalia and in the fight to retain the former Italian colony of Eritrea.

Somalia, on the Horn of Africa, owes its prominence to its strategic position at the exit of the Red Sea and the western flank of the Indian Ocean. This thinly populated, poor country has long been of considerable interest to the great maritime powers. The government in Mogadishu, the capital, has played on the advantages of its location until this day. Economically, Somalia, with its many, often still nomadic, tribes, offers few opportunities even in the long term.

Conditions are different in the Sudan. This is one of the large states of Africa with almost 20 million inhabitants. Its cotton fields and a few industrial—especially textile—capacities show hopeful economic beginnings. But twenty-year-old Hoechst Sudan finds it difficult, apart from a modest industrial business, to penetrate the vast interior of the country. The military and political chaos will have to be replaced by more stable conditions before new opportunities are likely to arise.

The capital, Khartoum, was one of the last stops on our trip. It is the intersection between the black south and the Arab Moslem north of Africa. City life is characterized by the deep-dark, tall Nilotic tribes of the Nuans, Dinkas, and Shiluks as well as by the Arab Nubians of North Africa, who dominate both politically and economically. They look extraordinarily picturesque in their long, white djellabas and white turbans. We took the view that in spite of all the problems, the market of this country with its large population should not be neglected. For this reason, we established a pharmaceutical agency together with a Sudanese partner, Shibeika Bros., which is still looking after the business today. It was expanded by an office for industrial products, which is, however, leading a rather difficult existence.

The Maghreb: Independence and French Tradition

It is often forgotten that hardly twenty years have passed since the Maghreb rid itself of French rule. The freedom fight of Algeria, Tunisia, and Morocco at first ensured that certain common characteristics of these three states would survive. But soon after they had achieved sovereignty, the differences in state form and divergent foreign policies became evident. Since then, each has gone its own way. All that remains is a strong economic and cultural orientation toward France, which has provided a home for many people from these areas.

Algeria—the Will to Self-Sufficiency

As the greatest sufferer in the Maghreb's war of liberation, Algeria has also moved furthest away from its former mother country. In its quest for economic independence, Algeria has been helped by significant deposits of oil and natural gas. Nevertheless, in spite of all the stresses and disputes, Algeria and France still depend upon each other in many respects. More than 600,000 Algerians work in France and they are often the providers for their large families back home.

The government of Boumediène successfully converted the economic power of the country, based on its raw material resources, into international political influence. As the meeting place for the nonaligned nations and as a fervent protagonist of developing countries in the United Nations, Algeria has gained a great deal of attention. In the

Arab camp, Algeria has always been among those inciting opposition to Israel. The successors to the late Colonel Boumediène seem to take a more realistic stand with regard to Arab affairs as well as to their relationship with the West.

According to present calculations, the oil reserves of Algeria will last scarcely beyond the end of this century. Therefore, the country has far greater expectations of its natural gas reserves which, allegedly, are exceeded only by those of the Soviet Union. Liquefaction plants of huge dimensions ensure an export business that is to include the United States, now that the technical problems of transporting the gas over great distances have been largely solved.

Algeria's natural gas and oil reserves are, of course, factors favoring the establishment of a chemical industry. With the extension of refinery capacities, crude oil exporting will be reduced in favor of processing in the country itself. The realization of these ambitious plans could result in capacities handling some two-thirds of future annual output. At Skidka alone, 15 million tons of refinery capacity per year will provide the basis for petrochemical activities.

Man-made fibers and plastics, already a field of worldwide overcapacities, are to be included in the secondary production program. Algerian industrialization policy in the chemical field is a striking example of the problems that arise when ample raw material resources and domestic processing reach dimensions that can remain economic only through exporting. Together with DIAG (German industry and plant society), Hoechst has occasionally participated as an adviser in planning the country's industrialization. The socialist economic system in national guise admits private ownership only in commerce and small trading. Industry is the domain of state-owned companies.

As with most business visitors, my knowledge of the country has been confined to the capital. In Algiers, there is clear architectural evidence of where the French used to live. Today, only a few French people are encountered, most of whom are development assistants under intergovernmental agreements. In the center of Algiers is the Kasbah, the old Arab city, which was the site of passionate fighting during the Algerian uprising.

Libya: Oil and Revolution

The Kingdom of Libya was formed in 1951 as the result of a dispute between the major powers over the political form of this huge desert area. Its feudal monarchy was removed in 1969 by a revolt of military officers while King Idris was on a trip abroad. Since then, the country's 2.6 million inhabitants have been tightly governed by Colonel Qaddafi. A foreign policy difficult for us to understand, with strong Islamic and Pan-Arab nationalistic elements in which left-wing socialist tendencies are also embedded, has provided the world with many surprises.

The force of this policy is based on oil. The usable reserves in the Libyan desert sand are estimated at 3.5 million tons, or 4 percent of world oil reserves. The high quality of Libyan oil, which has low sulphur content, and the short transport routes to Europe have strongly favored the sales of Libyan oil at high prices. When I first got to know Libya in the late 1950s, the oil was already flowing. Libya had become an Italian colony in 1934, and the traces of Italian occupation were beginning to vanish. The Italian plantations had been continued only on a minor scale because agricultural interest had greatly diminished; the chief earning potential lies in oil, the new money source. Indeed, like many other oil countries, Libya has a problem in finding enough specialists to implement the ambitious program that can easily be realized from the new wealth. All that has remained of Italian rule are the good roads, but much else has been constructed. The changes that have been made since the feudal system of King Idris came to an end are very extensive indeed. At any rate, things looked quite different during my first visit. There was even concern as to what should be done with the vast amount of money that was being earned. Nobody had yet thought of investing it abroad.

The situation changed overnight when the military regime gained power. Surprisingly, the new regime did not adopt a communist form of socialism. Rather, the revolutionary council of President Qaddafi aimed to revive Islam, to return the country to pure beliefs, and to subject the daily lives of the people, who had begun to move away from religious principles as their well-being increased, to the laws of the Koran once more.

This had many surprising effects. For example, the offices of foreign companies in Libya, when they are still admitted, must conduct their correspondence and documentation in Arabic, the language of the country. Passports of foreigners who have temporary permission to work and live in Libya must be familiar with the language even if— as is usually the case—the authorities of their native countries, and indeed the owners of the papers themselves, can neither read nor understand the statements in their passports. It is of course understandable that the oil wealth has strengthened Libya's national consciousness and that Western discrimination, which without doubt existed for a long period, is being more than balanced out.

The Difficulty of Commercial Contacts

The rapid development of the oil industry caught the early attention of the chemical industry. We tried to get into the business with chemical auxiliaries used for oil drilling. In the beginning, we appeared to be doing well. In 1964, the head of the sales department responsible for these products traveled to Tripoli in order to demonstrate to us his successes in this field. Unfortunately, on the way from the airport to the hotel, he had to tell us that our partners had become bankrupt. Again, we had the bitter experience of learning that a chemicals business related to oil can be conducted satisfactorily only through specialized service companies. The nationalist government of Colonel Qaddafi has raised many obstacles to the operation of foreign companies. We therefore maintain only a technical advisory bureau in Libya; it looks after the pharmaceutical business and the transactions covering the rest of our product line.

Libya's oil exports provide about 3 times as much income as the country needs for its imports. What is it doing with the rest? The country is a direct neighbor of Egypt, and it has been estimated that in spite of numerous confrontations, more than 100,000 Egyptians work in Libya as teachers, civil servants, and in other vocations. Their replacement would be difficult. Nevertheless, it is not easy to foretell how the relationship between the two countries will develop, since Colonel Qaddafi has a pronounced sense of mission and lays claim to being the leader of the Arab world. Our subsidiary Uhde has constructed large installations in Libya for ammonia and urea, and new projects are being discussed. The industrial capacities being erected far exceed the needs not only of the country itself, but those of the neighboring states. In what form these large capacities will be used and whether, for example, they will help to reduce the poverty of the developing countries lacking oil, remain open questions. I fear that many upsets in world trade will result if products, manufactured according to Western technology and adequate in quality for modern requirements, are sold in markets that have so far been supplied by the West.

Colonel Qaddafi provided the Western world with yet another sensation when he acquired a minority holding of 9.5 percent in the

Italian Fiat concern. This investment should be regarded not only as an expression of a newly awakened self-confidence of the Libyan people and its leaders. Colonel Qaddafi also seems to have recognized the need to build bridges to nations with which Libya was once closely linked for many centuries and which should remain partners in the future.

Many extraordinary monuments testify to the historic past of the country. It was not only a granary of the Roman Empire, but also had close cultural bonds with Rome and, earlier, with Greece and Phoenicia. There are many well-preserved ruins, especially those of Leptis Magna and Sabrata, which date to the most flourishing period of Greek-Roman art and which form a particularly stunning contrast to the harsh expanse of this desert country.

Morocco—Rich in Crude Phosphate

What oil and natural gas are to Algeria, crude phosphate is to Morocco. The estimated crude phosphate reserves of this country are 50 to 60 billion tons, or about 60 to 75 percent of all the world's currently known reserves. Only 10 percent of the output is processed on site into phosphoric acid and fertilizers, although this proportion will be increased to 30 percent.

The crude phosphate reserves at Youssoufia and Khouribja currently are being exploited to the level of 20 million tons per year, or 0.5 percent of the estimated reserves. Even at that level, Morocco is the world's largest phosphate exporter. With a share of 40 percent, it even leads the United States and the Soviet Union. In 1974, these three countries increased crude phosphate prices at a stroke from US$14 to US$68. At first, this appeared to be a parallel to the OPEC policies. However, such a high price could not be maintained for long.

The United States and the Soviet Union, so far the most important phosphate producers, have increasing requirements. Yet their crude phosphate reserves soon will no longer be sufficient for their needs. Therefore, Morocco has excluded long-term supply contracts for crude phosphoric acids with these countries. The industrialization policy of Morocco is founded on this almost inexhaustible supply of crude phosphate, unmatched in these quantities anywhere else in the Third World. Hoechst has contributed much to this development through Uhde, its engineering subsidiary. Considerable Hoechst know-how is incorporated in the new installation in the Safi area which forms part of the two Uhde projects, Maroc Phosphor I and II, that produce exclusively for exports.

Leaving Casablanca by the main road to the country's capital, Rabat, one passes the modern plant of Polymedic S.A., which has been cleverly integrated with the landscape. In the course of the last ten years, Hoechst, together with Moroccan partners who replaced the former French owners, has expanded this company into the greatest pharmaceutical plant in the country. Alongside it are the administrative offices of Hoechst Maroc, which were skillfully designed by the same German architect, Ewerth. Hoechst Maroc is operated together with two Moroccan partners, Karim Lamrani and Bensalem Guessous, with whom we have worked closely for many years. This Moroccan organization, like those in Tunis and Algiers, was developed by Hoechst France, which continues to exercise a mediator function between the companies and its partners in Morocco.

Apart from the pharmaceutical plant, which produces for other prominent companies as well as Hoechst, Hoechst Maroc owns a dispersion plant which is located farther away on the outskirts. When I spent my 1976 Easter leave with my wife and children in this country, whose landscape and culture are so attractive, the cornerstone for this factory was just being laid. The ceremony consisted of François Donnay, Bensalem Guessous, Abderrahmnan Guerraoni, and myself planting a palm tree. No doubt, the next time I visit Casablanca it will have grown considerably.

The constitutional monarchy of Morocco, formed in 1972 from the sultanates that had been French and Spanish protectorates, brought stability to the country in spite of several attempts at revolt. Together with a Western-oriented Tunisian foreign policy, the constitution ensures that the Maghreb is today linked by special agreement with the European Community. Also, the economic interests of Tunisia and Morocco call for links with the West.

The relationships between Morocco and Algeria were never very close. Conflicts in the western Sahara, promoted by Polisario, a guerrilla-style liberation movement, are again straining relations between the two countries. Strategic interests, access to the Atlantic, and valuable raw material deposits have rendered an otherwise useless strip of desert into a bone of contention.

For many months in the year, Morocco has a moderate climate. With the Oriental magic of the old residential cities of Marrakesh, Meknes, and Fes, and with the hospitality of its people, it is a particularly engaging tourist country. Modern hotels have been built, although tourism has not yet reached the dimensions achieved in Tunisia. We loved to stay in the old Mamounia hotel, located not far from the mosque in the center of Marrakesh, in the middle of flower gardens. In the Marché des Fous, the stranger can watch conjurors, snake charmers, or storytellers. Behind the market are the convoluted narrow lanes of the souks, or bazaars, where handicrafts are sold. Mosques, Koran schools, and palaces alternate in colorful sequence. Moroccan cuisine, regarded as one of the best in the world, often begins with a pastilla, a pastry filled with pigeon meat, almonds, and raisins, and powdered with sugar and cinnamon. *Majouri,* or grilled mutton, the braised dishes, called *tajine,* and *couscous* are known far beyond the borders of Morocco. Dessert usually consists of peppermint tea accompanied by biscuits made from almond dough and macaroons.

Hassan II is popular among the people. The frontier conflicts with Algeria have raised the national consciousness of the Moroccans. Since 1977, representation from the opposition parties, including the Istiqlal, is included in the political life. In the long run, this will probably contribute to a stabilization of conditions.

Tunisia: Between Tourism and Tradition

Tunisia is practically sandwiched between Algeria and Libya. It has 6 million inhabitants and is, economically, a well-developed Maghreb state. Industrialization, promoted with inducements for foreign investors, and flourishing tourism have contributed to its growth. Like neighboring Algeria, Tunisia has oil and natural gas. However, in view of the ambitious plans for their processing, these reserves will hardly last longer than fifteen to twenty years. Oil currently accounts for about one-third of the export yield of Tunisia and it is supplied largely to French refineries. Only a small part of the annual output remains in the country for further processing. The refineries, operated jointly by the Italian company, Ente Nazionale Idrocarburi (ENI) and the Tunisian state, are in Gabès, the future center of a petrochemical complex.

Hoechst Tunisia has a somewhat complex development history because foreign companies in Tunisia are subject to many complicated regulations and formalities. For many years we shared the same roof with the Bayer agency in a pleasant association. Later, we founded our own company with Tunisian partners. This company does not have an easy time because, apart from the high-quality carpet industry, industrialization is developing only slowly. Tourism is a major pillar of the Tunisian economy and the Germans play a considerable role in it. Foreigners are enticed not only by beautiful sandy beaches lined picturesquely with palms, but also by fairly well-preserved Roman ruins which bear witness to the long association of the country with the Roman Empire. The state-owned tourist organization has erected many modern hotels reflecting the architectural style of the country.

Politically, Tunisia has maintained a surprising continuity in the postwar era. The country has been headed for more than twenty years by Habib Bourgiba, who, as president for life, has extraordinary powers. He has resisted the political advances of his eastern neighbor with much circumspection and adroitness. Although a union with Libya, suggested by Tripoli two years before, was declared in 1974, Tunisia canceled the agreement only three days later. The big question is whether Tunisia will be able to maintain its political stability once President Bourgiba has gone. It is highly desirable that Tunisia maintain its independent position in the Maghreb world.

South Africa—Gold and Racial Problems

I had an early interest in getting to know South Africa, and not only because the country promised a great economic future with its wealth of gold, uranium, and other rare minerals, such as manganese, chromium, vanadium, and titanium. There had been many conflicting views concerning the apartheid policy of the Boer government and the sometimes grotesque consequences this policy of racial segregation had in daily life. In 1959, I combined my first sojourn in the country with a three-week vacation, journeying from Johannesburg to Cape Town, visiting several game preserves, and viewing the magnificent Victoria Falls in neighboring Rhodesia.

At the time, Hoechst was represented by the company of a Dutch business friend, Otto Schoemaker, who accompanied us for part of our journey. Like almost all the white South Africans we met, he believed that the policy of racial segregation and the creation of tribal areas for black Africans was the only way to maintain the country's peace and economic leadership, ensuring that the black population would achieve a living standard far above that in the adjacent countries.

While I agreed with his view, I was disturbed by the racial discrimination evident everywhere in public life. Called the "petty apartheid," it specified, for example, that in parks or on public transport, whites and blacks had to sit in separate areas. They had to stand apart when applying to post office counters and other public services. Even the driver of a bus for white passengers could not be a black. In addition, there were the "job reservations," dating back fifty years, when the higher positions in industry and administration were reserved for a large number of poor white immigrants.

Black Supervisors and Foremen

However, these negative concomitants of the apartheid policy have been gradually eliminated as the result of the economic boom in South Africa. White labor was in short supply and blacks had to take their place. In the factories, they advanced to supervisors and foremen. The whites were forced to rely on the labor potential of the black population if the momentum of progress was to be maintained. No doubt, some forms of discrimination still remain. But many companies, especially the foreign ones, have nevertheless aligned the wages of their black workers with those of the whites, often far beyond the officially permitted limits.

Gold mining and agricultural production facilitated the economic rise of the country before 1970 solely on the basis of low wages for black workers. With increasing industrialization, which took place mainly in the seventies, a major reorientation has occurred. Today, the aim of the country's wage policy is not to maintain a low wage level but to create a high demand backed by considerable purchasing power. At the present stage of South Africa's development, low wages would inhibit the growth of the economy and thus retard technological progress. High wages, on the other hand, would permit increasing utilization of the advantages of mass production so that the country could join the ranks of the highly developed industrial nations.

There have been considerable improvements in wages in the South African economy since 1970. For example, in gold mining, the wages of the black workers were raised 500 percent by 1977. In all other branches of industry, real wages have at least doubled. Visitors to South Africa can see for themselves that the standard of living of the black people, having been remarkably improved in the past decade, is now far above that of black workers in many other African countries.

Furthermore, there has also been a leveling of black and white incomes since 1970. The previous racially based wage structure for the same jobs has been eliminated. Through the wide use of job-evaluation methods, progressive companies have taken into account such factors as work loads, training, responsibility, and degree of danger when fixing salary levels. As a result, many jobs which, only a few years ago, were reserved for whites either by law or through tradition, are today occupied by blacks; examples include technical assistants, nurses, office workers, bookkeepers, sales personnel, machine-tenders. The labor market has therefore succeeded in true integration where the political area failed. The large multinational companies are very much concerned in accelerating this process. It was much more than a symbolic gesture when Arno L. Baltzer, the head of Hoechst South Africa Ltd., removed all signs saying "Whites Only" from the pharmaceutical factory in Johannesburg.

The Republic of South Africa owes its high level of economic development to more than the relatively cheap labor provided by its black population. However important such labor may be, the extraordinary mineral wealth of the country is without doubt the main reason for South Africa's progress. Among the many valuable minerals, gold forms the backbone of the economy, providing more than one-third of the total export value. However, in global economic and political importance, it is being displaced by uranium. Even less spectacular minerals, such as manganese, chromium, vanadium, and titanium, occur in sizable amounts and provide South Africa with a dominating, and in some cases almost a monopoly, position.

This unique raw material wealth has attracted many foreign investors who provided an impetus for industrial development. South Africa can be compared, in its economic structure, to an industrial country with a large agricultural sector. With a 20 percent share in the gross national product, the processing industries are still underdeveloped. The comparable figure for the Federal Republic of Germany is more than 40 percent. Diversification, linked to the government's politically motivated ambition for self-sufficiency, is given priority.

However, South Africa has no oil. In the face of the almost explosive rise in the cost of oil, the country has become conscious overnight of the seriousness of the energy problem. Mining requires more energy than all the rest of industry together. Nuclear energy as a supply source has just begun. The first nuclear power station, erected with French help near Cape Town, will eventually have the respectable output of 1800 megawatts. The hydroelectric power station of Carbora Bassa in Mozambique is already exporting a volume of 1400 megawatts destined for South Africa. Yet, for some time, none of these provisions will change the unique position of coal, which provides 90 percent of the energy required in the country.

Coal Liquefaction in Sasolburg

Some years ago I was very impressed by a visit to Sasolburg where the coal liquefaction process, developed in Germany during the war, is being employed. Sasolburg offers the important advantage that the coal is obtained on site by surface mining. As a result, it has become the largest energy source in the country, and many chemical industries have been established in the vicinity. Hoechst is participating in a factory in which Hostalen and polypropylene are produced from the output of a cracker in Sasolburg. Safripol is the name of the company, in which the South African Boer Sentrachem Company has a majority holding.

A second coal liquefaction plant, Sasol II, is being constructed in the Eastern Transvaal. When complete, the two installations together will supply about a third of the domestic gasoline requirement. Sasol II is expected to go on stream in 1981.

The earlier small trading company of Dutch origin has in the meantime become a rather comprehensive Hoechst company, also producing pharmaceuticals and dispersions and employing a considerable number of staff. A Trevira plant has been erected in the vicinity of Cape Town, the center of the textile industry. This sector at times causes much concern even in South Africa. There would be no fiber factories if the government were not pursuing a deliberate policy of autonomy. The present political conditions have necessarily resulted in reduced investments in the country. In the past, investments were to a large extent from abroad, especially from England, the United States, Germany, France, and Japan. However, the big problem continues to be the lack of oil.

Nobody can forecast the political future of the country. In addition to Rhodesia and Namibia, South Africa has become the target for attacks from all directions. Spearheading these attacks are not only African politicians, who have not exactly distinguished themselves through racial tolerance in their own countries, but also countries of the Eastern bloc. Now the churches and the Western powers, led by the United States, have joined in. The South African government is much concerned about providing its millions of simple people with a better and happier existence, including material requirements as well as education, health care, and social emancipation. Unfortunately, Pretoria is not always a clever interpreter of its own intentions. This weakness has contributed much to the negative image of South Africa in the Western world. However, the country is a vastly important raw material reservoir, which explains the great involvement of foreign industries. But this wealth in itself cannot suffice to calm the many emotions that exist in respect to South Africa. The state itself must tackle the problem of racial discrimination. Throughout Africa there is so far no black government that is sharing responsibility with a white minority. But this is not sufficient reason for South Africa's not taking tentative steps in this direction.

Developments Need Time to Mature

It is probably too early to draw up an intermediate balance sheet to show which forms of independence, following colonial rule, have yielded the best results for the people in the various areas of Africa. The process started only fifteen years ago and still continues. We shall probably have to accept the likelihood that the peoples between the Sahara and the Zambesi must experience the development phases that the European Continent had to suffer before the particularism at the end of the Middle Ages gave way to the nation states of the nineteenth century.

However, Africa is facing a far more difficult task. This third largest continent is a unit only geographically. Its area is larger than that of the whole of Europe, including the Soviet Union. More than three-quarters of the continent is situated in the tropical zone, where the climate is a serious handicap for economic development. Tropical diseases—yellow fever, malaria, sleeping sickness—are the eternal plagues of its inhabitants. Trypanosomiasis, the scourge transmitted by the tsetse fly, continually decimates the animal population. For many centuries, animals simply could not be used for transport, so business had to be conducted literally on the backs of the people. Thus, geography and climate have determined the colonization process in Africa. In contrast with the North American policy of the colonization of large areas, the colonial powers in Africa avoided the interior and confined themselves to trade from locations along the coast.

From Barter Economy to Industry

A condition for the desired economic independence of the young African state is the establishment of an independent industrial structure that meets the population's needs and provides for exports from its mineral wealth and agrarian products. There is no other way to end the practice of primitive barter among the many states whose annual gross product per capita is less than the weekly wage of an unskilled worker in the West.

Development aid and the policy pursued by supranational institutions aim at reducing the prosperity gradient between northern and southern areas. The decrease in prosperity from north to south is a well-known phenomenon in some countries. It is very pronounced in Italy, but France and Great Britain are also familiar with it.

The phenomenon is essentially a climate-based condition with all its consequences for the biological substance of a population, its mode of living, and the soil it occupies. In Africa, this gradient is further accented by the differences in the degree of education of the various ethnic groups and the gaps in political and economic maturity. Western development aid has so far barely taken account of these conditions that vary so greatly from region to region. Political considerations or a special foreign policy constellation has created a situation in which more than three-quarters of the members of the United Nations are regarded as developing countries. This is hardly in a tolerable ratio to the abilities of the more advanced countries that are supposed to help. While this situation continues, the laborious dialogue between north and south, or more precisely, between west and south, will not progress very rapidly.

Project-Based Development Aid

The Eastern bloc has always selected its protégés on the basis of their potential strategic usefulness. Where a country had no such usefulness, the Eastern bloc assumed the role of the uninvolved since development aid, according to its vision, was only a kind of restitution for colonial exploitation.

Western governments are tying their support more and more to concrete projects that might bring industrial progress to the country concerned. In spite of many negative criticisms, the considerable efforts that international groups are making to establish industrial organizations in African countries are surely more successful as a development contribution, both in their effect and in the long term, than aid funds that often are diverted to the wrong channels.

The policy of many a developing country toward foreign investors is highly controversial. Concern with excessive numbers of foreigners and with alleged exploitation of raw materials often blocks even the much-invoked cooperation. As far as the presence of too many foreigners is concerned, there is some substance in these worries, since control of prospering economic groups is—for the time being—primarily in European hands. Black entrepreneurs in Africa will have to play a much larger role in commerce and industry if they want to gain self-confidence and reduce the degree of white superiority. This is clearly shown by the example of Nigeria and some other countries.

The future of those African countries that are not blessed with oil or mineral wealth must be judged with some reserve. Tribal thinking still largely rules the political landscape. Most of the states are large in size and therefore difficult to assess. Moreover, there is a substantial population flight from the rural land into the alleged prosperous areas, the cities, where social problems become aggravated.

There is indeed a catalog of discrepancies that bar the progress of Africa toward a happier future. To play a decisive role in helping to solve these problems may well be one of the major tasks of the United Nations, provided its members take a less dogmatic attitude toward the problems.

Concessions Step by Step

Over the last few years, Hoechst has tried to find its place on the "Black Continent" within the framework of the new developments. The statement has been frequently made that the countries of the Third World are planning to supply a quarter of world industrial production from the year 2000 onward. Even if this goal may not be quite realistic, it is obvious that investments in these countries are both sensible and have a future. Western industries can open up new markets at the same time.

However much the developing countries may be interested in Western management experience and technology, their investment policies not only vary greatly from one country to another but, at times, are extraordinarily difficult to follow. Where a government is particularly interested in certain industries, and where a company submits an attractive proposal and offers substantial investments, it will be easier

In Nigeria, Chief Ben Olnwole welcomes Professor Winnacker, Kurt Lanz, and the German Ambassador Dr. Harald Graf von Posadowski-Wehner.

verleaf: Suspended bridge, Ivory Coast, Africa.

oechst's Research Farm in Malelane, South africa.

A Factory in West Africa.

With subsidiaries and plants in almost every country of the world, Hoechst places special emphasis on the training of native employees.

Modern household items made of Hoechst plastics are displayed in a Nigerian market.

A symbolic gesture at the
ground-breaking ceremony in
Morocco: Kurt Lanz plants a
palm tree.

The Hoechst office building in
Casablanca, Morocco.

In Egypt, people still suffer from a disease that has caused the death of millions of people since the times of the pharaohs: "The Bilharziose" biliary cirrhosis.

At Hoechst Orient in Cairo, researchers are working on a vaccine against bilharzia, a parasitic disease that has plagued the area since the days of the pharaohs.

to find a way of modifying inhibitory passages of such policies in terms of a concrete case.

Therefore, there should be mutual rapprochement in small steps. Since policies are frequently altered overnight, much toughness and negotiating skill, quite apart from infinite patience, will be needed to maintain, as well as to consolidate, an established position. The variability with which governments of the Third World at times interpret the modalities that they themselves have laid down is strikingly illustrated in an example from India. IBM was required to turn over to Indian nationals 60 percent of its capital; Coca Cola was faced by the government's demand that it reveal its jealously guarded manufacturing secret. No one will be surprised that both companies preferred to abandon their interests in India. Such extreme cases, however, are the exception. On the whole, early experience teaches that cooperation with countries of the Third World opens up new dimensions on a partnership basis and will thus be indispensable for both sides to ensure tomorrow's economic success.

19
The United States, Supermarket of Chemistry

A widely traveled acquaintance of mine once thought of writing his memoirs from the angle of the many hotels in which he stayed and which played a particular role in both his professional and private life. If I were to adopt this idea with regard to the United States, I would have to begin with the Plaza in New York. This noble hostelry cultivates a turn-of-the-century interior with such dedication that the marble splendor of the Palm Court still sports a dress-suited violinist. It was in this building, constructed in 1907 on the corner of Fifth Avenue and 59th Street as the "world's most luxurious hotel," that my discovery of America began in 1954 as a member of a small delegation headed by Professor Winnacker. At that time we stayed only a few days in the Plaza before setting out on a six-week tour of the States.

Gerstacker's *Regulators in Arkansas,* Cooper's *Leather-Stocking Tales,* and Mark Twain's *Tom Sawyer* and *Huckleberry Finn* provided me with my first images of the "land of the unlimited possibilities." In the thirties, Manfred Hausmann's *Kleine Liebe zu Amerika* was something of a best-seller. The interesting anecdotal manner in which Hausmann described America and the Americans greatly impressed me. In my first visit, my own experience confirmed many of the attractive qualities about which I had read in the book, especially the widely praised openness towards both neighbors and strangers. This candor greatly facilitates discussion and negotiation in business life. The camaraderie and friendliness with which the Americans meet one another strikes me as truly remarkable even if this cordiality does not always run very deep.

In the early postwar years, a German traveling through the United States for the first time did so with mixed feelings. We had been defeated by the Allies in 1945 and our country had been occupied by them. On the other hand, they had freed us from a hopeless situation. Each one of us carried conflicting memories of this period; but no nation should be judged solely by its army of occupation. During my first visit to America, I found few traces of this recent past. With changing political conditions, the American view of the Germans as opponents changed into that of Germans as allies. The formation of the European Community had been expressly promoted by the United States.

Dwight D. Eisenhower, who enjoyed immense popularity among the American people as the victorious commander in World War II, assumed the Presidency of the United States in 1952. During his administration, and with the influence of Secretary of State John Foster Dulles, political relations with the Federal Republic of Germany became increasingly closer. The growing mutual esteem, not only between the two governments but also between large groups of the population, was reflected in the election of Federal Chancellor Konrad Adanauer as "Man of the Year" by the editorial board of *Time,* the American newsweekly which featured his picture on its cover in January 1954.

Looking at a Giant Market

It was high time for me to have a look at a market which in many industrial fields, such as chemistry, pharmaceuticals, and man-made fibers, consumed about half the production of the Western world. In spite of World War II, or perhaps because of it, America had become a giant in economic, technical, and industrial respects. Its chemical industry, in particular, had achieved such a great lead that it was soon to become essential for us to catch up with it.

The history of the American chemical industry covers some two centuries. Small chemical undertakings were formed some fifteen to twenty years after the arrival of the Pilgrim Fathers. By the late nineteenth century, there were numerous small chemical companies serving local markets. Some later grew into chemical giants. The establishment in 1892 of the Midland Chemical Company by Herbert Dow, who as a young college graduate found a new method of bromine extraction, laid the foundation of Dow Chemical Company, today one of the largest chemical companies in the world.

Before the turn of the century, another small company was engaged in the manufacture of carbon electrodes used for the street lighting then coming into vogue. It was one of the companies which later constituted Union Carbide in 1917, another of the world's largest concerns. A comparatively early start was also made by the Monsanto Company, founded in 1901 by John F. Queeny with a modest starting capital of US$5,000. The Celanese Corporation was formed in 1918.

Prior to World War I, the American chemical industry lacked the scientific and technological significance that might have frightened its competitors. It concentrated on relatively simple processes. In the field of more sophisticated production, as demanded in the manufacture of dyestuffs and pharmaceuticals and the related organic intermediate and starting products, it had little capability. These areas were accepted as the domain of the German chemical industry, in those years acknowledged, if not admired, as the center of world chemistry.

Lessons from the Crisis Years

The extent of this German dominance of the American market was demonstrated during World War I and the resulting maritime blockade. Dyestuffs and, even more important, pharmaceuticals were suddenly in short supply in America because the British fleet had cut off the overseas supply lines from Germany. Before America joined the war in 1917, German commercial submarines had been able to supply the United States with at least some of the most essential medicines and dyestuffs. From then on, however, America went ahead with an energetic and systematic development of all those branches of the chemical industry whose vital importance had been revealed so dramatically.

Financing was no problem because of the boom conditions generated by the war. Practically unlimited sums of money, together with scientific and technological knowledge, partly gained through the confiscation of the German patents, produced an almost meteoric growth in the American chemical industry.

When the war ended, setbacks occurred. Much had happened too quickly and too hectically. Many a company that had sprung up overnight without the appropriate commercial and technical background disappeared with equal rapidity. Moreover, German competition was back in action more quickly and more comprehensively than had been expected. In order to avoid unduly intense competition, both sides of the Atlantic sought to divide the world into spheres of interest, ushering in a basically unhealthy era of monopolies and syndicates.

Some of the most important chemical companies had learned that the successes of the German companies and I. G. Farbenindustrie were based on fundamental comprehensive scientific research and intensive cooperation with the universities and polytechnic institutes. In the twenties, Du Pont, a company which had earned billions of dollars during the war by supplying munitions, was no longer content to enlarge its production program by merely acquiring other companies. The product line already comprised paints and lacquers, dyestuffs and pigments, acids and heavy chemicals, cellulosic plastics, coated textiles and—under license from the French Comptoir des Textiles Artificiels —rayon and cellophane.

Expanding Dimensions

Shortly before World War II, the chemical industry in the United States had annual sales of more than US$4 billion. Factories and processes were the most modern in the world. Over 10,000 scientific staff, more than in any other branch of industry, were proof that the nation's chemical industry had realized the unique chance that it had been offered.

A comparison with the conditions in Germany during the last prewar years is of great interest. The annual sales of the I.G. in 1939 were almost RM2 billion, and the number of scientific staff was 1200. However, production and process-engineering knowhow was hardly less advanced than that in America. Between 1935 and 1944, the production of synthetic fuel alone was increased to 3.5 million tons a year. However, such comparisons are not entirely valid, since German industry was working from an entirely different starting position because of the political conditions of the time.

Saving US$140 Billion

What was to happen, however, to the enormous capacities that had been created to meet possible war requirements? Would there be setbacks similar to those after 1918? American industrialists have often told me that they were greatly concerned with these prospects. During the war, the Americans had saved US$140 billion when they were unable to buy many things. A large part of this money was now dissipated on a spending spree. Whatever items were involved, whether lipsticks, curtains, or Cadillacs, chemical products usually were involved. Surprisingly, during the "fabulous fifties," a period currently creating much nostalgia in the United States and elsewhere, the dollar remained stable and prices increased only a slight 2 to 3 percent per year.

We were astonished and impressed by this "supermarket of chemistry" which we were able to study during our first American reconnaissance. We no longer had any doubt that Hoechst had played its cards right by deciding to back mass-produced plastics and fibers. It was clear that these products had great future prospects. We had no exact idea how far the Americans were ahead of us. The country was a giant domestic market with a population that not only demanded ever higher standards of living but that, in considerable contrast to Europe, also had the required purchasing power. In 1954, no more than 4 to 5 percent of the output of the American chemical industry was exported. Further, American industry had available almost unlimited resources of raw materials and energy.

It was most impressive to watch how rapidly the United States effected the changeover from traditional coal to petrochemistry. While acetylene chemistry still remained in the forefront in Europe, with ethylene obtained from coal and acetylene, the oil-rich America chose the direct route of ethylene from oil. In 1954, 1 million tons of ethylene was produced from oil and natural gas. There was no doubt that this trend toward petrochemistry would intensify. For example, in the two years between 1963 and 1965, domestic ethylene production rose from 3.5 to almost 4.5 million tons. The nation was also far ahead in the production of aromatic hydrocarbons by the petrochemical route. In 1959, for example, the United States produced some 60 percent of its total benzene requirements. Four years later, in 1963, the amount of petrochemically produced benzene in West Germany was only 25 percent of the total requirements.

Naturally, we also saw just how far ahead of us the American companies were in respect to sales and profit. In 1953, Hoechst had achieved sales of just about DM1.2 billion (US$300 million), a sum of which we were fairly proud—not without reason, from our point of view. In the area of profits, the differences were even greater. In contrast with European, and especially German, competitors, the American companies had a high ratio of equity to debt. With companies like Du Pont, Union Carbide, Allied Chemicals, and others, the ratio was about 70 to 80 percent on average, making their interest burden very much smaller than ours. Further, United States productivity was much higher than in the Federal Republic.

One of the few respects in which we were better off than the Americans in those postwar years was the economic flexibility which Konrad Adenauer and Ludwig Erhard allowed the large companies. As the result of much arbitrary action during the early years of American industry, antitrust laws and regulations limited the potential of U.S. companies to merge and cooperate with one another.

From Small Beginnings

Our immediate beginning in the United States was of almost dwarf-like dimension: four modest offices on the eighty-third floor of the Empire State Building, New York City. They were occupied by Intercontinental Chemical Corporation, the forerunner of the present American Hoechst Corporation. Max E. Klee, who had been active for I.G. in the thirties, was its first president. Our modest activities in the early 1950s centered on dyestuffs. Apart from pharmaceuticals, these had been the most important exports of I.G. before the war. A large part of them had gone to America, although, after World War I, that country had begun to build up its own dyestuff industry.

The German dyestuffs chemistry had lost its dominant position after the First World War but succeeded in regaining some ground between the wars. The I.G. even found it worthwhile to erect its own dyestuffs production facilities in the United States. It established the General Aniline and Film Corporation (GAF), which achieved annual sales of US$50 million, a sum not to be sniffed at in those days. As the company's name indicates, the proceeds of the film and photographic side were contained in this figure. During World War II, along with all the German patents, this company was requisitioned as enemy property. The United States eventually turned it over to private hands, and today GAF continues to be a renowned concern.

Nevertheless, there was every justification for restarting our United States business with the traditional dyestuffs even if, as in so many other fields, we had to start from zero. The beginning was made one year before I set foot in America for the first time. In April 1953, we had acquired the Progressive Color and Chemical Company in New York from Adolf Kuhl, a German-born businessman. It had been in existence since 1923, acting solely as a trading company for dyestuffs.

One year later, literally on the first day of our visit, we concluded negotiations for the acquisition of Metro Dyestuff Corporation in Coventry, Rhode Island, which provided us with production facilities. This company had also belonged to a former German, Dr. Harry Grimmel, who had originally worked as a chemist for Bayer in Leverkusen. After the war, Grimmel decided to apply his expert knowledge to the establishment of his own company. He found American partners, who provided him with the necessary capital; acquired land and buildings in a small former textile factory; and began the manufacture of dyestuffs. For a person familiar with this field, such an undertaking provided a sound existence at that time. Grimmel operated on a modest scale, and his machines and other equipment were not the most modern. Nevertheless, he achieved sales of US$1.2 million in 1954.

The third trading company we acquired was Carbic Color and Chemical Company. We thus had a solid basis for our own dyestuff business in America, and we expanded slowly. The range was extended to include more sophisticated classes of dyestuffs and the organic intermediates, which serve as starting materials not only for dyestuffs but also for pharmaceuticals and agricultural chemicals. These developments led to the present Rhode Island Works, which represent one of the important divisions of the American Hoechst Corporation.

The Beginning of Our Pharmaceutical Activities

Of course, we could not live from the dyestuffs business alone. From the beginning, my goal was to set up Hoechst in the United States as a pharmaceutical producer. As early as 1950, when Hoechst was still under American administration, we estabished our first overseas contacts. With the help of the American I.G. Control Office, we were able to conclude a license agreement with Merck in Rahway, New Jersey, which permitted us to manufacture penicillin, the famous antibiotic previously so rare in Europe. At the official opening of the new penicillin plant in Hoechst, one of the first speakers was John J. McCloy, then High Commissioner for the Federal Republic. Both during his period of office and afterward, McCloy contributed decisively to the normalization of relations between the United States and West Germany.

At about the same time, we concluded a general license agreement with the Upjohn Company of Kalamazoo, Michigan. Under this agreement, Upjohn acquired the exclusive right to exploit new Hoechst pharmaceutical developments in the American market. Upjohn continues to be one of the most important American pharmaceutical companies still in family ownership. It was founded in 1885 by a young doctor, William Erastus Upjohn, who lived in Kalamazoo, the Indian word for "boiling water."

In that year, Dr. Upjohn acquired a patent for a process that produced pills in a form that could easily be crushed with the thumb. This was a remarkable advance over many competitive products that were so hard that they passed practically unchanged through the digestive tract. A pharmacist customer told Dr. Upjohn that some competitive preparations became so hard after a period of time that they could be nailed to a board without causing them any great damage. After Upjohn had tested this claim, he sent a small board and samples of these competitive pills to several thousand doctors and asked them to try the experiment—a truly striking promotional idea.

When we made contact with Upjohn, the company was already supplying more than 500 different preparations. Apart from antibiotics, other drugs, and vitamins, the company's special preoccupation was with synthetic hormones and hormone intermediates. For example, Upjohn was one of the first companies to market a suprarenal gland hormone preparation.

The year 1955 was a big one for Upjohn. Hoechst research workers had discovered Rastinon, the first oral antidiabetic, which represented a milestone in postwar pharmaceutical history. Millions of diabetics were now freed from the need for their daily insulin injection. Upjohn took over the new preparation for the American market, conducting even further investigations into the new preparation. Twelve million tablets were sent to some 3000 American doctors, who prescribed the antidiabetic to some 20,000 patients. In diabetic cases suitable for this medication, the results were as outstanding as they had been in Germany. In 1957, under the name Orinase, it was approved by the U.S. Food and Drug Administration (FDA). It became one of Upjohn's most successful preparations and, by 1970, had achieved a turnover of U.S.$20 million. Hoechst collected the appropriate license fees.

Such fees were one of the factors that made it difficult for Hoechst later on to decide whether to continue such lucrative and reliable cooperation or to build its own, initially high-risk production for the American pharmaceutical market. In the fifties, this market had an almost compulsive fascination for every European chemical company. Its total turnover increased from US$1 billion in 1950 to approximately US$2 billion only ten years later. Profits were more than 10 percent on average, and the return on investment was more favorable than in any other branch of industry.

On the other hand, the risks were also very high, particularly with respect to the high research expenditures. This applied especially to a newcomer like ourselves who, though backed by an experienced and creative research team, did not have available the necessary comprehensive marketing resources to compete with the large American companies.

In the fifties, there were no less than 1000 pharmaceutical companies operating in the United States. From our offices in the Empire State Building, the view encompassed the American headquarters of more than a dozen world-famous pharmaceutical concerns, then truly an incomparable "empire" in this field.

In the Shadow of the Kefauver Hearings

The American pharmaceutical market represented a force with powerful financial backing and tremendous research capacity. It was highly doubtful whether we would be able to succeed against such strong and experienced competition. Our intention to gain a foothold in the American pharmaceutical industry coincided with a development that greatly tarnished the hitherto bright image of the pharmaceutical industry: the sensational "Kefauver Hearings" which, in 1959, began to investigate the pharmaceutical industry through a committee of the United States Senate. Very serious charges had been leveled: that a huge margin existed between the production and selling costs of pharmaceuticals, and that the efficacy of the preparations was tested inadequately, if at all.

Carey Estes Kefauver, Democratic Senator from Tennessee, and later, opponent to Richard M. Nixon in the race for the vice-presidency, headed this investigation. Kefauver had gained fame through the publicity that he had achieved in 1952 as head of a Senate committee investigating organized crime.

The pharmaceutical industry was totally surprised by these investigations. Only a few years earlier, it had enjoyed general admiration for its development of the new antibiotics, the corticosteroids, the oral antidiabetics, and many other preparations that had achieved dramatic progress in the treatment of disease.

All now seemed suddenly forgotten, with talk of profits of 1000 or more percent. To make matters worse, cases of birth defects were reported following the administration of thalidomide. The product was never sold in the United States, although samples were distributed to hundreds of physicians. Without thalidomide, Kefauver might never have succeeded in passing his proposed legislation. Under these circumstances, however, the Kefauver investigations resulted in the 1962 amendments to the Federal Food, Drug, and Cosmetics Act which provided the FDA with new and extensive powers.

The affair had a positive side, however. Far-reaching scientific reforms were carried out within the pharmaceutical industry. It was recognized that however important the production of pharmaceuticals might be, research, long-term toxicity tests, and above all, quality control and continuous information for the physician were no less indispensable.

Hoechst Pharmaceuticals Are Produced in the United States

In this atmosphere, it was difficult to decide if we should actively enter the American pharmaceutical market and build up our own production. There was a strong faction in the Hoechst board of management which regarded the risk as very high and argued that we should be satisfied with licenses.

In the end, we did decide to go ahead. The United States was, after all, the largest pharmaceutical market in the world, accounting for 30 percent of world pharmaceutical consumption. With regard to preparations against "civilization diseases," the figure goes as high as 50 percent. Moreover, from a medical standpoint, America was and is the most progressive market.

Finally, marketing methods in the United States were far more developed than in Germany. For example, medical representatives were not doctors but superbly trained lay people—at that time, a totally unimaginable situation for the Federal Republic. The number of medical advisers was very large, and with it, of course, so was the opportunity for providing comprehensive and intensive information.

We could not afford to play a passive role in this market. In 1960, we asked Max P. Tiefenbacher, then the head of Hoechst's pharmaceutical exports, to go to America to reconnoiter the market and carry out a systematic search for projects that looked suitable for our plans. The first question was whether we ought to establish a new company or acquire an existing one. After thorough consideration, we decided upon the latter course and purchased Lloyd Brothers of Cincinnati for US$4 million. The Midwestern firm had been established in 1870 as a family concern for the preparation of botanical medicines, yet had failed to keep in touch with modern pharmaceutical industry research. Our main purpose was not so much to have the building and production facilities as to take over Lloyd's sales organization and to sell its products. In this way, we were able to bridge the gap until we entered the market with our own pharmaceuticals.

At the time of acquisition by Hoechst, Lloyd Brothers was achieving annual sales of almost US$3 million with a rather old-fashioned drug range. Each extension to the range proved extraordinarily difficult. It takes a great deal of stoicism and patience to meet the requirements for new drug approval by the Food and Drug Administration. However, drug safety justifies the most stringent tests.

Breakthrough in the Seventies

Our first real success in America was achieved with Lasix, the Hoechst diuretic. It was highly doubtful at one time whether we would be able to launch this product under our own banner. Our license agreement with Upjohn provided that, following the introduction of Orinase, our partner would have an option to take over the American sales of all other Hoechst pharmaceutical novelties. While we were no longer happy with this agreement, we nevertheless offered Lasix to Upjohn. Surprisingly, Upjohn was not interested and did not exercise its option, thus missing out on a major opportunity. Lasix was a best-seller on doctors' prescription pads, not only in the United States, but throughout the world.

Nevertheless, the progress of Lasix tried our nerves. Clinical tests had begun in 1963. Two years later, thousands of pages of clinical reports were overwhelming the newly formed clinical research group. We had to call for reinforcements from Hoechst and, as a result, Dr. Hubert S. Huckel, who today is board chairman of our pharmaceutical company in the United States, arrived at our plant on the Ohio River.

Many a medium-sized company would probably not have survived the long period of clinical evaluation and government approval which, in the end, amounted to three and a half years. It cost us many millions of dollars. Eventually, our patience and the outstanding properties of our new diuretic won the day. Lasix was first marketed in the United States in July 1966, and the extraordinary success which it achieved during the succeeding ten years has surely justified the efforts that were invested in this project. Lasix had the immense advantage that the American competition was unable to offer doctors a comparable product.

Please, Who Is Hoechst?

In those first years, we were continually surprised to find how few Americans knew the name of Hoechst which, at one time, had enjoyed a considerable reputation in their country. Yet, since the foundation of I.G. in the mid-twenties, the well-known Hoechst drugs had been marketed throughout the world only under the Bayer cross.

Max Tiefenbacher once told us a tale from the period when he began to work in the United States. He had a talk with the chief buyer of McKesson & Robbins, the most important pharmaceutical wholesaler. The buyer listened for a while to Tiefenbacher's remarks and then asked courteously: "Mr. Tiefenbacher, could you please explain to me who is Hoechst?" He did not know even Hoechst's home country.

Happily, we never encountered any animosity toward Hoechst or other German companies during those early years of building our business in America. I well remember the day when, together with Tiefenbacher, I attended my first meeting of the Pharmaceutical Manufacturers Association. We met the chiefs of the large companies, such as Merck, Squibb, Pfizer, Upjohn, and Eli Lilly. When we entered the conference room, Jack Powers of Pfizer approached us and warmly and demonstratively greeted us with the words: "We welcome you as a competitor; let's compete with one another!"

This was not empty talk but an expression of the American business mentality, which regards competition as natural. "If we go with our products into Europe," it was said, "then it's only right that the Europeans, and therefore also the Germans, should compete with us in the U.S."

As it turned out, the American companies have never been hostile toward us. On the contrary, they frequently helped us when, as comparative novices, we grappled with particular problems.

Introducing Behring Diagnostics

At the beginning, we had little luck with our attempts to introduce veterinary products from Marburg and Hoechst into the United States market. We acquired two small companies in this field, including National Laboratories in Kansas City, which had been leaders in the field of hog cholera vaccines. But with the elimination of hog cholera in 1967, these laboratories were closed. Although we continued to market the remaining veterinary range, we did not achieve any real success and therefore developed a partnership with Ralston-Purina in St. Louis, Missouri, an important producer of animal feedstuffs. The partnership did not last long because the business showed too little profit, and after two years we went our separate ways. Since then we have operated the veterinary program as part of our other activities in New Jersey.

In 1958, we began to transfer to the United States the clinical diagnostics products of the Behring Diagnostics Division, a modern installation in Marburg. These diagnostics already enjoyed a high reputation in the progressive American medical climate. It appeared to us a rewarding task to exploit this position, and to participate in its scientific development and economic growth. Far-reaching developments are especially likely in the field of modern diagnostics, and they certainly will have impact on both continents. In 1978, American Hoechst acquired the Calbiochem Laboratories in La Jolla, California. They produced a supplementary line of diagnostic reagents which Behring will sell worldwide. Furthermore, our potential for exchanging new scientific developments was again strengthened by this transaction.

Hoechst today is one of the major research-intensive pharmaceutical companies in the United States.

The Trevira Era

Even more difficult than the issue of our own pharmaceutical production in the United States was the decision to erect a Trevira plant. As far back as 1954, Hoechst had acquired from Imperial Chemical Industries (ICI) a production and sales license for this promising product, but we had had to confine our polyester activities initially to Germany and a few Eastern countries.

During the late sixties, the relevant ICI patents gradually expired. Every producer, so far as financially able to do so, was now at liberty to compete with ICI worldwide. Hoechst decided to join in this competition, which promised rich rewards. Although such worldwide planning required much anxious thought, it was necessary to make quick decisions. The truism "first come, first served" was proved true over and over again.

Within a few years, we erected Trevira factories in Austria, Chile, and even in the United Kingdom, the home country of polyester itself. However, the most promising market, because it was the largest, was once again in the United States.

For such a large project, it was advisable to enter into partnership with an American company. After a close look at the "marriage market," we settled on the Hercules Company, previously known as Hercules Powder because it originally manufactured gunpowder. We had previously engaged in a mutually fruitful exchange of knowhow concerning plastics, especially polypropylene and low-pressure polyethylene. Our experience during this cooperation had been excellent. The leading people of the two groups had moved closer to one another in human terms. Most important, Hercules was one of the leading American producers of dimethylterephthalate (DMT), the starting product for polyester.

Our decisions had been preceded by intense preliminary work. At that time, nine companies in the country already were producing polyester. Du Pont, for example, had operated an experimental plant since 1950, and had gone into large-scale production in 1953, marketing its product under the trade name of Dacron. Among the other companies engaged in the manufacture of polyester were such powerful firms as Celanese, Eastman, Monsanto, Philipps, and Dow. When we first began to explore this market, annual polyester production in the United States already amounted to 360,000 tons.

Initial negotiations on the establishment of a joint company for the production of Trevira fibers took place in 1965. We soon achieved agreement with Hercules and jointly founded Hystron Fibers. By 1967, we were producing 12,000 tons of staple fiber in a new plant in Spartanburg, South Carolina. Prior to that, we had sold small quantities, imported from Germany, to prepare the market for Trevira. But it was clear to us from the start that we would never achieve a significant position in the American fiber business on the basis of imported material.

It soon appeared appropriate to expand our capacity in the Spartanburg plant. We did this in 1968 and 1969 by including the manufacture of filament in addition to staple fibers, which soon gave us a satisfactory share of the staple fiber market. Out of 1.1 million tons per year, 125,000 are provided by our plant. Over the years, Hoechst has become the fourth largest polyester manufacturer in the United States.

The Spartanburg site offered two surprising advantages. This region in South Carolina had become the focal point of the American textile industry.

Thus, we were able to enjoy the advantage of close proximity to the customer. In the case of bulk goods, short transport distances are a special economic advantage in a country where what Europeans would regard as a long distance is considered no more than a stone's throw. Also, Spartanburg has an on-site DMT plant which Hercules had previously constructed.

We knew at the outset that it would not be easy to get Trevira accepted in the American market. It would take considerable effort and many advertising millions over a long period to achieve a breakthrough with Trevira, and we were not mistaken. Even in its third year, Hystron did not succeed in making a profit.

In view of the painfully slow progress and the high losses, Hercules preferred to withdraw from the joint venture and to concentrate on its earlier activity, the production of DMT and its starting product terephthalic acid. This decision in no way affected the friendly feelings on either side. We fully appreciated our partner's decision to abandon what was to Hercules an entirely new field, and to concentrate on its area of expertise.

Hoechst, on the other hand, had accumulated much valuable experience with synthetic fibers. We remained convinced that our project in Spartanburg would soon be out of the woods and that Trevira would be launched on the highway to success. Consequently, Hoechst acquired Hercules's share in the joint venture, including the DMT plant which our partner had once brought into Hystron as a "dowry." The plant continued under the new name Hoechst Fibers Industries.

Our confidence in the future opportunities of Trevira in the United States was fully justified. With the aid of an intensive advertising campaign announcing the arrival of the "Trevira Era," we marketed

Trevira not as an unbranded raw material for cheap, mass-produced textiles, but as the material for haute couture. The most talented European fashion designers created exquisite models from Trevira. These fabrics gained the status of having a particular flair with the "European look," a wonderful vantage point from which to attempt the expansion of the business. As we had expected, Hoechst Fibers was soon in the black, and the American Trevira business enjoyed a period of immense success.

Black Days for Trevira

This happy situation changed in 1975 when polyester producers suffered a worldwide crisis. Among the many factors causing this setback was a reduction in consumption in America and Western Europe. The full extent of this shift could not possibly have been foreseen. Another important reason was the overcapacities that resulted not only from extensions and new plants but also, paradoxically, from continuous improvements in technology. Progress in this field meant that the productivity of many plants rose by leaps and bounds. Continually increasing imports of textiles from the Third World were another factor.

The "black days" of the polyester business throughout the world had begun. The decline was particularly serious in the United States, and our competitors fared no better. The American market is a shining example of the regulatory mechanisms of a free economy. Under the compulsion of intense competition, a fundamental process of selection soon took place. This had bitter consequences for some competitors that were no longer able to meet the requirements of the market. Yet, this weeding out is likely to again establish a proper balance between supply and demand in the foreseeable future.

Such a development would also be desirable for Western Europe because it would surely result in regulation of the market in many fields. State measures are rendering recovery difficult in many countries. It is not my intention to extol the *"laissez faire, laissez aller"* of the "nightwatchman" state of the last century, because the fabric of our social order requires some state intervention. Under extreme market conditions, it may be unavoidable temporarily to restrict the free play of market forces. Yet, state intervention must be confined to the absolute minimum. In many cases, such intrusion not only leads to clumsiness and more bureaucracy, but also prevents measures that would be urgently required for economic reasons. Instead of tackling the root of the evil, some states gloss over the difficulty with laws and regulations, so the real solution is continually postponed.

Hoechst Polystyrene

In 1974 we entered into a partnership—quite different from the one with Hercules—with Foster Grant. The Foster Grant Company was founded in 1919 by Samuel Foster, Jr. He started work in a small building in Leominster, Massachusetts. His first employee was Grace Goodall. She feared that, in view of the slow start of the company, her job might turn out to be rather short-lived. She stayed for forty-nine years.

The small family concern at first concerned itself primarily with the manufacture of combs and novelties, and then, in the late twenties, of sun glasses. Eventually it became the largest producer in the world. The first Foster Grant sunglasses, designed on a piece of brown wrapping paper, cost 10 cents each. A large part of the frames were originally made from a special plastic, polystyrene, which the company purchased. By 1948, Foster Grant was the largest polystyrene user in the world, and realized the need to be independent of subcontractors for this plastic. A factory with an annual capacity of 50,000 tons was constructed at the company's headquarters in Massachusetts. In 1954, a styrene plant, the polystyrene monomer, was established in Baton Rouge in Louisiana. Output grew in excess of the company's own needs, and polystyrene was offered on the open market with considerable success.

Sales were not confined to the United States. Although over sixty years old, Joe Foster, the son of the founder, was still full of entrepreneurial spirit. He began to sell his polystyrene also in Western Europe, and eventually decided to build a small plant from which to supply his European customers. This plant went on stream in Breda, Holland, in 1963. The French representative of the Breda factory then offered our Paris office the sales agency for Foster Grant polystyrene throughout France.

Hoechst France accepted the offer. Initially, the arrangement was intended as an experiment because polystyrene was one of the few plastics not yet part of the Hoechst line. Nothing could be lost, therefore, by investigating the possibilities of this product, using France as a test market. The project proved a complete success. As a result, we acquired a 50 percent share in the Breda factory in 1966 and launched a program of vigorous expansion. In 1968, our American partner withdrew entirely from the European polystyrene business, which was then taken over completely by Hoechst. Subsequently, the Breda capacity was increased to more than 200,000 tons of polystyrene a year. The product is marketed under the trademark Hostyren.

Meanwhile, Joe Foster went even a step further in America. In 1969, he increased production of styrene in Baton Rouge from 250,000 to 700,000 tons, setting up the most modern large-scale plant. This decision was a courageous one for a family concern, particularly since the 250,000 tons of styrene that could previously be produced annually in Baton Rouge, like the polystyrene, were largely destined for the general market.

Lock, Stock, and Barrel

The situation changed drastically when Joe Foster died in 1971. His family showed no particular interest in continuing the perfectly healthy company. At first, United Brands acquired a majority interest. However, it soon became known that both this concern and Joe Foster's heirs and partners were interested in selling the company outright.

At the end of 1974, Hoechst took over the company lock, stock, and barrel, for a price of approximately US$100 million. The deal included the highly efficient management with which Hoechst had maintained close technical cooperation for many years. This acquisition opened up new perspectives for Hoechst whose plastics business in America was still very weak.

That we should become one of the largest sunglasses manufacturers was not our original intention, but it has since proved a very profitable activity. However, the main importance of Foster Grant to American Hoechst lies in the plastics field, where polystyrene has provided us with a further springboard in America.

American Investments in Europe

In most cases, American chemical companies are anxious to manage their production plants in Europe themselves. They generally do not find the absence of European partners a difficulty, because their equity is frequently higher than that of comparable German companies. Americans are often able to finance extensive investments largely from their own pockets, while we usually have to draw on outside capital. In this way, the American chemical industry exploited the postwar European opportunities more intensively than did other branches of industry. It did not confine itself to selling American-produced goods, but built up production facilities on a considerable scale.

Almost all the important American chemical companies are represented in Europe with sizable manufacturing facilities. Du Pont, for example, has factories of various kinds in Germany, Belgium, Luxemburg, Holland, Spain, and England. Monsanto operates fiber plants in Lingen on Ems and in Coleraine, Northern Ireland, not far from our own Trevira factory in Limavady. It also has plants in Antwerp for the manufacture of ABS resins. The European production of Union Carbide Corporation is even more comprehensive, with plants in Germany, Great Britain, France, Spain, Belgium, Italy, Scandinavia, Greece, and Switzerland. In addition, it has a central European administration in Geneva. Among the products manufactured in Europe are high-density polyethylene, resins, laminates, and graphite electrodes.

The Americans have invested almost twice as much in Europe as Europeans have in the United States. No doubt, the rapidly growing postwar market and the earlier high dollar rate have been contributory factors. Among the European companies investing in the United States, German companies head the list. All the indications are that in the coming years, that country will be the preferred investment area of the West European chemical industry.

Dow's Energy Policy Is Different, Too

The problem of assuring raw material supplies, especially oil, makes a long-term energy program one of the most important tasks of any American government. The chemical industry, in particular, must plan carefully for the future. For this reason, more and more companies are following the tracks of Dow, acquiring oil and gas deposits and forging closer links with oil producers.

A return to our technological past—to using coal—may even be considered since it is a chemical raw material that is still available in ample quantities. Up to 1973, for example, Dow bought up lignite deposits in Texas and reserves of about 600 million tons, an amount that will ensure Dow's supplies over the next forty years. Dow already produces 80 percent of the energy needed by its factories, and its experts are convinced that current from lignite will be cheaper than that obtained from oil after 1982.

The Silent Spring

Whether using oil, natural gas, or coal, the chemical industry will always be caught in the cross fire of opinion in the debate about the environment and the responsibility of industry toward it. No one in the industry can afford to withdraw from discussion of this subject.

Environmental protection had not yet become the preoccupation of the public conscience and a valid topic of civic concern when Rachel Carson caused a sensation that extended far beyond the frontiers of the United States with the publication of her book *Silent Spring* (1962). This eloquent book, winner of eight awards, serialized in the *New Yorker* magazine, became enormously popular and controversial, alerting Americans to the dangers of DDT and other chemicals in the environment. The title derives from a "fable for tomorrow" with which the book opens, a cautionary tale of a town in the heart of America that once lived in harmony with nature. Then a strange blight crept over the area. Domestic animals died, new and mysterious human illnesses appeared, grasses and flowers withered, and wild birds disappeared. The few that remained trembled and could not fly. "On the mornings that had once throbbed with the dawn chorus . . . of bird voices, there was now no sound; only silence lay over the fields and woods and marsh."

The book helped stimulate a massive campaign against industry, and the chemical industry suffered its share of the attack. No one should deny that, in the years after World War II with the rapid growth of industry, many sins of omission were committed. No one was concerned with doing more than the usual to keep water and air clean in those heady years. Cities and towns acted with no less negligence than industry by not building enough new sewage plants or not modifying existing ones to meet the needs of an increasing populace.

The reaction sometimes reached the level of hysteria, occasionally of considerable intensity. The new problem became a popular subject for political and ideological exploitation. Little has changed in this respect to this day, although much has been done by the industrialized

countries of the Western world, especially the United States and West
Germany, to provide increased environmental protection.

Some 25 percent of the chemical industry investment budgets has
been spent in past years on environmental purposes. To this must be
added the considerable investments needed for the maintenance of
modern installations for the purification of both water and air and for
noise abatement. It is beyond question that much progress has been
made.

However, environmental protection, largely safeguarded through
laws and regulations, is itself threatened by the perfectionism of the
authorities. In this respect, both the United States and Germany have
a story to tell. The Environmental Protection Agency (EPA) has been
unduly harsh at times in dealing with industry. Decisions by this
authority have prevented investments in locations where they would
have been beneficial from a political and economic point of view.
Similar considerations apply to Germany. In the Federal Republic,
matters are not made any easier by the separation of responsibilities
into regions and subjects.

It is necessary to strongly warn against such perfectionism. That
civic consciousness for the creation of an environment worthy of its
people has been aroused and intensified is welcome and commendable.
But if the authorities delay or prevent investments, or render them
vastly more expensive because of impractical or excessive demands, it
will be the people who will have to pay the bill one day.

Bridgewater in New Jersey

In 1968, we moved almost all our divisions from New York to
Bridgewater, New Jersey, where we had acquired a complex of 44
hectares in the green, hilly landscape north of Somerville. By the time
construction was completed, we had available nine buildings with a
useful floor space of 37,000 square meters.

During the next two years, we concentrated most administrative
and sales functions in one complex in Bridgewater: the sales and
service departments for chemicals, plastics, dyestuffs, pigments, and
Hoechst Pharmaceuticals, until then in Cincinnati, including administration, research, finishing, and quality control. The move of Hoechst
Pharmaceuticals was an especially difficult venture. We experienced
all the problems associated with such relocation, problems that affect

American staff just as much as Europeans, in spite of their greater mobility. A sales center completed the complex, which had cost some US$40 million. Bringing together the various individual companies dealing with our activities in the United States had become one of my most important tasks during the last few years. It was indispensable for both the American market and the financial world. In the end, our pharmaceutical production, Hoechst Fibers, the former Hystron, and the Azoplate reprographic materials unit were joined to form American Hoechst.

New York—the Wonderful Catastrophe

Since I became chairman of the supervisory board of American Hoechst in 1968, I have visited America even more frequently than before. My headquarters on these occasions is New York City.

For me, New York has long been one of the most fascinating of the world's cities. It is the eternally pulsating commercial and financial center of the nation. Any company with any claim to importance, including every large chemical company, has had a presence there in one form or another. In New York, as in almost all other large cities of the industrialized countries, companies are transferring their administrative headquarters increasingly from the city to the suburbs or further afield. Nevertheless, the city remains the most important economic nerve center of the United States, especially insofar as contact with the European countries and their industries are concerned.

Naturally, such an agglomeration of people and interests often results in huge problems. Public services, for example, have frequently reached the verge of complete collapse. Sometimes it seems as if these problems can no longer be solved by conventional means. Nevertheless, like many citizens of New York, I take the view that the city's regenerative powers are sufficient to ensure its survival as a symbol of a young and optimistic nation.

Of course, New York, the "wonderful catastrophe," as Sabine Lietzman has described it, is not America. Indeed, in many respects, this gigantic concrete jungle is atypical of the spacious land with its countless geographical and human aspects.

America, and above all, New York, has always been described as the "melting pot of races and people." I am not so certain whether these winged words really apply everywhere in the country. Take the many streets, and even quarters, in which only Chinese or Germans, Puerto Ricans or Italians, Negroes or Jewish immigrants from Eastern Europe, live. Such closed enclaves do not seem to me to be a sign of people living happily together, or even alongside one another. On the other hand, one must not forget that much has been achieved in the last decades to bring about a better relationship between blacks and whites. The young people, for example, go to the same schools and universities; in the armed forces, as in sports, integration has long been complete. Black persons have become ministers, ambassadors, mayors, and every other type of professional. American democracy is well on the road to solving this old social problem.

Between Jazz and Hemingway

The Americans themselves know best how much human substance its black people have contributed to the life of the nation. How, for example, would the modern music of this country have evolved during the last fifty years without jazz, which has made American music famous throughout the world, from dance halls to concert halls? How many black writers have earned both national and international fame?

These cultural achievements in America have always attracted me, perhaps because I belong to a generation which, for political reasons, was cut off in its youth from most of the cultural life of the world. This isolation also applied to American literature. It was, therefore, not until well after the war that I was able to read the outstanding works of American literature from the twenties and thirties, such as Thomas Wolfe and Ernest Hemingway.

An invaluable mine of information at that time, for both me and most of my contemporaries, were the American Cultural Institutes, which were not exactly in short supply, especially in the United States zone of occupation. They could be found even in small provincial towns, and they were beautifully equipped institutes whose task, pursued with much enthusiasm, was to make the Germans familiar with American culture, technology, and democratic policy.

In the theater, there were interesting encounters with such important dramatists as Thornton Wilder of *Our Town,* Eugene O'Neill and his tragedies, and young Tennessee Williams, who achieved world acclaim with his first play *A Streetcar Named Desire,* a success repeated later by the film version.

New York: From the Musicals to the Museums

During my visits to New York, I frequently take the opportunity to pay a flying visit to those indigenous arts that cannot be exported as easily as books. I have made it a rule, even if I am in New York only for a few days, to visit a theater, musical, concert, or museum.

I usually visit a museum on weekends if the weather is not conducive to golf. The Metropolitan Museum, with its wealth of art treasures, always attracts me. The seventeen departments contain works from various periods and parts of the world. The range extends from ancient Egypt to Greek and Roman antiquity, from examples of ancient Oriental, Islamic, and Asiatic art to the primitive works of American Indians before their continent was discovered by Columbus. I also recall with pleasure many exhibitions in the Museum of Modern Art, particularly the works of Chagall and the late paintings, of Cézanne.

But my visits are not only to theaters and museums. There is another compulsory, though rather more profane, destination: the New York Athletic Club on 59th Street, only a few meters from the Park Lane Hotel. A few rounds in the club swimming pool or a sauna are the price of my well-being.

While I regard swimming as more of a health duty, golf has been my continuous inclination ever since my first American trip. That inclination and talent do not always go together has been amply proved in my case. However, I have made some progress and spent many an enjoyable weekend with friends on the beautifully designed golf courses around New York and on the West Coast. Most convention hotels in the country are surrounded by beautiful golf courses. Dozens of little carts scurry across them, replacing the caddies who have become redundant in this affluent society.

20

Canada: Large Cities and Lonely Forests

After a long night flight from London, I arrived in Montreal one day in the autumn of 1954. When business discussions had been completed, we drove to the northern copper and gold mines near Rouyn and Noranda, on the border between Quebec and Ontario. From there, we continued southward to Toronto via North Bay.

The contrast was remarkable between Montreal and Toronto, each with a million inhabitants, and the complete loneliness of the rest of our journey that was only occasionally interrupted by a small village. We spent a night in a tiny guest house with six rooms in the middle of Parc de la Verendrye, one of the large, beautiful wildlife preserves found in all Canadian provinces. There were bears, elks, and red deer nearby.

I was interested in the northern mining operations because they used our chemicals in ore flotation. Ekkehard Maurer had prepared everything very well. At that time he was looking after the Canadian interests of the Duisburger Kupferhütte, which was part of our organization. Our program included a 900-meter descent into the copper mine at Noranda, my first experience of this kind.

Toronto, which looked very British at that period, was our last stop. It was time to think about the impressions and experiences gained on this trip. For the first time, I had become really conscious of the tremendous expanse of the country. Yet, we had seen only a small part of it during our five days there.

Three Cities

Canada is 40 times as large as the Federal Republic with barely 40 percent of Germany's population. These statistics suddenly became reality because of personal experience. It is surprising how much of the population (23.5 million in 1977) lives in the three large cities: 2.8 million each in Montreal and Toronto, 1 million in Vancouver, accounting for almost 30 percent of the population. In Montreal, two-thirds of the inhabitants speak French, and it is therefore the second largest French-speaking city in the world.

Although Toronto is about 1000 kilometers from the sea, it is nevertheless a large seaport. The St. Lawrence River, which is crossed in Toronto by bridges up to 5 kilometers long, forms a huge stream. Seagoing vessels up to 30,000 tons can anchor in the harbor here.

When we began to develop a sales organization in 1953, we were faced by very strong international competition. Du Pont, Imperial Chemical Industries, (ICI), and Union Carbide in particular had long maintained large organizations in Canada. They imported some of their products from the parent company, but to a large extent they were producing in the country itself. At the time, Hoechst did not have the money to finance a starting-up period, so the Canadian organization had to stand on its own feet from the beginning.

After the war we sold small amounts of dyestuffs through a one-man company in Montreal. This belonged to Adolf Kuhl, a German-Canadian who had been active for the I.G. before the war. For a reasonable price we bought his firm, the Progressive Color and Chemical Company, as well as a small dyestuffs warehouse. These units were the embryo of our own organization. The company's name was soon changed to Canadian Hoechst Ltd.

We chose a different route to establish our chemicals business. In the beginning, there were not enough products to meet the costs of a proper sales organization and regular customer visits, even if we were to confine those activities to the cities of Montreal and Toronto.

We therefore linked up with a young English chemicals trading company. It had founded a Canadian subsidiary, Kingsley & Keith (Canada) Ltd., and was on the lookout for further sales products.

We agreed that Kingsley & Keith would at the same time become one of the chemicals departments within Canadian Hoechst, too. In this way, the same company sold products of different origin under the Kingsley & Keith name and our products under the name of Canadian Hoechst. Our cooperation developed very satisfactorily. However, after six years, both the Hoechst and the non-Hoechst business reached a sales level that justified independent organizations. Therefore we separated amicably. Kingsley & Keith (Canada) Ltd. is continuing to operate successfully in its fields.

Our Route into the Pharmaceutical Market

One of Hoechst's strengths is the variety of its product line. This diversity leads to the interesting task of shaping the sales organization to the products being sold.

We realized that our start in the North American pharmaceutical sector would have to be launched with a major product of above-average interest. This prerequisite was fulfilled in 1955 when our research succeeded in developing the first oral antidiabetic in the world. I have already described the reasons why we agreed that Upjohn should sell this antidiabetic, called Rastinon in Germany and Orinase in Canada and the United States—under license in the latter country.

The question was whether to pursue the same route in Canada. Integrating these preparations into the Upjohn organization, which was highly efficient also in Canada, would probably have resulted quickly in high, no-risk license fees. But the establishment of our own pharmaceutical organization, into which we could introduce further pharmaceuticals later, would thereby have been postponed for many years, in all probability.

We therefore decided not to go in for licensing but to found a joint venture with Upjohn under the name Hoechst Pharmaceuticals of Canada Ltd. Upjohn and Hoechst had equal shares in this concern. Hoechst contributed products and scientific results, while Upjohn provided management with experience and knowledge of pharmaceutical requirements in Canada.

It had been planned from the beginning that one of the partners should be free to take over the shares of the other at some later date.

That this would probably be Hoechst was indicated by the company name. Apart from that, Upjohn was already operating its own pharmaceutical company in Canada under its own name. This joint enterprise of Hoechst and Upjohn operated successfully. Although there never was any conflict of interests, it became clear in the course of time that this possibility could not be entirely excluded in the future. We therefore discussed with Upjohn a takeover of its shares by Hoechst. Our partner showed full understanding. In 1963, Hoechst Pharmaceuticals of Canada Ltd. was acquired by Hoechst in its entirety and integrated as a pharmaceuticals division into Canadian Hoechst Ltd., thus gathering all Hoechst activities in this organization.

In a country as important as Canada, our sales organization could not be based on imports alone. The difficulties of establishing manufacturing facilities in the country itself derived not only from the immediate vicinity of large U.S. plants. In Canada, efficiently operating chemical companies had been set up by our international competitors during World War II and in the postwar years. In 1963, we started with the construction of a polymerization plant whose capacity has been greatly increased since then. Other plants have been added in Varennes, Quebec; in Cambridge, Ontario; and a pharmaceutical plant has also been set up.

While the output from these plants is destined essentially for the Canadian market, a subsidiary of our associate Süddeutsche Kalkstickstoff-Werke in Becancours, 130 kilometers downstream from Montreal, constructed a large plant for the manufacture of ferrosilicon and silicon-metal. In the main, the output of this plant is destined for the world market. Its high energy requirements are met by the ample supply of electricity available in Quebec at reasonable prices.

Twenty-one years after my first visit, which was followed by several other Canadian trips, I traveled to western Canada in 1975 to see an entirely different part of the country: the Athabasca oil sands near Fort MacMurray in the northern part of Alberta province. It is only after leaving the cities that one perceives the full expanse and magnitude of Canada. Athabasca also gives one some idea of the major developments that may take place in the coming decades.

On the journey from the airport to the oil sands processing plants, I became aware of the rigors of living and working in these surroundings. Our plane had landed the morning after the first night-frost of the year had arrived. The snow and ice clearance had not begun yet and the hilly street was completely glazed over. I was glad when, after only one descent into a ditch, we arrived two hours late at the works of Great Canadian Oil Sands. We saw the open-cast mining of oil sand and the separation of the oil from the sand. The amount of oil in the sand of this surface area of about 25,000 square kilometers is hardly imaginable. It has been estimated that some 6 billion tons of oil can be produced here by open-pit mining. The amount that might be obtained through underground mining is claimed to be ten times as high.

Athabasca is one of the world's great oil reserves. Its present problems are the steadily rising construction cost estimates for future extraction plants as well as a steady increase in operating costs. Because of these, and in spite of the high price of oil from the Middle East, the economic exploitation of these oil sands is by no means assured.

Canada also has large reserves of natural gas. They have been either discovered or assumed to be along the Arctic coast and on Canada's Arctic islands. Here, too, the question of cost is decisive—not so much the cost of production, but the cost of transportation to the user. The business executive from the chemical industry, looking at Canada's development, will, of course, assess these factors and raw materials in terms of their potential usefulness to the chemical industry.

The landscape of western Canada is particularly spectacular. There are extensive forests which are used to only a small extent for wood industry purposes. Picturesque bays, islands, and lakes make it an ideal vacation area. There is not much industry, and that is why I have managed only twice so far to visit Vancouver and its wonderful surroundings. There is little difference in the daily life of Canada and of the United States, although their historical development and their political institutions vary significantly.

Mentioned in Passing

All I learned in my history lessons about the year 1763 was that at the end of the Seven Years' War Prussia acquired the Province of Silesia. In the same year, England won most of the vast French colonial empire in North America. Defeated in the French and Indian War, France ceded Canada and all the lands lying east of the Mississippi—except New Orleans— to Great Britain. Her ally, Spain, lost Florida, in return for which France gave Spain New Orleans and Louisiana. But this conquest was obviously regarded as unimportant for our lessons, which were oriented entirely toward Europe.

When one thinks of the terrible loss of life not only in the two World Wars but also in the Thirty Years' War and the Seven Years' War, it is all the more remarkable that England succeeded in conquering this vast area of North America with a loss of life of only about 1500 English people. Of the former French colonies, only Quebec has retained its French character. One reason was the extraordinary care with which the French language was protected by English Canadian legislation. Another reason was the great vitality of the Franco-Canadian population. In this connection, Alain Peyrefitte said in his book *Le Mal Francais* that the 60,000 population of 1760 had increased to 6 million in 200 years without any significant French-speaking immigration.

Out of 23.5 million Canadians, 6 million are French-Canadians, 5 million of whom live in Quebec province, which has a total of 6.5 million inhabitants. The other 1.5 million are English-Canadians or immigrants who do not speak French and have attached themselves to the English-speaking group. In contrast with this, the melting pot of the United States has produced very uniform customs and habits, based on a common language. But in Canada, tradition and links with the past are fostered, with all the advantages and disadvantages that adherence to tradition brings. It is surely a good thing when one's present-day life is rooted in a strong and beautiful past. But it is dangerous to let the past overshadow the present, as seems to be happening in Quebec these days.

To say that the current dispute is simply a conflict between French-Canadians and English-Canadians would be only partially right. The protagonists are the prime minister, who is fighting for the unity of Canada, and the head of the Parti Québécois and current premier of Quebec province, whose party demands a separate Quebec.

The former prime minister, Pierre Trudeau, the premier of Quebec, René Levésque, and the governor general, Jules Léger, all happen to be French-Canadians from Quebec. The head of the commission established by the federal government to examine suitable measures for maintaining the unity of Canada is Jean-Luc Pépin, another French-Canadian from Quebec.

Among the French-Canadians of Quebec, there is a large group which regards the whole of Canada as its country and Quebec province as its immediate homestead. But there are others who do not feel linked to Canada, who believe that Quebec, with its 80 percent French-speaking population, can maintain its language, tradition, and culture only as a sovereign state.

When one studies the Canada-Quebec problem, it appears that there is no suppression of French-Canadians in politics. The government of Quebec province is entirely French-Canadian, and even in the federal government the most important positions are held by French-Canadians. In the past, too, French-Canadians have played a decisive role in political developments. In Canada's 110 years of federal government, its most important office of prime minister was held by French-Canadians for 33 years.

The situation is different in language and commerce. For a European, it is remarkable that many English-speaking Canadians in Quebec, whose ancestors have in some cases lived there for generations, have not thought it necessary to learn and speak French. That attitude is only now beginning to change.

In the business area, it is surprising that the number of French-Canadians in influential positions was and is very limited. In part, this is due to the French-speaking school and university system. This was developed by the church, largely along classical lines. Instruction in the sciences was intensified only after World War II. The Université de Montreal, the largest French-speaking university of the province, had its first nonclerical rector only in 1965. Thereafter, however, instruction in the natural sciences was rapidly expanded. Only now is training being given to a French-speaking generation which really has a chance to advance into leading positions in the economy.

Another reason can probably be found in the development of Canadian industry, which was given a decisive impetus during World War II when British and American companies established Canadian subsidiaries. Having the same language and a similar life-style as the Anglo-Canadians often resulted in closer links with them than with the French-Canadians.

Quebec's industry showed a strange income pyramid. The base of the pyramid was almost entirely made up of the French-speaking people, but the nearer the apex, the smaller was the percentage of French-Canadians. The top was almost without exception English-speaking. Many heads of large companies could talk to their employees only through interpreters, which made for misunderstandings and other problems. These factors should be borne in mind when one investigates the reasons for the present conflict in Canada.

I am convinced that any fragmentation of Canada, whether wholly or in part, would have an extraordinarily unfavorable effect on its economic development. Although Canada is an important market with its 23.5 million people, it has only 10 percent of the population of its

Headquarters of American Hoechst Corporation in Bridgewater, N.J. AHC manages a dozen U.S. companies with close to 10,000 employees.

Overleaf: American Hoechst Corporation started small but high. The forerunner of AHC began modestly with four rooms on the 82nd floor of the Empire State Building in New York.

Cattle-breeding plays an important role on the agricultural scene. With vete medical preparations, Hoechst makes an important contribution to animal health.

Headquarters of American Hoechst Corporation in Bridgewater, N.J. AHC manages a dozen U.S. companies with close to 10,000 employees.

Overleaf: American Hoechst Corporation started small but high. The forerunner of AHC began modestly with four rooms on the 82nd floor of the Empire State Building in New York.

Cattle-breeding plays an important role on the agricultural scene. With vete. medical preparations, Hoechst makes an important contribution to animal health.

The West Warwick, R.I. paint facility —the first Hoechst production site in the United States.

The TREVIRA® ERA

begins here in Spartanburg

Intensive advertising of "the Trevira era" introduced polyester fiber to the American consumer. The campaign successfully positioned Trevira in the fashion market rather than as a fiber for mass-produced merchandise.

A scene inside the Trevira plant in Spartan, S.C. Hoechst Fibers has become one of the largest polyester fiber producers in the world.

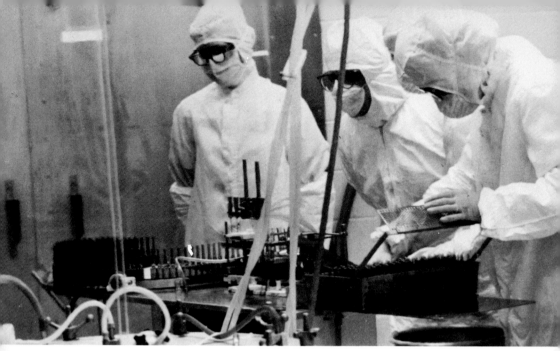

Originally Lloyd Brothers in Cincinnati, the pharmaceutical division of American Hoechst has been centered in Bridgewater under the name Hoechst-Roussel Pharmaceuticals, Inc. since 1970. Research at this location includes the development of effective antirheumatics and analgesics.

Calbiochem, of La Jolla, Calif., a maker of diagnostic was acquired by American Hoechst Corporation in 197 and combined with AHC's Behring Diagnostics to become Calbiochem-Behring.

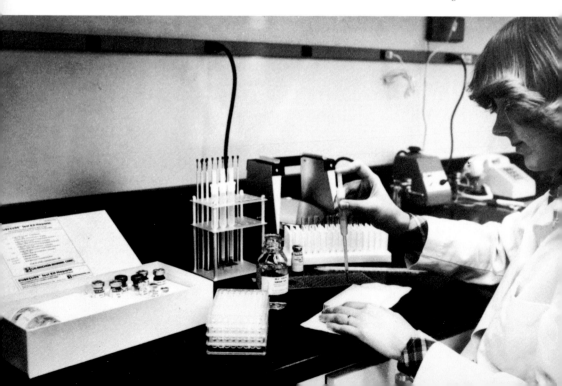

Eocom, a recent addition to American Hoechst, manufactures Laserite 100-E printing systems. Research at Eocom has been supplemented by the work of two other Hoechst companies involved with reprography—Azoplate in the United States and Kalle in Germany.

With the acquisition of Foster Grant, Hoechst became the world's largest manufacturer of sunglasses. Foster Grant activities in plastics and petrochemicals have since been absorbed by AHC. Sunglasses remain under the well-known Foster Grant name.

The $20 million U.S. corporate headquarters was dedicated on October 14, 1970. An audience of 400 heard addresses by William T. Cahill, governor of New Jersey (3rd from left), Professor Rolf Sammet, Kurt Lanz, and John G. Brookhuis.

ssor Winnacker and Mr. Lanz at a press conference in New York City.

The Eocom Laserite system 100-E received praise in the United States and was acclaimed among thousands of competitors as a prestigious "R & D product of the year."

The headquarters of Hoechst Canada, Inc. in Montreal. In a country as important as Canada sales could not succeed solely on imports. So a series of plants has been built, including facilities for plastics and pharmaceuticals.

In 1972, then Finance Minister Helmut Schmidt of West Germany paid a visit to the Hoechst Fibers installation in Spartanburg, S.C. Schmidt, now Chancellor, is third from the left.

Two production plants in Bayport, Tex. are scheduled to start operation in 1980, producing styrene monomer and high-density polyethylene.

neighbor, the United States. This is an important consideration in investing. Separation into two areas with populations of 17 million and 6.5 million respectively would have serious consequences for both sides.

We know from Europe's experience how long it took and how difficult it was to create the Common Market from several sovereign states. We know what difficulties the further development of this European Community will present. It is very unlikely that such matters would be easier in Canada. As an advocate of a politically unified Europe, I hope sincerely that Canada will be able to solve its problems and remain a unified country. Perhaps its people have taken its unity too much for granted in the past; perhaps the realization that this unity is not cast in stone may result in a stronger feeling of responsibility and obligation of each individual toward this beautiful and unusual country. In fact, I like to think that the development of such sentiments has already begun.

21
Latin America—Continent of Contrasts

Alexander von Humboldt, a friend of Schiller and Goethe, and himself one of the greatest men of his time, was thirty years old when one of his dreams was fulfilled. In June 1799 he boarded the Spanish frigate *Pizzaro* in order to explore and describe Latin America, which had a magic fascination for him. When he returned to Europe only five years later, his travel stories, his later scientific discoveries, and his correspondence with the great names of his time soon made him world-famous.

I still enjoy thumbing through the writings of this historian, natural scientist, and humanist who gave his work the epigraph, "Man must attempt the good and the great. The rest depends upon fate." The superficial reader might gain the impression from Humboldt's travel descriptions that Latin America is a continent with ubiquitous common features and few differences between the individual countries. In reality, Latin America is a world of vivid contrasts. Its disparities became clear to me on my many trips from Venezuela to Tierra del Fuego.

Even at first sight, one notices that there are distances, and not only in kilometers, between the countries of the east coast that had been settled by large waves of European immigrants, and the states of the west coast, among whom Mexico and the nations of Central America may also be counted, which all look back to a great Indian past. Except for Portuguese-speaking Brazil, the Latin American nations have, as a common feature, the Spanish language, which became their own during centuries of dependence on their Spanish conquerors. Roman law and the European system of education were essential prerequisites for the special development of Latin America.

These European influences have undoubtedly changed the surface characteristics of the Latin American people. But the temperamental and fun-loving South Americans of the Atlantic and Pacific coasts are encountered only rarely in the Andes regions, however. The population there, mostly of pure Indian descent, still suffers from the heavy burdens that were imposed on their ancestors in pre-Columbian times.

Europe, and especially West Germany, has realized only in the last few decades that Latin America has produced literature of international rank. This was made very clear to us in the sixties and seventies when Miguel Angel Asturias and Pablo Neruda were awarded Nobel prizes for literature. They represent roughly a dozen South American authors whose work was influenced by the political stresses of their modern environment. Their writing therefore also captures the interest of the European reader who is trying to get an understanding of this particular world.

The music of small countries like Paraguay and Ecuador is no less popular than that of its big neighbors such as Mexico and Brazil. This music reveals even to the lay listener the cultural diversity of the people from Mexico to the Straits of Magellan.

The countries under the Spanish yoke won their release, under the liberals' leadership, in stages around 1820. Since then, they have pursued widely varying political courses right up to the present day. There have also been many disputes among the countries, especially over border questions.

Soccer: The Common Passion

Although I claimed that the continent was full of contrasts, there are various common features that transcend all frontiers. In the first place, there is a feeling of community vis-à-vis the rest of the world—a feeling that even our delegates from the parent company have developed and are likely to retain after they return from several years in South America.

Then there is the great sports passion, soccer. On Bolivia's Altiplano, 4300 meters above sea level, or on the Copacabana beach in Rio, everywhere and everyone, children and grown-ups alike, whatever their race or color, are passionately devoted to soccer. It is no accident that the largest soccer stadiums in the world are in Rio de Janeiro, Mexico City, and Buenos Aires. It is probably the dream of every young boy to become as famous, and perhaps also as rich, as Pelé. The soccer heroes have long since won what the Latin American nations have not fully achieved so far in economic and political respects, namely, to be among the world leaders and to have a major impact on events.

What are the economic problems of Latin America? There is a yearly population increase of more than 3 percent and unemployment is high. The prices for monoculture products vary, and the lack of foreign exchange is chronic. All this dictates industrialization. But this requires the establishment or improvement of the infrastructure. This, in turn, requires much capital, which in turn—except for the oil countries—means further indebtedness abroad.

It is somewhat paradoxical that Latin America's most important economic partner, the United States, whose investments and far-reaching financial aid make the largest contribution to the development of these countries, is rather unpopular in those very countries. Even the Alliance for Progress (Alianza para el Progreso), set up in 1961 by President Kennedy with much goodwill and at great expense, did not change anything in this respect. It appears as though Latin America wishes to abolish its dependence on the United States, whatever the price. This dependence is felt in technological and economic as well as political and military matters. It is not likely that, by now, some Latin American countries may have recognized that full detachment from this dependence will be possible only in the long term, if at all desirable.

Certain initiatives, such as the creation of a 200-mile offshore zone, originally suggested by Peru, or the formation of the Organization of Petroleum Exporting Countries (OPEC), advocated for more than ten years by Venezuela, have been internationally accepted. But the essential goal of becoming independent through the economic integration of the countries in the Asociacion Latinoamericand de libre comercio (ALALC; Latin American Free-Trade Association) has not been achieved. There is a lack of improved communication; and the idea of seeking a unilateral advantage from such a merger serves the goal of independence as little as the economic imbalance between the member states.

The Andes group, whose proposal was only belatedly accepted by the other ALALC members, has also failed to solve the problem in spite of more favorable beginnings. Most progress has been made by the small Central American common market (Mercado Común Centroamericano), to which Guatemala, El Salvador, Nicaragua, Honduras, and Costa Rica belong.

Latin Americans and Germans have always been linked by mutual sympathy. For this reason, we tried to rebuild our contacts with these markets soon after the war. However, the road that led to our fairly extensive involvement today was not free from worries and disappointments. But that is something we may also have to live with in the future on this continent of deep social contrasts and vast economic opportunities.

Mexico: Democracy in the Single-Party State

I had romantic visions of Mexico even in my early youth. There were the fine civilizations of the Mayas and Aztecs, the adventurous and bloody conquest of the country by a few Spanish adventurers who came largely from the impoverished landed gentry and were led by Hernando Cortés. *The White Gods,* by Eduard Stucken, was a best-seller during my school days. It was based on the assumption that the Indians had surrendered to the Spanish conquerors as the result of a prophecy. But it is more probable that both the Aztec and Maya empires had at that time passed the zenith of their power and glory. There is now authentic information about this period, for instance, in the diary of the monk Francisco López de Ganara which was found in a monastery library in Spain.

Mexico does not disappoint the visitor. The country is rich with cultural monuments that have been retrieved from the primeval forests or the soil. They go back to the golden age of the Mayas, Mixtecs, Toltecs, and Aztecs. There was scarcely a trip to Mexico on which I did not take the opportunity to spend a weekend viewing new discoveries. Chichén Itzá and Uxmal, in the south near Mérida in the province of Yucatán, were the beginning and climax of these archeological excursions.

There never was a Mexican nation until independence was achieved in 1821. There were, however, many Indian tribes separated by mountains and rivers, and often at war with one another. After the Spanish conquest, the governors appointed by the Spanish Crown aimed only to consolidate their rule, exploit their subjects, and above all, increase their wealth. As a result, many parts of the empire, which under Cortés extended from Mexico through Texas and New Mexico to California, were lost through internal strife.

After the 1910–1917 revolution, which claimed more than a million victims and destroyed most political, economic, and social institutions, Mexico was so weak that foreign capital from the United States was soon able to gain a dominant position. Those events have left the Mexican people with a pronounced fear of dependency, such as one encounters in many developing countries. The great problems of the country are not blamed on faults at home or on Mexico's technological or cultural backwardness, but on the large powers that, it is said, have granted the developing nation only a subordinate and dependent role in the international system.

The excessive nationalism, the inherent mistrust of most Mexicans, and their aversion to foreign influences must be seen against this background. Most of this antagonism is directed against their northern neighbor. The poor and the rich seldom make good neighbors, unfortunately.

Rio Grande: More than a Frontier

The Rio Grande, the frontier river between Mexico and the United States, separates two worlds of entirely different mentality: the Anglo-Puritan in the north and the Latin-Catholic in the south. One notices it as soon as one crosses the frontier. On one side, order, cleanliness, and prosperity; a few yards further, improvisation, poverty, and misery.

More than in any other American country, the early breakaway from the Spanish Crown and the conflict over the new national constitution in the north left their traces on the mentality of the people. The fact that as early as 1938 President Cárdenas expropriated and nationalized foreign oil companies, especially the large companies from the United States and England, shows what a change had taken place after decades of internal unrest.

Mexico achieved its political equilibrium only gradually in the wake of conflicts in which liberal and extreme left elements fought for power. A presidential democracy emerged. The central political power continues to be the Revolutionary Institutional Party (Partido Revolucionario Institucional). The president is virtually appointed by this party and serves for six years. For the term of his office, he is the almost undisputed ruler of the country.

Single-party rule brings with it all the negative facets that such a "democratic" form of government implies. The question is whether, on a longer run, the present system, which has produced presidents of varying belief and political fortune, will succeed in mastering the problems of the country.

Our Start in Mexico

Hoechst was back in the Mexican market soon after the war. In 1950, Farcol S.A. was founded as our official agency. Its manager, who also provided most of the capital, was Dr. Kurt Schneevoigt, who had been a representative of AEG in Mexico before the war. His collaborator was a former I.G. dyestuffs technician, Dr. Georg Sandor, a native Hungarian. Together these two men built up our industrial business, especially our business with the textile industry.

We had to overcome difficult early years also in Mexico because the means available to the two men were limited. We finally acquired Farcol and renamed it Química Hoechst de Mexico in 1957. Eventually, Dr. Jürgen Brand, who had been with the company some time and had spent his early years in Peru, was sent to Mexico in order to reinforce the management there. This appointment proved very successful. We were now at last in a position to pursue the recovery of our pharmaceutical trademarks. As everywhere else, this effort required many and complex negotiations and numerous interventions.

In 1953 we finally achieved our goal. The trademarks were returned to us, and we were able to take over the pharmaceutical business which had been inadequately handled by a nationalized company. In addition, we were able to acquire the building and grounds of the former Instituto Behring in the San Angel quarter of Mexico City.

This proved a stroke of good fortune. Although the installations were prewar and fairly antiquated, the grounds measured almost 20,000 square meters and offered sufficient space for a modern pharmaceutical plant. In addition, we modernized the Instituto Behring, and finally we also erected a small, attractive administrative building for our company.

We were very glad when we were able to inaugurate the new complex with due ceremony in 1959. We had demanded a great deal from our staff in the meantime. Mexico was a country that did not make things easy for the importer. The domestic industry had developed, and the customs laws were aimed at protecting domestic industry from foreign competition. For this reason, Hoechst was forced to make industrial investments. This was indicated for pharmaceuticals in any case, and as already mentioned, the site in San Angel offered us this opportunity.

For our industrial business we acquired ground in Santa Clara, an industrial suburb of Mexico City, where we erected a plant for Mowilith dispersions and organic pigments. To this we later added facilities for the production of textile dyestuffs.

Together with Mexican majority partners, the Banco de Comercio, we founded a subsidiary, the Sociedad Mexicana de Química Industrial S.A., on the same land and within the framework of the Mexican laws for foreign investments. This plant produces a range similar to that in other overseas countries: pharmaceutical raw materials, dyestuffs and intermediates, surfactants, etc.

Self-Sufficiency Written in Large Letters

Many industrial projects are much more difficult for non-Mexicans to realize. The country has an extensive policy of self-sufficiency. As a result, basic industry is mainly state-owned. Many other economic activities are reserved for purely Mexican capital or for a Mexican-capital majority.

An impressive example of this policy is the Pemex oil company. It is responsible for the entire oil industry. There is now enough oil not only for the requirements of the country but also increasingly for exports. New reserves have been discovered recently which will make Mexico a leading oil exporter. The relevant laws were passed in the early post-war years and make the production of petrochemical base materials and the first stage of further processing a state monopoly.

Like other German competitors, we have only a small share in the big chemical industry of Mexico. Besides chemical raw materials and fertilizers, in which the country is now self-sufficient, it produces all the base chemicals for petrochemistry. Man-made fibers, too, are produced in Mexico in such volume that there are scarcely any import opportunities.

Nevertheless, there were enough challenging possibilities for our plant in Santa Clara to extend its range and to become more sophisicated. Parallel with our policy in West Germany, we started to become involved in paint manufacture in 1972. The company concerned is Barnices Aislantes S.A. (BASA), on the road to Toluca from Mexico City.

Mexico's Industrialists Have Caught Up

Mexico City is one of the most dynamic capitals of Latin America. It is growing at such a rate that strangers believe they can perceive changes even at intervals of two to three years.

Most of our staff from Germany become acclimatized very quickly. Once they have overcome their initial isolation, they soon establish good relationships with Mexican colleagues and friends.

Mexican industrialists, usually trained in the United States, are dynamic and successful. With their help, the economy has gained a great deal of ground. Although Mexico is dependent on foreign technology, it knows how to tailor it to the country's needs and develop it independently. Because of some shortage of capital, Mexico's private banks and financial institutions are much favored as partners for industrial companies.

The frequently overestimated capacity of the Mexican market is nonetheless large enough, despite the self-sufficiency policy, to generate fairly lively competition. Industry looks increasingly toward exports in order to sell its excess production.

In the pharmaceutical sector, there is a trend toward state intervention that is typical in Latin America, but all too noticeable in the rest of the world as well. We shall be able to cope with this only if we continue to provide the market with the latest results of research, thus keeping one jump ahead of local industry, which confines itself mainly to imitations and licensed preparations. One unsolved problem is Mexico's almost nonexistent patent protection.

As far as I can tell from my own observations, Mexico's big problems are the population explosion and education. There is a growing proletariat in the large cities. Mexico City and its suburbs have 12 million inhabitants. The meaning of this density of population is difficult to appreciate for those who have not seen the conditions at first hand. Industry and the administration not only cannot absorb all the young people who come from the universities. What is far more serious, the people who flee the countryside, thinking they can find an existence in the cities, get stuck in the suburban slums. In his book *The Children of Sanchez,* published in 1962, and based on tape recordings, the American anthropologist Oscar Lewis has presented the story of such a Mexican family in the most moving manner. The process of integration is long and complex, especially since the people who come from the country are frequently illiterate and lack the necessary mental agility for existing in a large city or following an industrial occupation.

More serious still is the fact that the agricultural population has no share in the industrial upsurge of the country, in the wealth which tourism, especially from the United States, provides year after year. It is estimated that about 20 percent of the population live under conditions that do not differ greatly from those at the time of the Spanish conquest. This, of course, is political dynamite. There can be no doubt (and one would expect it in a one-party system) that there is an extraparliamentary opposition, although it has little opportunity to appear in public. The government's most important task will be to let the rural population and the urban proletarian participate, through suitable reforms, in the great upsurge experience of the country as a whole.

No doubt Mexico's large deposits of oil and gas will play an important role in the future. It is claimed that the income from oil and natural gas exports will enable the Mexican government to solve the major problems of the country: food and education for its rapidly growing population. Half of Mexico's 64 million inhabitants are less than fifteen years old. Some 900,000 new jobs need to be created each year. One wonders how this is to be achieved. Oil and gas alone, even if they are processed in Mexico itself, will offer only a limited number of new jobs.

Of course, many a visitor who is enjoying life in one of the beautiful modern hotels will hardly perceive these problems. Charmed by the special atmosphere of the country, the magic of its architecture, and its hospitality, the visitor will usually just wish to have a good time.

We often discuss the country's complex problems when I am together with people from our company. Dishes typical of the Mexican cuisine provide the background for such evenings. They are among the irresistible attractions of this country. Mexico fascinates the occasional visitor no less than those colleagues of mine who agreed to go there for a few years and have stayed for the rest of their lives.

Ecuador: Bananas, Oil, and Cocoa

I visited Ecuador for the first time in the mid-fifties. On that trip, I had a few hours' stopover in Guayaquil. I used this time to visit our pharmaceutical branch in this tropical port. With almost a million inhabitants, Guayaquil is the largest city in the country. However, the capital, Quito, in the highlands of the Andes, is scenically much more attractive and has a much better climate because of its altitude (2900 meters) and its proximity to the equator. But I got to know it only two years later.

What impressed me greatly at the time was the fact that the two medical representatives employed by Hoechst Pharmaceuticals made their daily rounds by bicycle. But this was absolutely in line with the thriftiness of the European merchants there. This thrift, apart from many other qualities, also distinguished Alfredo Zeller, who hailed from Hamburg and was the chief of our former agency, Eteco S.A. When we founded Hoechst Eteco in 1961, our small office, with its staff of half a dozen, was on an upper floor of the rented Eteco building in the center of Quito. Our partners were concerned mainly with the import of machinery, which was exhibited in the large windows of the ground floor. Alfredo Zeller had a good eye for efficient young people. His right-hand man was U. J. Thomsen, who took over the management of our Peruvian business in 1967.

Though governments changed frequently in Ecuador, alternating without bloodshed as a rule between the military and the politicians, the country was progressing well economically. And yet, the condition of the population (10 percent white, 40 percent Indian, 40 percent Mestizo, and 10 percent African and Asian), the pronounced prosperity gradient, and above all, the backwardness of the rural population offered enough points of conflict.

The boom in the banana business, which started in 1950 after a plant disease had paralyzed production in several Central American states, resulted in the first upswing. Ten years later, the country discovered oil. Its production was first left to international companies in which the state had an interest. Then the income from this oil began flowing in such quantity that oil-based industries were established. The infrastructure and the agrarian economy, too, have been provided with new momentum. According to the ideas of the military junta of Guillermo Rodriguez Lara, which governed the country from 1972 to 1976, this development was to take place within a mixed economy that was attractive enough to lure foreign capital. As a result, much Latin

American money has gone into Ecuador in the last few years and has produced additional prosperity.

In this connection, I recall a discussion I had in Quito in 1974 with Coronel, Ecuador's minister for mineral wealth. Our free exchange of ideas about his country's plans and the prospects of the Andes Pact was unconventionally but refreshingly accompanied by folk music from a transistor radio on his desk. The wide-ranging development projects of all the OPEC countries were a timely subject also in Ecuador. Our discussion showed me that there was some concern with how to manage the problem of sudden wealth, where and how it had best be invested, and how the organizational difficulties could be mastered. There are, unfortunately, various places in the world, and I am not including Ecuador, where oil millions have been largely wasted. What is more, the huge monetary blessing has often had an adverse effect on the work morale of the population.

Hoechst has developed well in Ecuador. Our company now has modern offices in Quito and Guayaquil. A well-equipped pharmaceutical plant that also produces cosmetics has been established in Quito. A plant for the manufacture of plastics dispersions is under construction in Guayaquil. This plant also increases employment, which the country with its 7 million inhabitants badly needs. During my last visit, I joined its staff of 300 on an outing at a large farm in the interior. The attraction of the day was a bullfight in a small arena that belonged to the farm. In line with the peaceful character of the people, the bulls were not killed. But the contest is not without its risks, as one of our employees found out. He went into the arena as a voluntary *torero*. Since then, he has firmly resolved never to do so again.

Peru: El Dorado of the Conquest

I know no other place where the Spanish colonial history of Latin America is as vivid as in front of the baroque cathedral of Lima. The magnificent equestrian statue of Francisco Pizarro, who conquered and destroyed the highly developed empire of the Incas in 1531, impressed me deeply with its pride and cruelty. It reminds me of the artistically superior equestrian statue of Bartolommeo Colleoni in the square in front of the gothic church of SS. Giovanni e Paolo in Venice. Both statues glorify bandits who carried out their raids largely for their own benefit. Incidentally, in contrast with Peru, Mexico erected no statues to its conqueror. Instead, the Mexicans honor the memory of the unhappy, defeated Aztec emperor Montezuma and his courageous nephew Cuauthémoc.

Only in recent years has the Peruvian military tried to restore the picture of the Indian freedom-fighter Tupac Amaru, executed in 1780, and thus to do away with any symbols of foreign conquest. His name was then promptly misappropriated by the Uruguayan terrorists who called themselves Tupamaros.

A visit to the Gold Museum in Lima or to one of the other exhibitions of Peruvian gold gives one an idea of the art treasures found by the Spanish conquerors. They melted them down and sent the gold to Spain in order to cover the chronic deficit in the state coffers of the "Reyes Catolicos," and also to ensure their own bloody rule. Even more than in Mexico, the visitor to Peru encounters signs of the conquered past of its 17 million people.

In Cuzco, the old capital of the Inca Empire, I learned that the people, 50 percent of them pure Indian, have practically no share in the economic and political life of the country. The military government that came to power in 1968, replacing the rule of the whites, the élite, and the great families, has not made any fundamental changes in the political and social structure up to now. When Hoechst looked at Peru after the war, it found that the very tolerant feudal rule of the whites was uncontested in principle.

Clemens A. Ostendorf, the second-generation Peruvian from a merchant family in Hamburg, a large landowner, and the owner of the Peruvian Volkswagen Company, offered us the services of his agency, Transmares S.A. With the help of a handful of efficient young people from Hoechst, who worked closely with a well-qualified domestic team, our basis was established in a few years.

The jointly founded Hoechst Peruana consolidated our activities in 1962. Ostendorf, then a spartan bachelor, had the gift of delegating responsibility to talented young employees. Thus U. J. Thomsen, after his apprenticeship in Ecuador and Mexico, soon became an independent manager, to whom Ostendorf, thirty years his senior, was adviser and paternal friend.

Our business flourished, and the spectacular pharmaceutical plant was followed in the second half of the 1960s by the usual middle stages of local manufacture: auxiliaries for the textile industry, plastic dispersions, pigment preparation, and a cosmetics plant.

The Military and Socialism

In 1968, a military junta gained power. It attempted to integrate the people—a task that could be completed, if at all, only over a long period of time. The new masters introduced left-wing socialist ideas, expropriated large plots of the vast, privately owned land, and nationalized the basic industries: minerals, oil, fish meal, iron, steel, base chemicals, and all newspapers. In addition, there were the new industrial foundations resembling the Yugoslav model, the Empresas de Propiedad Social. The founding of the so-called *communidades industriales* has also had an adverse effect on the investment climate. Under this system, all companies every year had to pay their employees 15 percent of profits before tax in the form of shares, until the employees held 50 percent of the capital. There were additional payments which benefited either the work force or the state.

As a result of those provisions, the Hoechst plants and our trading company were at the point of losing their majority to native Peruvians and of coming under state control. Fortunately, the regulations that threatened to strangle any kind of private initiative have been abandoned under the more pragmatically oriented government of Morales Bermúdez. Rigorous application of the antiforeigner provisions of the Andes Pact has now given way to a more realistic approach. Peru's rulers have realized that foreign capital and Western technology are necessary for exploiting the country's mineral wealth, such as copper, zinc, oil, and gas, and for establishing processing industries.

The country is rich in raw materials and agricultural resources, both capable of extensive development. I hope that the current pragmatic economic policy will last. A relapse into the methods of the "people's democracy" would postpone the desired prosperity for a long time.

For the culturally and archeologically interested visitor, Peru offers a wealth of magnificent impressions. About Cuzco, one of the highlights of our excursion, German Arciniegas writes in his cultural history of Latin America: "Cuzco, capital of the Incas, built in the Andes 3000 meters high with cyclopean walls, a model of architecture, where the palaces and temples are faced with giant blocks of dark stone, with terraced gardens at their feet. Cuzco is the center of the star from which roads lead to the four corners of the empire."

We traveled through the high valley of Cuzco to the ruined Inca city of Machu Picchu, which was wrested from the primeval forest only a few decades ago. Those massive, moss-covered structures, shaped like pyramids, exude a whiff of the old Inca Empire's tremendous power. We made our trip by means of a narrow-gage railway. Shortly after leaving Cuzco, we visited a nitrogenous fertilizer factory constructed by our engineering subsidiary Uhde and financed with money from the development aid program. After some years, however, this plant was reequipped for the production of ammonium nitrate for explosives production. This is a typical example of state misplanning. The market for fertilizers, which was to be created in this region within a few years, does not exist even today.

On another trip through the highlands around Lake Titicaca, I attended a colorful Indian festival, the *chonguinada,* in a small village. We stopped in order to admire the beautifully embroidered shawls of the Indian women and to watch the dancers, who had unfortunately become a little unsteady because of their intake of alcohol. Since strangers seldom come into this region, and stop there even less often, we were eyed from all sides until one of the notables addressed himself to us. It was the *curandero,* the healer, who in that remote part of the world is often the only help for the sick. After the *curandero* learned that we were pharmaceutical producers, and thus colleagues in a sense, we got into some interesting shop talk. He told us something about syndromes and his particular form of therapy, and especially about the herbs he used. He emphasized, however, that without the amulet he personally prepares for each individual case, no treatment can be successful. Such amulets, small vials that are hung around the neck of the sick person, can be bought in the Indian market in Cuzco. They often have gruesome contents, but are no doubt prescribed in the knowledge that the sick person, apart from physical treatment, also needs psychological support—a placebo.

Finally we had to take our leave, but not before we had given the *curandero* a doctor's sample of one of our preparations for headaches, from which his wife was obviously suffering.

In 1972, we brought U. J. Thomsen, the chief of Hoechst in Peru, back to the Hoechst head office. In the central sales department, he was able to pass on some of his ample overseas experience to the young people who wanted to spend part of their career abroad. Thomsen himself knew how necessary it is for the central office to have a section that occupies itself precisely with the problems of foreign assignments.

Colombia: Resources on the Equator

For many years, my visits to Colombia were confined to the medium-sized industrial and agricultural center of Cali. Surrounded by tropical vegetation, the city is built in the Spanish *cuadra* (block) style. It offered few attractions. The only possible hotel for foreigners was the Alferez Real, a humble establishment that did not invite one to linger.

Cali had little industry—only paper, metalworking, and a few chemicals. With its 2 million inhabitants, Bogotá, the seat of government, economic authorities, and important industries, would have been much more attractive. So would Medellín, the center of the textile industry. Why then had Hoechst chosen the provincial town of Cali as its headquarters?

The story is as follows. Two employees of the former I. G. Farbenindustrie agency—the textile merchant Günther Seyd, who had trained with the I.G. in the Grüneburg, and his colleague J. Klötzner from the colorants sector—were interned in Cali during the war. Immediately afterward, they founded a medium-sized company in Cali. Under the name of Inquinal Ltd., it produced auxiliaries for the well-developed textile industry of the country. In addition, it manufactured a number of chemotechnical products. To round off the business, the two men then established an import business for dyestuffs and chemicals, which developed into a base for Hoechst. As early as 1948, we concluded an agency agreement. In 1958, after many development stages, the operation became Hoechst Colombiana. That company built a number of manufacturing plants for pharmaceuticals, agrochemicals (an important field in Colombia, where more than 40 percent of the inhabitants work in agriculture), dispersions, and pigment preparations. This took place in close cooperation with Inquinal, our friends' company, which provided help and also became an important customer.

In Colombia all roads lead to Bogotá, not to Cali. I have always been delighted when, after hours of waiting, I have been able to board a rickety, crowded Avianca Super Constellation for the two-hour flight from Cali to Bogotá. The land route was too risky in the early fifties. The *violencia* ruled wide areas of the interior. Every day there were reports of new attacks by this originally politically motivated guerilla struggle.

Reasonably stable conditions have existed in Colombia only since the beginning of the sixties. At that time, traditional major parties, the Liberals and the Conservatives, formed a coalition. It lasted for sixteen years. The two parties took turns at filling the presidency, each turn lasting for four years.

Economically, much was done during this period to open the country, which is so blessed with mineral wealth and hydroelectric power. But land reform, which was urgently needed, did not get much beyond some modest beginnings. Corruption and smuggling have continued unchecked. In 1974, the pact between the two parties was dissolved. This introduced unrest into the political scene once more. Although newly won raw materials and oil and gas production have greatly improved the standard of living, the country has not been spared a high inflation rate that was the result of the state's growth policy.

Dogmatism of the Andes Pact

Colombia became a member of the Andes Pact, signed in 1969. It appears as though Colombia intends to apply Article 24 of this pact, which often frightens off foreign investment by its dogmatism. This and the state's growing investment control will in no way accelerate an economic upswing. The development of the chemical industry will also be impeded by these measures. This is regrettable because, quite apart from the base materials, there are already some remarkable manufacturing facilities for man-made fibers, plastics, and fertilizers. It is likely that Colombia's gross national product will continue to rise faster than that of its neighbors.

Today, Cali is no longer a sleepy provincial town. But the actual power center is still Bogotá. For that reason, Hoechst decided to move its administrative headquarters to the capital. The new building, conveniently near the airport for the hurried visitor, was inaugurated in June 1978. The production plants will remain in Cali and will, I hope, continue to develop in step with the general growth of industry and agriculture.

Is God a Brazilian?

In 1951, when I was still working in the Hoechst staff departments, I received marching orders that immediately awakened in me the most attractive expectations. I was to fly to Rio de Janeiro together with Dr. Engelbertz, a white-haired giant who represented the production plants on the Hoechst board of management. It was time, I was told, to have a look at the business situation in Brazil, where half the inhabitants of South America live.

At that time, more than twenty exhausting hours of flight with three intermediate stops separated Frankfurt and Rio, the dream city of the world. What a difference from my flight in 1977, when the Concorde took us from Paris to Rio in precisely five and a half hours.

Our job in 1952 was to prepare a recommendation to the Hoechst board on whether, and how, Hoechst should become active in Brazil. We could tackle this question only because, even at that time, we had a close relationship with Willy Kurz and his company. Kurz, like his partner Melchior Müller, came from the old I.G. Both had lived in Brazil since 1923. Since they had detached themselves from I.G. and become independent at the beginning of the war, they had amassed a sizable fortune. Nevertheless, it was a matter of honor for both of them to represent one of the I.G. successors. Pontosan S.A. had been founded in 1949. Hoechst took it over step by step until, in 1957, it became Hoechst do Brasil. Günter Maulshagen later transferred our center of activities to São Paulo.

Then as now, Rio was of breathtaking beauty. But the infrastructure of the giant city, which had grown too quickly, showed many defects. Traffic was chaotic and consisted mostly of large cars made in the United States. One of the standard jokes of the time was this: *Question:* What is the difference between a rich and a poor Brazilian? *Answer:* The poor Brazilian has to wash his Cadillac himself. In succeeding years, the Cadillacs were largely replaced by Volkswagen Beetles, which conquered almost two-thirds of the market.

Even at that time, I was greatly moved by the poverty of the *favelas,* those shantytowns, often in close proximity to the villas of the rich, with huts nailed together from boards, corrugated iron, and tar paper. They were a depressing attempt to house the continuous flow of penniless agricultural workers, mainly from Brazil's Northeast.

Black Orpheus, an Italian-French film that gave Marcel Camus fame as director some fifteen years ago, is set against a background of the *favela.* The film also illustrates another aspect of these simple people's lives: the *macumba,* that secret combination of ancient African rites and Catholic liturgy. Indians, Africans, and Portuguese in this Latin American melting pot have merged into a group that includes many intelligent, diligent, and good-natured people who are not given to violent reactions, who are lively, but with, at times, a touch of melancholy.

I have pleasant recollections of a trip to Teresopolis in the mountains near Rio. This town owes its existence to Pedro II, Brazil's last emperor, whose desire to abolish slavery led to his abdication in 1889. Pedro built Teresopolis more than a hundred years ago for his wife. For himself, he erected a summer residence at Petropolis not far away. Both locations have gradually developed into exclusive living and recreation centers. What touches a German like me particularly in Petropolos is that here, at Gonçalves Dias 34, in a house on a quiet street with flowering gardens, the writer Stefan Zweig and his second wife committed suicide in February 1942. The Brazilians obviously knew whom they were sheltering, because they gave Stefan Zweig a state funeral. Thousands of mourners, many of them ordinary folk, lined the streets through which six officers carried his coffin.

The counterpart to Petropolis is Teresopolis. In this town there is now a research institute of Behringwerke which works on the development of a vaccine against the Chagas disease, so widely encountered in certain parts of Brazil. This disease is named after the Brazilian Dr. Carlos Chagas. Its carriers are flagellates called *Trypanosoma cruzi.* The "cruzi" commemorates Oswaldo Cruz, who waged war on malaria and yellow fever at the turn of the century.

The first medical research institute in Brazil with whom Behring wished to cooperate carries the name of Oswaldo Cruz. Chagas disease manifests itself in fever, swellings, inflammation of liver and lymph glands, and heart damage. Behring's small team in Teresopolis has already been trying out a suitable vaccine in animal experiments, and is confident that it will soon be able to provide Brazilian doctors with a live vaccine to conquer this disease.

Not Starting from Zero

Back to our first excursion in 1952. Even then a certain amount of industry had already been established. The first large steel works had been erected in Volta Redonda. Construction in the industrial triangle of Rio–São Paulo–Santos was proceeding apace. Petrochemical projects to complement the Santos refinery were being discussed.

Dr. Engelbertz and I found it difficult to recommend a start from zero. The following year Hoechst therefore acquired an interest in Fongra, which had settled in Suzano not far from São Paulo. Hoechst took over the interest of the Brazilian cofounder, Fontoura. The other partner was the American company W. R. Grace. Our partnership with Grace turned out to be unsatisfactory, and we dissolved it after some years. Today, the plant at Suzano, which produces base chemicals, dyestuffs, pigments, organics, and pharmaceuticals, is the largest installation of Hoechst do Brazil.

Further manufacturing facilities in the country have in the meantime partly replaced and partly supplemented the increasingly difficult import business. In 1970, Hoechst entered the man-made fiber business. We acquired a majority holding in Cia. Brasileira de Sintéticos (CBS), which was greatly expanded. Two years later we acquired an interest in the Oxford paint factory. This, too, was soon expanded.

The happy times when entrepreneurs in Brazil could hardly fail to get rich overnight ended decades ago. Brazil is a market with intense competition from all directions. The large industrial companies of the world are represented and manufacture at least part of their products in the country. After World War II, starting from the core of the Brazilian textile industry, other processing industries have also been established. They were eventually joined by a substantial growth of basic industry. Brazil's political conditions, especially its attempt to establish a democracy on the Western multiparty model, created conditions that might have become explosive. Therefore the military intervened in 1964 and established a government of army officers. This brought some reality to *Ordem e Progresso* (Order and Progress), the motto the country chose when it obtained its independence 150 years ago.

Much has been written about the Brazilian military regime, often very negatively and also objectively wrong. I must say that the picture of Latin America painted by the European media often strikes me as greatly distorted. Like every authoritarian system, the Brazilian administration too has many weak spots. Compared with dictatorships of different origin, however, it has the advantage that at least the freedom of both the press and the individual has been largely maintained.

The military regime which, after the presidencies of Castello Branco, Arturo da Costa e Silva, Emilio Garrastazu Medici, and General Ernesto Geisel, now followed by General Figueiredo, has introduced a two-party system, giving the population an opportunity, in free and secret elections, to report agreement or rejection of the regime. Ernesto Geisel had been the head of the state-owned oil company Petrobras and had given it a new lease on life. As the country's president, he was seriously concerned with bringing about some political liberalization. He and his successors deserve our best wishes in this endeavor.

The second phase of Brazilian economic progress, the years between

1965 and 1973, in which the gross national product increased by 8 percent annually, was sharply interrupted by the oil-price shock in 1974. Nationalistic motives had led Brazil at an early date to declare oil a national patrimony and to forbid all foreign participation in oil production.

Brazilian Oil

Necessarily, this policy led to a slower development of this resource. Even today, Brazil still has to import 80 percent of the oil it needs. Its natural gas deposits are better utilized, so that the two products together can supply some 50 percent of the energy the country requires. The rest is supplied by hydraulic power. This, however, does not alter the fact that when the price of imported oil increased four- and five-fold overnight, it completely upset the balance of payments, which had just been achieved with a great deal of effort. There also was substantial inflation because the increase in oil prices meant that imports also became more expensive. The inflation rate that had been reduced to less than 20 percent now climbed to twice that figure. Today, inflation remains the biggest problem of the Brazilian economy.

The government has conceived a new economic plan that will promote agriculture more than in past years. But above all, structural changes are to be carried out so as to reduce dependence upon foreign oil. To foster production and speed up exploitation, further contracts have been signed with foreign companies, especially for offshore exploration.

The classic Brazilian export continues to be coffee. Apart from this, much effort is being directed to utilizing other raw materials, especially iron ore, which is available in vast quantities in processed products for export.

Wage development is indexed and most prices are subject to state control. The prime aim is to reduce the inflation to a tolerable level. Every effort is made to raise the real income of the people at large. Also, the middle class, which gradually came into being over the last twenty years, is to have an opportunity to develop further and thus become a stabilizing factor.

Under such difficult circumstances, it was not surprising that foreign companies were obliged to put up with many obstacles. We, too, had to do so in our enterprises, especially with regard to profitability.

Foreign capital in Brazil continues to be both welcome and necessary. But there is pressure to ensure that this capital becomes active in partnership with Brazilian capital. This is not always simple. One's partners often have very different ideas of the time when profits should be realized. In addition, capital in Brazil is in short supply, and the financial requirements of foreign companies must often be met from hard-currency credits. This is an unreliable business, since the currency is continually being adjusted to the dollar, that is to say, permanently devalued.

Effective rate-of-exchange protection is not possible. Furthermore, Brazil discriminates against foreign companies by not allowing those with a foreign capital majority to use domestic capital for long-term credits.

Among the German companies in Brazil, Volkswagen has the largest direct investment and has achieved considerable success. Volkswagen does what has become increasingly necessary for all foreign investors: it exports a significant part of its output.

Another example is Mannesmann do Brazil, which operates by far the most important pipe manufacture in the country. Some years ago, it got into the limelight because of some financial maneuvers by the local management. Even in German economic circles, there was doubt that the company would be able to get itself out of those difficulties. But, due to the patience, persistence, and optimism of Dr. Egon Overbeck, the German head of the company, the scandal was resolved without too much noise, without political complications, and above all, to the satisfaction of all concerned. Mannesmann do Brazil is flourishing once again and has enlarged its capacities.

Regaining the Trademarks

Our sister companies were no less successful and active than Hoechst. We are in intense but successful competition with domestic as well as foreign companies. For the pharmaceutical industry, the situation was difficult from the beginning. During the war, all German trademarks were confiscated. There probably was no other country in which it was as difficult and troublesome to regain these trademarks as Bayer and Hoechst found it to be in Brazil.

State intervention and the enlistment of prominent lawyers were without success. Eventually, Bayer's former pharmaceutical manager, J. K. Ahrens, who had meanwhile joined our company, turned to one of his old friends from the Ministry of Health, the lawyer Saboja Lima. With his aid we finally succeeded in recovering the trademarks without undue expense. When we asked Saboja Lima whether we could do him a favor, he let me know that he would be very pleased if a mass were to be held on his behalf in Cologne Cathedral. This was arranged. We were all deeply impressed by this man's attitude and his friendship.

A Bundle of Worries

We have rebuilt a large pharmaceutical business in Hoechst do Brazil, but it has remained a problem child. The country's tremendous expanse demands a large organization. Naturally, the pharmaceuticals have to be manufactured in the country, as well as the essential raw materials. All these were conditions we could fulfill and had learned to regard as normal. But it was difficult to get the state price-control authorities to agree to adjustments of pharmaceutical prices, although the rate of inflation sometimes reached 90 percent. This subject is political dynamite in Brazil. Furthermore, most of the pharmaceutical companies are in foreign hands, a fact that at times led to some fairly unpleasant press campaigns. Such campaigns hardly induce the authorities to show any indulgence for the approval of reasonable prices.

We are still facing this problem, and it is linked to others, such as the transfer of royalties and the recognition of patent rights. We know that in this respect we still have difficult times ahead of us. But we are confident that the famous Brazilian realism will gain the upper hand.

"So Deus quem sabe" (Only God knows) is the Brazilian reply to a question for which a proper answer is not readily at hand. I hope for the sake of Brazil and its industrious and intelligent people that the saying, *"Deus e Brasileiro"* (God is a Brazilian), will prove to be true.

Japan's Avant-Garde: Vegetables and Fruit

It is interesting to note the extent to which the Japanese have become engaged in Brazil. The beginning of this involvement was quite a simple one. Japanese farmers were invited to Brazil in order to start vegetable and fruit cultivation in the vicinity of the large cities, above all near São Paulo and the new capital of Brasilia, which is about 1000 kilometers away in the interior. In a few years, this project brought a fundamental change to the eating habits of the Brazilians. But Japanese activities have not been confined to agriculture. Major Japanese industries followed and now play an important role in Brazil in all the the fields in which they lead. They include petrochemicals, electronics, and many other items in which Japan is prominent as a regular supplier not only in Southeast Asia but also in Latin America.

As far as Hoechst is concerned, it should be mentioned that, as the result of the takeover of Foster Grant in the United States, we also acquired an interest in Foster Grant's Brazilian project, Estireno do Nordeste. A capacity for styrene monomer, to be followed by one for polystyrene, is being erected in the new petrochemical center of Bahía.

Foster Grant has a one-third share in this project and has provided the technology. At the same time, Hoechst has led long negotiations with Petroquisa, the petrochemical subsidiary of the state-owned oil company Petrobras. These discussions are to establish whether, and under what conditions, Hoechst technology for the polyolefins—in this case low-pressure polyethylene—can be contributed. We made this contribution conditional on our obtaining its selling rights for Hoechst do Brazil, and we prevailed after long negotiations. When these plants go into operation in the early eighties, Hoechst will also be a factor in the plastics business of Brazil.

Carnival under Sugarloaf Mountain

When I was in Rio in May 1977, our arrival coincided with the Carnival weekend. We therefore spent Saturday and Sunday with friends outside Rio on the paradisiacally beautiful coastal strip between Rio and Santos. However, we returned in good time to witness the unique spectacle of the samba dance schools. Contrary to what usually happens when Rio celebrates its carnival, it was not raining. We sat with tens of thousands of other spectators on the bleachers that lined a broad street. Between late afternoon and early next morning, nine samba schools presented themselves. Some of them had as many as 2000 participants. While they paraded in samba rhythm, the groups reached a kind of ecstasy in which rhythm, music, atmosphere, and some alcohol were decisive factors.

Everything is on show, from the magnificently colored and precious glittering costumes worthy of an ice revue to the barest mini-bikini. Close rapport between the dancers and the spectators who follow the rhythm is quickly established. It is probably more comfortable to watch this giant show, from which some of our group returned to the hotel at 8 o'clock in the morning, on television. But one would be missing the really electric atmosphere that makes such an evening unforgettable.

Until 1963 Hoechst's main office was in Rio, but it was never effectively equipped. Sarsa, the Brazilian Roussel-Uclaf subsidiary, has retained its office in Rio, where it also manufactures its pharmaceuticals in a modern plant. Roussel has succeeded in advancing into the front line of Brazil's pharmaceutical producers with preparations that were carefully designed for the needs of the country. In spite of all the difficulties, Roussel is fighting hard to maintain this position. This will mean additional investments and research and basic manufacture. In all these respects, we cooperate closely and successfully with Roussel.

The new capital of Brasilia, which has already been mentioned, is a major hit. It needed immense faith in the nation's unlimited possibilities not only to tackle such a project but also to bring it to a successful conclusion. The croaking that was heard at the time throughout the world and in the country itself has long since died away. Anyone who takes in the inspiration and beauty of modern architecture as created in Brasilia by the German-Brazilian Oscar Niemeyer, starting from scratch, so to speak, has to admit that a truly courageous step into the future has been taken here. The subsequent governments of the country have consistently followed a policy of transferring administrative centers to Brasilia, a policy that has greatly promoted aviation.

No European visitor to São Paulo should omit a visit to the snake farm of the Butantan Institute. Professor Buckel presented me with the dried tail end of a rattlesnake. If it is carried in one's purse, says the legend, it will ensure that the purse is always full of money. Unfortunately, I have not been able to check this out. My purse with its rattlesnake skin somehow got lost the next day on the Santos beach. Since then I have carried paper money in a money clip.

Argentina: Once the Richest Country of Latin America

One of the largest concerns of Argentina, Bunger & Born, started as a family undertaking of Dutch-German origin. The company's founders came to the Argentine three generations ago. They began in the grain trade. This activity was later augmented by flour-milling plants, a field in which Bunger & Born became the leaders. Later, the company diversified. Textile plants, paint factories, and finally, chemical plants enlarged the operation. Its chemical interests were concentrated in Companía Química, which gained a leading position in the early post-war years. In petrochemistry, too, it was evaluating joint ventures.

In 1955, when I visited the senior chief of this company in Buenos Aires, he invited me to lunch at his home after our business discussions. His house was at the edge of the city, with a beautiful view of the wide plain over which Buenos Aires is spread. In contrast with the company's sober, unpretentious offices, Jorge Born's house was modern and grand. With its valuable French paintings, it offered a stylish background for the head of the concern who gave the impression of a *grand seigneur* but is personally a very sober and modest man.

Even at that time, Hoechst for some years had had commercial links with Companía Qúimica. Jorge Born and his chemical adviser Günther Lewinsky were looking for even closer links that might extend into the other countries of Latin America in which the B & B Group was active. That ambition of this far-sighted businessman could not materialize. Both sides had already mapped out their positions in too much detail. We did, however, establish close collaboration in Argentina. In 1957, we founded Química Hoechst, in which Bunger & Born had a 50 percent interest. The company's co-managing director, Gustavo Nachmann, stayed on as co-manager even after the partnership was dissolved in 1968. We continued to be partners in Anilsud, a dyestuffs factory founded at the beginning of the sixties by Bayer and Hoechst together with Companía Química.

Perón and the Consequences

I came to know Argentina in the final phase of the Perón regime. The trade unions, Perón's base of power, had gradually become the rulers of the state. The dictator granted them ever new concessions, far exceeding all rises in productivity. His beautiful wife, Evita Perón, was the guardian angel for the poorer sections of the population. She was revered by large masses of the poor even after her early death. Unfortunately, this aspect of Peronism exploited the confidence and love of simple people who did not understand the actual conditions. In addition, as always with such a regime, there were those who did not want to see the real facts. After the war, this rich agricultural country, with its 25 million inhabitants, was one of the wealthiest Latin American countries. But all its riches were wasted and gambled away in the Perón years. When Perón finally had to go into exile in Madrid in 1955, he left behind him an impoverished economy and the beginning of an industry that was viable only behind high protective-tariff walls. Politically, there were no forces that could have turned Argentina into a democracy.

It was a very hectic period. Between 1955 and 1974, Argentina underwent no fewer than ten changes of government. The military governments, excepting that of General Ongania, did not reap much political and economic fortune. They were unable to bring the country back into the world economy, and internally, they did not try to establish a more stable structure of society.

Meatless in the Land of Milk and Honey

Nevertheless, during that period of political instability and continuous changes in economic policy, Argentina continued to develop. New processing plants were established. Agriculture had degenerated to such an extent that a country that traditionally exported large quantities of meat had to decree one meatless day per week.

In spite of all the political and economic problems, the Hoechst group made good progress. In 1969, we acquired the majority in Indur, a synthetic resin factory. The connection had been established by Chemische Werke Albert, the Hoechst subsidiary that had granted the production license.

In 1971, our relationships with the Villa Aufricht company had progressed so well that we were able to acquire a majority interest in this family company. Today it operates under the name of Sudamfos. Here, Dr. Hirsch and his partners with great courage and ingenuity undertook the manufacture of phosphoric acid and phosphorous derivatives, as well as nonprescription drugs and hospital supplies. The phosphorous component fitted well into the Knapsack program and supplemented our industrial program in Argentina.

To introduce or produce Hoechst pharmaceuticals was not easy in Argentina. Química Bayer, which had been selling the Hoechst products, had been taken over by the state, which operated it in a mediocre manner.

For a long time we negotiated with different partners. Eventually Dr. Carlos Piscione, the last trustee-manager of Química Bayer, came to the conclusion that it would be better for business and employment to link up once more with the former owners. In this way, when the former Bayer property was divided in 1958, we were able to regain both the trademarks and the Instituto Behring in San Isidro. This is a suburb of Buenos Aires where we then established a new pharmaceutical plant. The grounds on which Behring had kept horses for serum were large enough to put up our Ozalid plant also. Following our tested principles, the pharmaceutical area was also incorporated into Química Hoechst.

Apart from the basic chemicals industry, a petrochemicals industry developed in Argentina, but only hesitantly. Except for a few minor oil drillings, its only raw material basis was natural gas, which had to be brought from the remote south of the country but was, at the beginning, adequate for the relatively modest capacities. Recently, however, further projects for petrochemical plants have been considered. There is every prospect that they will be realized in the coming years, especially now that Argentina is meeting some 95 percent of its energy requirements increasingly through hydroelectric and atomic power stations, so that it can utilize more of its oil reserves for petrochemistry.

Between Left-Wing and Right-Wing Extremism

At the beginning of the seventies, political conditions became more and more chaotic. People began to call for Perón's return from Spanish exile. Although outsiders could at first not understand this recall at all, Perón did indeed return in 1973. Old and sick as he was, he took over the government. Recalling the great support that Evita had given him, he appointed his young second wife, Isabelita, as his deputy.

After Perón's death in 1974, Isabelita became President. Perón had not been able to solve the country's political and economic problems, but complete turmoil ensued under Isabelita's government. Left-wing extremists and right-wing extreme syndicalists fought one another, and bloody assassinations and kidnappings were the order of the day. Terrorism became a common phenomenon.

There was general insecurity in the country. The leading people of Química Hoechst had every reason to feel threatened, too. For months, like many other people in industry, they moved about only in the company of bodyguards. Eventually, an end was put to the unbearable situation. The military, who, after the luckless regimes of Generals Ongania and Lanusse, at first had little inclination to assume power once more, eventually felt compelled to do so.

In the meantime, the myth of Peronism had largely disintegrated—

one of the prerequisites for a return to normal conditions. General Videla, the chief of the new military government, avoided all draconian measures. He combatted internal insecurity and the abysmal corruption of public administration with determination, but without exaggerated publicity maneuvers. The government countered the inflation, which had reached unbearable heights, with the classic measures of economic retrenchment. At times this approach led to liquidity difficulties in both industry and commerce. But the beneficial effects of this realistic economic policy can gradually be seen. Although the inflation rate is still high, it has been brought under greater control.

The economy of the country is on the road to recovery. Argentina can draw on its inner reserves, especially in grain cultivation and animal husbandry. But both basic and processing industries, whose growing share in the gross national product already amounts to 37 percent, are being expanded. This is accelerated by the fact that foreign companies are showing renewed willingness to invest in the country.

For some years, Argentina looked with some envy at Brazil, whose economy took a major upturn in a relatively short time. After the most recent developments, however, Argentina no longer has much reason for such envy. The country offers all the necessary conditions for becoming a strong and politically stable member of the Latin American community. Its per capita income is the second highest in South America, and substantial export excesses and foreign currency reserves are being accumulated. The confidence of the outside world has also grown greatly, as evidenced not only by the investment projects but also by the speed with which the recent state loans were subscribed to.

The Argentine people have had to go through many hardships. It is my belief that they will now devote their energies to revitalizing their economy and to establishing a sound position for themselves among the countries of Latin America.

La Petite Différence

Among the culinary attractions of Buenos Aires there is the famous restaurant La Cabaña. Its speciality is baby beef, a steak that spreads beyond one's plate and tastes delicious. To start with, there is a consommé from fresh meat. As I mentioned, rationing was introduced, but there has never been a lack of good-quality mutton or pork.

Among the most sought-after delicacies in Latin America are *criadillas.* I was offered some once at an official dinner for Hoechst visitors. When I asked the lady on my right what I was eating, she answered with perfect diplomacy, "You are being offered the difference between a bull and an ox."

Chile Comes to Grips with Itself

It's Santiago, Plaza de la Constitución, on a sunny Sunday morning in 1955. It is the "changing of the guard" of the police. Their appearance surprises us, not only because of the Prussian-sounding march music but also because of the German-type steel helmets worn by the soldiers. Indeed, we were told, both things are remnants of Chile's military cooperation with Germany in the years between the two World Wars.

Military cooperation between Germany and Chile began in the final decades of the last century. After the nitrate war between Chile and Peru from 1879 to 1883, Chile reorganized its army on the German model with help and instruction from German officers. The parade uniform of the cadet school is still like the uniform of the German Cuirassier Guards, with red and white plumes on their helmets. By the late 1890s, Chile had the best-organized army in South America.

It used to be said that the military played no role in politics, and that there was neither the risk nor the chance of the military intervening in order to sustain the fabric of a middle-class democracy, which was not very stable even at that time. One example of the Chileans' democratic outlook was President Jorge Alessandri, who succeeded Carlos Ibáñez. He went on foot every day from his house to the Government Palace, right across the the center of town and accompanied, if at all, only by a personal friend. Since then, unfortunately, much has changed in Chile.

Hoechst started its postwar activities through the trading house Vorwerk y Cía. This company, from Hamburg, was established in Chile 130 years ago. It was the oldest and for a long time the most important German undertaking in Chile. Its former senior chief, Paul-Joachim Crasemann, received me in his spartan office, still equipped with standing desks. He was the majority owner of a sheep farm of 25,000 hectares in the north (expropriated under President Frei) and a pine plantation of 1500 hectares in the south.

Georg Mosel, who looked after the Hoechst business, came to Chile before the war as a former I.G. trainee. His sister was distantly related to the Crasemann family. But this did not alter the fact that under the tight regime of his chief, Mosel had little flexibility in developing the business as the market demanded. With its more than 10 million white inhabitants, its wealthy agriculture, its textile industry, and its enormous output of minerals, especially copper, Chile had again become an important trading partner for Germany at that time.

In 1955 Química Hoechst Chile was founded. Since then, Chile has become one of our promising markets in Latin America. Hoechst's first pharmaceutical operation in South America was established there in 1957. We had acquired an empty factory building, called "the railway station" because of its arched roof.

When we inaugurated this pharmaceutical plant, the second government of General Carlos Ibañez was in office. Within six years, he changed his ministers more than a hundred times. The minister of health, who had been invited to the inauguration, had only just been appointed and was hardly known. He arrived unattended, and when he entered the room where the festivities were taking place, no one took any notice of him at first. Eventually he introduced himself with the words, "I am the Minister of Health."

The Hoechst organization developed quickly. The textile industry offered a sizable market. Under the management of an able chemist, Dieter Gevert, we were soon producing textile auxiliaries. After we acquired the appropriate domestic plants, our output reached quite some volume.

Since polyester fibers were in great demand, and since domestic production was under tariff protection, we decided to build a Trevira plant, which went on stream in 1968. Our partners were the Deutsche Entwicklungs gesellschaft (DEG, or German Development Corporation) and the Adela Investment Co., a finance institute concerned specifically with the promotion of new industries in South America. Everything therefore seemed to be proceeding well, and the German staff was happy in the beautiful city of Santiago, picturesquely situated at the foot of the Andes. The people were friendly and the Germans were popular; in fact, many German families that came to Chile generations ago still speak German.

This friendship has not suffered from the fact that Chile lost its great nitrate income as the result of the German Haber-Bosch process for producing synthetic nitrogen. Since then, Chile's nitrate trade has been replaced by the copper trade, which more than compensates for it. The export value of nitrate production was about US$40 million, or one-thirtieth the value of the copper trade.

Company Car in No-Man's-Land

Here is an example of the Chileans' very friendly attitude toward foreigners even in times of crisis. Under the government of Ibañez, there were serious street demonstrations in Santiago. It was the first major attempt in Chile, and also in other South American countries including Colombia, of international communism to overthrow weak governments and gain power through terror.

At the time of the unrest, Mr. Bangert was a senior Hoechst official in Chile. He had a great privilege, a company car. It probably was the most valuable asset the company owned. By the time Bangert remembered that he had parked this car outside the Government Palace, the unrest was already at its height. But come what might, Bangert was determined to rescue the car. If he did not, thrifty Crasemann would never forgive him.

Together with Georg Mosel, Bangert got to the government building by a roundabout route. On one side of the street was the angry crowd which was getting ready to storm the building; on the other side was a small but heavily armed guard that had been ordered to defend the palace and had taken up its position. In the middle, in no-man's-land, there stood the company car alone and forlorn. When Bangert and Mosel declared that all they wanted was to retrieve this car, both the crowd and the guard allowed them to pass and rescue their vehicle.

The insurrection was defeated after bloody fighting that lasted several days. Then came the great surprise. The middle-class parties, split among themselves, lost their majority in 1970. Under the leadership of Allende, a socialist minority government was elected with 36 percent of the votes.

Under Allende's regime, the economy deteriorated at a rate that can simply not be imagined. Although he attempted to observe democratic rules, he was systematically undercut by the left wing. As a result, apart from the official nationalization of industry and agriculture, there were continuous illegal occupations of land and factories, which were, however, tolerated by Allende and his government.

The Hoechst factories were occupied—and released—four times. This sequence can be regarded as lucky, since other companies were taken over by the state after the first, or at most the second, occupation. Hoechst profited from the fact that the occupations were badly timed and coincided several times with the debt renegotiations then underway with West Germany. In order not to prejudice the discussions, Allende's government was forced to back off.

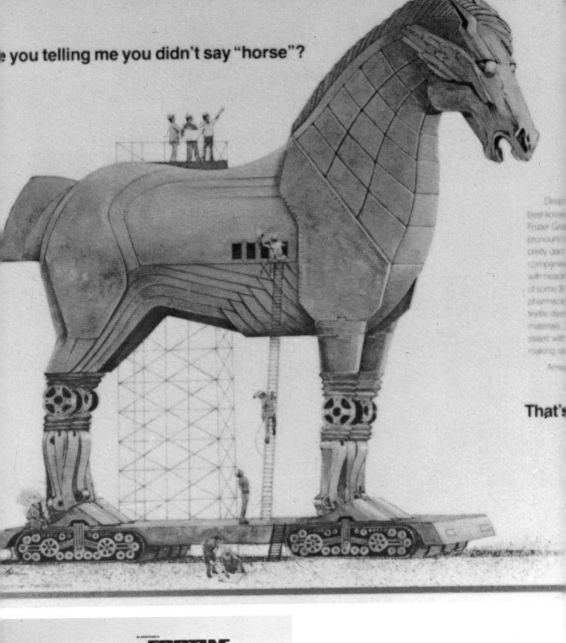

e you telling me you didn't say "horse"?

That's

as advertised in
BusinessWeek and **FORTUNE**

Are you telling me you didn't say "horse"?

That's Hoechst, pronounced success.

Hoechst

Good health is our most valuable asset.
Hoechst helps protect it.

A man protected from disease and epidemics can look forward to a longer, richer life. And contribute more to the welfare of our country.

Hoechst helps protect health in many different ways.

- Through sera and vaccines proven worldwide which immunize the body to all sorts of disease germs.
- Through diagnostic aids which enable the physician to detect diseases, thus paving the way for effective cures.
- Through a vast array of effective medicines to combat acute and chronic diseases.

Working for Hoechst are not only thousands scientists who develop such medicines, but consultants who keep our doctors informed of latest developments. So that they can provide still better protection, better help.

Hoechst thinks ahead.

Nigerian Hoechst Ltd.
P. O. Box 261
Ikeja

Hoechst

Overleaf: The corporate advertising campaign conducted by American Hoechst centers on the problem of pronouncing the company name. "If you say 'Herkst', you come pretty close," according to the ad.

POLYES

Hoechst care is woven into the fabric of Britain

So many things that touch our lives owe something to the care of Hoechst.

Take clothing: Trevira polyester fibre has brought a lot to the world of fashion. From crisp classics and soft knits for women to stylish suits for men. Made from Trevira alone or in blends.

But Trevira not only goes into clothes. For example, it's also used for retractable covers so that you can use stadiums and swimming pools in all weathers. And for household textiles too, like curtains, carpets and upholstery.

Hoechst takes care of you in lots of other ways too. Have a look around you. There's almost certain to be a product associated with Hoechst.

Famous names like Optrex eye care, Corimist hair products, Hostalen plastics and many more all owe something to the care of Hoechst.

Hoechst care also helps farmers to grow cleaner, more productive crops. And Hoechst cares about the future too.

Every day over half a million spent on research for the produ To help make your world a better. In Britain, Hoechst employ people and it has offices, plants throughout the country.

For more information abou say 'Herkst') and what it stands Hoechst, Salisbury Road, Hounsl phone 01-570 7712 ext. 3169.

در مورد مشکلات خود با متخصصین
هوخست گفتگو فرمائید.

شما میتوانید مشکلات خود را با همکاران
متخصص هوخست در میان بگذارید و از
دانشها و تجارب دانشمندان بر جسته
بهره گیرید. آنان دانش خود را در مورد
مواد شیمیائی معدنی ، تولیدات فنی ،
مواد محرك لاسول ، مواد پروتئین زا ،
مواد شیمیائی آلی ،مشتقات بازل و خوشبو
کننده ، کود شیمیائی ، سموم دفع آفات ،
مواد علوفه برای علوفه، رنگ، رنگکنهای
آلی ، بهرختت و مواد تعاونی برای
تکمیل محصولات نساجی ،شامه ملولری،
نخ و الیاف مصنوعی ، ردین مصنوعی ، مواد
اولیه رنگکاری ، مواد پلاستیک، واکس، فیلم
وفوایذ طراحی دارو، مصرف داروئی،
مقادیر بالارجواب سوال

هوخست در زمینه های ذیل فعالیت
دارد :

شرکت سهامی خاص ایران هوخست
خ... تهران

วิชาการด้านเทคนิค
จำเป็นสำหรับอุตสาหกรรมสิ่งทอ
เฮ็กซ์อยู่ที่นี่เพื่อช่วย

เฮ็กซ์ ค้นคว้านำหน้าเสมอ

เฮ็กซ์

Care of Hoechst

Hoechst

Cuando sea mayor
seré médico y te curaré

Ojalá sea así,
pero ¿tendrá los medicamentos adecuados?

Hoechst, investigación responsable.

Hoechst

'n Ramp als in '53 mag nooit meer gebeuren.

Sinds eeuwen moeten de lage landen zich met zware dijken tegen de zee beschermen. Bij de stormramp van 1953 bleken de maatregelen onvoldoende. En nu nog, als de wind tot orkaankracht aanwakkert, denken we aan dat jaar terug. Zoiets mag nooit meer gebeuren.

We hebben gelukkig niet stilgestaan. Ook Hoechst heeft 'n bijdrage geleverd aan de zekerheid van vandaag. Hostalen® -kunststofpennen die mijlenlen betonblokken voor dijkfundamenten op hun plaats houden; dijkversteviging van Trevira® spunbond. Ook was het een idee van Hoechst om restproduktenvan de fabriek in Vlissingen te benutten voor o.a. de onderbouw van de dijken. De hiervoor gebruikte overslakken zijn erg goedkoop, snel ter plaatse en milieutechnisch gezien is het tevens een uitstekende bestemming.

Het zijn resultaten van samenwerking tussen waterbouwkundigen, chemici en techniici. Zij kunnen niets veranderen aan de immense krachten van wind en water. Wel hebben ze voor de mensen achter de dijken het leven veiliger gemaakt.

Wilt u meer over ons weten, vraag dan om toezending van de interessante brochure "Hoechst denkt vooruit".

Hoechst Holland N.V.
Afd. V.x-richting, Postbus 284, Amsterdam

Hoechst

بار محصول خود را افزایش دهید.
هوخست بشما کمک خواهد کرد.

هوخست بآینده می‌اندیشد

هوخست

Hoechst Singapore...
yes, I connect you with
a specialist in this field.

Because if there is a better, faster, cheaper way to do the job, one of our specialists will find it for you.

Hoechst has experts for:
Chemicals, aerosol propellents, refrigerants, radiochemicals, fertilizers, pesticides, feedstuff additives, dyes, surfactants, textile finishing agents, synthetic filaments and fibres, synthetic resin dispersions, synthetic resins, paint raw materials, plastics, waxes, films and sheeting, reprographic equipment, pharmaceuticals, veterinary preparations, cryogenic installations, welding equipment, design and construction of chemical plant.

Hoechst Singapore Pte. Ltd.
Soon Wing Industrial Building
2, Soon Wing Road
P.O. Box 89
Singapore 13
Tel.: 283 26 22 (8 lines)

Hoechst

Centralvarme
fra verdensrummet

Den kendsgerning, at olie-, gas- og kulreserverne ikke er uudtømmelige, har tvet videnskabsmænd og teknikere til at rette søgelyset mod jordens største energikilde: solen.

Overraskende hurtigt er der i løbet af de senere år udviklet systemer, der kan udnytte solvarme til boligopvarmning, de såkaldte solfang af forskellige typer. Forskere fra Hoechst har med succes bidraget til denne nye teknik. Ved at udvikle helt specielle plastmaterialer har de skabt forudsætningerne for særlig rationelle konstruktioner.

Varmepumpen er et andet system til udnyttelse af de eksisterende energireserver. Den omsætter grundvandets eller jordens varme til centralvarme. Også her har Hoechst ydet et afgørende bidrag - nemlig Frigen®.

Når de to opvarmningssystemer kombineres, kan et hus nærsten gøres uafhængigt af andre former for energi.

Hoechst Danmark A/S
Islevdalvej 110, 2610 Rødovre
Telefon (02) 91 26 22

Hoechst

Wird die Karies besiegt sein, wenn diese Kinder in die Schule kommen?

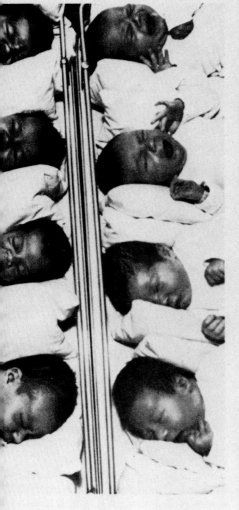

Die Volksseuche Karies kostet die österreichischen Krankenkassen jedes Jahr Milliarden Schilling. Ganz davon zu schweigen, was sie an Angst und Schmerzen kostet.

Forscher der mit Hoechst verbundenen Behringwerke sind dabei, einen Impfstoff gegen Karies zu entwickeln. Er soll Kinderzähne, die noch nicht von der Zahnfäule befallen sind (im Vorschulalter), immun machen.

Der Impfstoff gegen Karies wäre ein weiteres Beispiel für den vorbeugenden Schutz vor Krankheiten. Durch Impfungen konnten so gefürchtete Krankheiten wie Diphtherie, Keuchhusten, Wundstarrkrampf und Kinderlähmung unter Kontrolle gebracht werden.

Forscher der Behringwerke leisteten zu dieser Entwicklung wichtige Beiträge.

Hoechst denkt weiter.

Hoechst Austria AG
Altmannsdorfer Str. 104
1120 Wien

Hoe

dicono che tuo figlio vedrà il duemila in buona compagnia: 6 miliardi di uomini

Nel 2000 saremo in 6 miliardi. Scoraggiarsi non serve. Sono problemi che vanno affrontati con impegno serio!

Nei prossimi 25 anni la produzione di alimenti dovrà essere triplicata per il fabbisogno dell'umanità.

Per vivere in questa dimensione "duemila" occorre una operosa immaginazione.

Nuove tecnologie, nuovi piani di sviluppo: dalla salute alla casa, dall'alimentazione all'abbigliamento.

Con questo spirito lavora la Hoechst con i suoi 179.000 collaboratori in oltre 120 paesi.

Esperti Hoechst aiutano l'agricoltore ad aumentare i raccolti, a difendere il patrimonio zootecnico dai parassiti. Affiancano il medico per proteggere la salute, curare le malattie, prolungare la vita. Studiano come vestirci con più funzionalità ed eleganza. Progettano materiali da costruzione per una più vasta e moderna abitabilità.

In sintesi, grazie alla Hoechst, la speranza si è fatta ricerca.

Hoechst soluzioni per l'uomo.

Hoechst

伝統からの出発……❹

温故知新

父はライバル

未来こそヘキストの目標です

Hoechst
ヘキスト ジャパン株式会社
本社　東京都港区赤坂8-10-16　〒107

The Ni...

Proteja Ud. sus plantas
contra enfermedades y parásitos.
Hoechst ayuda.

Gran parte de su cosecha
es destruida cada año por
enfermedades y parásitos. Ya en
la próxima cosecha podría
Ud. limitar estas pérdidas. Con
productos fitosanitarios de
Hoechst.

Hoechst ofrece un amplio
surtido de productos
fitosanitarios de gran eficacia,
ecológicamente sin objeción y
acertados para cada caso. Así
hay medios pulverizadores
específicos para determinadas
enfermedades de plantas ó
determinados parásitos. Con
baja dosificación dan un
máximo de protección.

Hoechst es una de las
empresas químicas piloto. Miles
de científicos trabajan en el
desarrollo de los productos.
Para asegurar su empleo
apropiado están los asesores
de Hoechst. Pregunte a nuestro
asesor ¿que debe hacer para
cosechar más y mejor?

Hoechst Elteco S.A.
Casilla 1468
Quito
Tel. 23 90 93

Hoechst

Nossa telefon
conseguiu

A modern textile industry makes our country more independent.

Hoechst helps to build it.

As living standards rise, there's a rising demand for better, more modern clothing. That's why textiles rank among our major industries. Hoechst is making an important contribution toward expansion of the textile industry.

As one of the leading manufacturers of dyes and synthetic fibres, Hoechst supplies a complete line of products for the textile industry: dyes, fibres and textile auxiliaries in a coordinated, integrated system.

In addition to products and technical aids, Hoechst supplies comprehensive know-how through experienced Hoechst experts. Hoechst experts who are intimately familiar with the specific problems facing our country because Nigeria is their home.

Hoechst thinks ahead.

Nigerian Hoechst Ltd.
P.O. Box 261
Ikeja

Hoechst

April 19, 1978

m gênio: em apenas 15 dias, ela reitinho o nome da empresa.

"Hoechst do Brasil, bom dia". Essa é a nossa telefonista, uma das poucas pessoas que conseguem falar direitinho o nosso nome.

Peça para ela ensinar você: enquanto isso, vamos falar um pouco sobre nós mesmos. A Hoechst é uma das maiores indústrias químicas e farmacêuticas do mundo, com fábricas e laboratórios em mais de 120 países.

Inclusive aqui no Brasil, onde a Hoechst tem 3 fábricas, 8 escritórios e filiais, e mais quase uma dezena de empresas associadas, que seguem a mesma filosofia da Hoechst mundial: pesquisar, desenvolver e aperfeiçoar produtos e métodos.

E esse trabalho está facilitando a vida de engenheiros, químicos, médicos, veterinários, pecuaristas, agricultores, gráficos, fotógrafos, modistas, milhões e milhões de pessoas.

Agora você já sabe alguma coisa sobre a Hoechst. E mesmo que não consiga falar o nosso nome, não faz mal: nossa telefonista sabe, o que já é mais do que suficiente.

Hoechst do Brasil
Química e Farmacêutica S.A.
Caixa Postal 7333
01000 São Paulo - SP

Hoechst

Vor Kinderlähmung können wir sie schütze
Vor Krebs noch nicht.

Diese Kinder haben ein langes Leber
70 Jahre und mehr. Noch um die Jah
dertwende wurden die Menschen im
schnitt nur 38 Jahre alt.

Mit Arzneimitteln haben die Pharma-F
und Ärzte Geißeln der Menschheit wi
Diphtherie, Pocken und Typhus besie
Doch das Forschen für das Leben m
gehen. Denn trotz aller Erfolge könne
Ärzte bei den meisten Krankheiten m
vorhandenen Medikamenten nur die
Beschwerden lindern.

Um auch diese Leiden heilen zu könn
ist noch viel zäher Forschungswille, a
Geld notwendig.

Pharma-Forscher von Hoechst arbeit
allen Erdteilen. Für die Suche nach n
besseren Arzneimitteln setzt Hoechs
Jahr über 400 Millionen DM ein.
Damit die Zukunft der Kinder sichere
lebenswert ist.

Ihr Kind braucht u Forschung.

Dieses Motiv erhalten Sie kostenlos
als farbigen Poster R 10 407
Hoechst AG, Abt. VFW, 6230 Frankfurt am Main 80

Hoechst

One of a series of corporate advertisements appearing in West Germany. Focusing on the needs of the future, this ad says: "Your child needs our research."

In this fertile country, foodstuffs and everyday consumer goods increased in price day by day. The inflation rate exceeded 1000 percent. The patience of the population was at an end. It is claimed, probably not without reason, that the desperation and anger of the Chilean women contributed much to the revolution. I was told that the wives of the Chilean politicians and the high military resorted to the well-tried Lysistrata method, the separation from bed and table, in order to force their menfolk to intervene in the grotesque political situation.

Economic Policy with Question Marks

In September 1973, a military junta under General Augusto Pinochet took over the government—and also a grave inheritance. Politically, the country was split, and economically, it was facing ruin. Later generations will be in a better position to judge whether there might have been other and possibly better forces for bringing the country's life back to normal. For the start, at any rate, the people were prepared to tolerate the austerity policy which Pinochet's economic advisers prescribed. Since then, industry and agriculture have recovered reasonably well. But whether Chile will be able to maintain the economic policy of Milton Friedman's disciples is open to doubt. Without a minimum of protective measures, it will not be possible to pursue industrialization effectively in so small a market. Already, we have had to close our attractive Trevira factory because it did not enjoy any tariff protection. And this example is by no means an exception.

The new investment law seems to offer hope of attracting foreign capital to the country once more. It was precisely because of the antiforeign investment policy and the high protective tariffs of the Andes Pact countries that Chile withdrew from these agreements more than a year ago, causing a great deal of publicity.

Apart from modified investment legislation, the government will have to create confidence if foreign capital is to be invested in this country, which needs it greatly for the exploitation of its mineral wealth. This, incidentally, includes oil and natural gas. Following the establishment of Chile's two refineries, small initial capacities for a petrochemicals industry were erected. Monomers and plastics are being produced, and there are ambitious plans for future expansion.

The first visible success is the recent investment by Exxon Mineral Co., which bought a copper mine in the vicinity of Santiago for more than US$100 million and is contemplating further investments. Chile seems to be typical of the developing countries and the extraordinary fluctuations in their economic conditions that place investments under great risks. Within a decade, Chile's economic rules were reversed three times. Under President Frei, a private economy was controlled by the state and protected by high tariffs. Under Allende, the pendulum swung over to a totalitarian state economy that excluded private enterprise. And now, the latest military government is trying to promote a free private economy open to foreign investors.

I am confident that the friendly, industrious people of Chile, which is so richly endowed by nature, are well on their way to gaining their share of Latin America's progress and prosperity.

Montevideo

When I first visited Uruguay at the beginning of the fifties, the country had 2.5 million inhabitants and was regarded as one of the most stable and prosperous on the continent. The capital, Montevideo, where half the country's population lives, is situated in the delta of the Rio de la Plata. For my generation, it is associated with Curt Goetz's charming comedy *The House in Montevideo.*

Our beginnings there were somewhat idyllic. Henry Kahl, a respected merchant from Hamburg, had settled in Uruguay and was operating a modest import company with the help of his three daughters. He started to do business with Hoechst, and gradually a regular agency relationship developed. After his early death, Hoechst acquired an interest in the company. Rodolfo Pies, of German origin and married to an attractive Chilean woman, had been helping Henry Kahl, and was now put in charge of the business. Everything proceeded satisfactorily. In 1970, a small pharmaceutical plant that had become necessary was inaugurated.

Compared with the other South American countries, Uruguay's social legislation was progressive and benefited all layers of the population, particularly the important, well-to-do middle class. Because of this environment, and also because Uruguay's population is almost exclusively European in origin, there were no serious social problems.

But even in Montevideo, the world was not to stay as it was. Democratic governments by one or the other major party—the Colorados tended toward socialism, the Blancos were rather conservative—tried to establish the perfect welfare state. Pensions were raised to an unrealistic level, public housing was erected on a lavish scale. Such policies were bound to lead a healthy country with no racial problems not only to the edge of bankruptcy, but also to internal unrest. The Tupamaros—urban guerrillas who committed crimes, allegedly with political motives—reduced Montevideo to an unbelievable state of insecurity. The economy, already burdened by the welfare programs, was further weakened by these circumstances. Inflation and corruption did the rest.

Then the inevitable happened. The military took over in 1972 and later elected Juan María Bordaberry, one of the large landowners, as president and civilian guardian. The government dealt with the Tupamaro terrorists, but commerce remained sluggish, and the rate of inflation remained high. Uruguay, an agrarian country, was no longer able to export enough of its essential commodities: meat, wool, and grain. As in other cases, the protectionist agrarian policy of the European Economic Community (EEC) was not exactly helpful.

There is a certain degree of rivalry between the two big neighbors, Brazil and Argentina, over their buffer state, Uruguay. This should help in the long run to attract investments into such projects as hydroelectric power stations which, in turn, might promote further industrialization. If the country remains orderly, the tourists will no doubt return to the beautiful beaches of Punta del Este, especially the Argentinians who found refuge there in time of crisis.

Guatemala and the Common Market of Central America

In the autumn of 1969 we came from Mexico to inaugurate our new plant in Guatemala City, which has an altitude of 1500 meters. A modern complex for administration and manufacture had been erected there near one of the exit routes from the city. At first sight, the size of the complex might seem excessive in a country with barely 6 million inhabitants. But Guatemala is an important member of the Mercado Común Centroamericano (the Central American common market). Costa Rica, El Salvador, Guatemala, Nicaragua, and for a time Honduras formed a successful customs union and introduced a joint external tariff. It was therefore quite reasonable for Hoechst to operate in those countries, using Guatamala as its center. It does so through a system that allows the individual companies in the various countries their independence and responsibility and permits rational decisions in personnel and technical matters.

Guatemala's President Mendez Montenegro attended the inauguration. In his speech he emphasized that his nation welcomed foreign
investors because they provided new jobs for the rapidly growing
population.

Many different factors are responsible for the strained atmosphere
in the country. Two-thirds of the population are Indians. They play
no part in political life, and only a minor part in economic matters.
There are more sharp social contrasts in Guatemala than in any other
country of Latin America. Within the white élite, made up mainly of
large landowners and the military, there are conflicts whose causes are
not readily comprehensible to the outsider. The former United Fruit
Company played a role that was not exactly fortunate for the United
States. Not only did it pursue monopolistic policies with regard to the
country's coffee and bananas, but it also concluded rather too many
"deals" with the various governments. No doubt, this is now a matter
of the past, at any rate in this form, and it was probably much exaggerated in the media. All the same, the country will have to travel a
long way before it achieves political stability.

Still, the economy is progressing well. Agriculture, which employs
about two-thirds of the population, has had excellent yields in recent
years. Apart from coffee, the major export product, cotton crops and
cattle raising play important roles. The processing industry, especially
for textiles, is increasing in importance. Mineral wealth has so far been
exploited only to a limited extent. Quite recently, a Canadian company has begun to mine nickel on a large scale near Lake Izabal. Initial,
successful drillings near the Belize (formerly British Honduras) border
seem to justify hopes of significant oil reserves.

The small conference to which Hoechst invited the heads of its
offices throughout Central America and the Caribbean provided an
opportunity to exchange ideas concerning the various problems of
these markets. On such occasions, one always meets young people
who are earning their laurels in difficult jobs abroad, and one should
never lose sight of the best of them.

Venezuela: Happiness from Oil?

Venezuela has been generously endowed by nature. It has the largest
oil reserves in Latin America, high-grade iron ore which can be readily
obtained by open-cast mining, bauxite, phosphates, and to top all
these resources, a vast reservoir of hydroelectric power for producing

cheap electricity. However, as was the case with its mother country, Spain, all those riches have not entirely been of benefit to Venezuela.

Just as the gold and silver from her colonies caused the industry of Spain to decay because the required goods could be bought abroad with the precious metals, so are easy riches undermining the initiative of the Venezuelans. The 13 million people in the country, which is about 3½ times the size of West Germany, ought to be doing well. And yet, here too, there is the stark contrast, so typical of Latin America, between rich and poor, although there also is a middle class.

The slums of Caracas, locally known as *ranchos,* extend into the center of the city. Under President Pérez Jiménez in the 1950s, the *ranchos* were replaced by public housing. But apartment houses proved to be no solution to the problem. Mothers have to go out to work but cannot leave their children alone in an apartment. Furthermore, it is impossible to raise chickens or vegetables in these apartments, and that is one more reason why many poor Venezuelans prefer to live in their primitive dwellings. They will acquire a huge American car and other modern machines and appliances before they will think of abandoning their *ranchos.*

The government's slum-clearance policy is extensively supported by state funds. But the large-scale construction of public housing cannot keep pace with the city's increasing population, which already amounts to 3 million people. The general population increase is estimated at more than 2 percent per year. In the cities, especially in Caracas, it is even higher. About 50 percent of the population is born out of wedlock.

When I visited Venezuela in 1977, I was surprised and impressed by the political stability the country had achieved since my previous visit. The guerrilla movement which had greatly complicated life in the cities and in the countryside early in the sixties appeared to be a thing of the past. However, there is said to be a high crime rate in Caracas.

When I visited Caracas for the first time in 1953, the town still presented a typically Spanish colonial aspect—uniform *cuadras* (blocks) of one- and two-story houses with inner countyards and small gardens. But dictator Pérez Jiménez had already begun to transform the city into a modern metropolis. Broad streets in the center were flanked by modern skyscapers for the government authorities.

Today, the central city is a jungle of high-rise buildings, and traffic is chaotic. In 1963, when Hoechst's pharmaceutical plant was inaugurated together with an administrative building in the suburb of La Trinidad, this complex was still on the outskirts of the city. Today, the city has spread beyond this former suburb. Our attractive building in La Trinidad, which had originally been ample for administration, pharmaceutical production, and warehousing, has long since become too small. Although there was enough space for further extensions, it seems to be quite impossible to get a permit to expand—not even just the pharmaceutical plant or just the warehouse—in this quarter, which has become part of the inner city. We must therefore consider moving to the new periphery of Caracas for the further expansion of our company.

It was a long road that led to the 1963 inauguration of our establishment in La Trinidad. In 1952, we had been visited one day at Hoechst by a nice young Swabian named Heinz Weisshaar, who wanted to form our agency in Venezuela. He told us that he and his friends had established an import business and had by now accumulated sufficient capital to look after the interests of a well-known German company like Hoechst. A little later on, I met his senior partner and promoter of this group, Dr. Enrique Hartung. Hartung's parents had lived in Venezuela for some time. He himself had completed his studies in Germany after the war, and had then returned to Venezuela.

These two young people seemed very suitable to us. Until then, Hoechst had only an insignificant dyestuffs and chemical business in Venezuela, which was looked after by import agents. Through Hartung and Weisshaar the agency gained a new momentum. Their business was conducted purposefully and dynamically, and very soon we concentrated both pharmaceutical and other activities in Remedia, a company in which we acquired a 50 percent interest in 1960.

When Hartung, who became a member of the Hoechst board of management, returned to Europe in 1967, we acquired complete ownership of the company. We had not only attached a pharmaceutical production facility to this company, but in 1966 also a factory in the little town of Maracay on Lake Valencia, twenty minutes away from Caracas. This factory produces textile auxiliaries, dispersions, pigment preparations, and similar products that suitably complement the imported range.

Venezuela and Petrochemistry

In an oil country like Venezuela, we of course also take an interest in petrochemical products, partly on behalf of our engineering subsidiary Uhde and partly in respect to our own projects as well. But in

Venezuela, our plans encountered severe limitations from the beginning.

The state-owned Instituto Venezolano de Petroquímica (IVP) held the monopoly for petrochemistry. It was allowed to take on foreign partners, but its bureaucracy was of such perfection that in spite of our many discussions and negotiations no concrete project was ever realized. The IVP was later integrated into the state-owned Petroven oil company. Since 1976, Petroven has taken over all nationalized, largely foreign companies. It was, however, wise enough to negotiate service contracts with some of these companies, so as to secure expert advisers for further oil drillings.

Petrochemistry concentrated on two sites: Morón on the Caribbean coast, where mainly fertilizers are produced; and El Tablazo on Lake Maracaibo, where a large cracker was erected. Basic chemicals are produced there and capacities for polyolefins have also been established. There are still many difficulties in El Tablazo, because the cracker has problems in going on stream.

We had long negotiations concerning a project for the production of low-pressure polyethylene. At long last, however, conditions did not strike us as attractive enough to participate. The oil euphoria of 1973 intensified the latent tendency of the authorities to buy only technology and export services from foreign countries. Foreign participation is limited to a minimum.

It is difficult not to suspect that in the long term such participation might not be desirable at all, at any rate not in the base industries. Venezuela believes that as the result of its oil wealth, it will always have enough capital of its own. It has found in the last few years, however, that its oil income is not adequate to fully finance the very ambitious industrialization program of its government. Venezuela was forced to fall back on foreign loans which it obtained, of course, without difficulty.

The oil income, which has been rising enormously since 1973, resulted in a vigorous economic boom. But some less desirable consequences of such sudden wealth were not long in coming. Inflation increased, and in spite of all attempts to confine it, it is still in the region of 15 to 20 percent. Unemployment has risen, and the working morale of the population has greatly deteriorated. These conditions are no doubt due in part to very high wages and also to an excessively high level of social compensation.

These impressions were confirmed when, during a stop in Ciudad Guayana in 1972, I visited a new industrial development on the lower reaches of the Orinoco. More than 1000 kilometers from the capital, a large industrial complex is being established. It consists mainly of steel and aluminum plants. Cheap electric energy is obtained from the nearby Guri Dam, which will be enlarged considerably.

The favorable energy situation induced our French colleagues from Nobel-Bozel Electrometallurgie to erect a plant in Ciudad Guayana for the production of ferro-silicon, an important aggregate for the steel industry. The factory, after a building period of about two years, is in operation. It offered a good example of the problems that arise in such new locations where the infrastructure is only partially available. Living conditions for the people who work there are not ideal. There are modern apartment buildings, a hotel, and a former Italian liner that has been converted into another hotel. But there is little else far and wide. Therefore the work force in the entire industrial complex consists essentially of new Venezuelans, immigrants from all over the world.

The Nobel-Bozel company, named Venbozel, employs about 180 people. It had to transfer no less than thirty skilled workers, foremen, and engineers, from France to Ciudad Guayana in order to get the factory started and to keep it going. Qualified domestic labor was not available.

It will probably take years before this situation changes. On the one hand, there is the big advantage of secure, cheap electricity; but on the other, there are all the complications that such a site generates. Environmental protection measures are taken on the same scale as anywhere else.

Problems of the Pharmaceutical Industry

For the pharmaceutical industry in Venezuela, matters are no easier. The state is much concerned with strict observation of good manufacturing practices. Government inspectors regularly visit the plants in order to demand changes, down to the smallest detail, wherever they find that the rules are not, or not yet, completely observed. No doubt the plants in foreign hands are subject to special scrutiny concerning the observance of regulations.

But there are more serious forms of discrimination. A large part of the pharmaceutical turnover is accounted for by the national health insurance program, the Seguro Social. Since 1976, pharmaceutical companies with foreign majorities have been excluded by government decree from supplying the Seguro Social with their products. In addition, there is rigorous price control. This affects not only us but also the Venezuelan establishment of our partners from Roussel-Uclaf, which had achieved a top position in the pharmaceutical industry as the result of a very flexible policy. We shall have to see how we can jointly continue to operate under these difficult circumstances, though each company will, of course, maintain its own trademarks.

Politically, because of its economic and industrial weight and also because of its oil wealth, Venezuela has become the leading power in the Andes Pact countries. This, no doubt, is one of the reasons why the antiforeign investment laws of the Andes Pact are applied particularly strictly. Venezuela will have to see how far it can realize all its plans from its own resources. After all, its oil reserves are not unlimited, and Arturo Uslar Pietri's slogan, *"Hay que sembrar el petroleo"* ("You have to sow the oil"), will not apply for ever. If the standard of living is to be maintained when the oil wells dry up in a few decades, it would be wise, before this happens, to create some other viable base industries and processing plants. In this respect, thanks to its rich mineral resources and its potential energy sources, Venezuela offers the right conditions.

The problem is the people, but they are bound to become increasingly more realistic in their assessments. Complexes they have without doubt, especially about the big neighbor to the north. These will have to be overcome. Then a balanced approach may be found for reasonable cooperation with the Western world.

Bolivia—Democracy in the Andes?

Originally I had intended to journey to Bolivia from Peru over the top of the Andes, thence across Lake Titicaca to La Paz. But the itinerary proved to be rather long and somewhat adventurous for a civilized traveler, particularly as I hoped to take my wife with me. In the end, we flew to La Paz, at about 4000 meters the highest capital city on earth. It is not the country's capital; that honor belongs to Sucre, smaller in size and not situated so high up. But it is the real economic center of a country numbering 5.5 million inhabitants, 60 percent of whom are Indians, 25 percent cholo or mixed, and the rest whites from all over the world.

Most of the Indians live in the lower regions of the country, tropically hot savannas which are put only to agricultural use and are hardly suitable for European visitors.

One of the curiosities of La Paz is the fact that its airport is situated several hundred meters higher than the city in a veritable lunar landscape. A rickety taxi finally gets you to La Paz. The city offers few architectural attractions and can be compared with a medium-sized Spanish mountain town. Some impressive structures from the colonial period have survived, especially the cathedral where one can witness the enduring deep piety of the Indians.

The economic heartland of Bolivia is the Altiplano, a high plateau, rich in minerals, in the southwest. A large part of the working population lives in this area. The Altiplano is bounded by the western Corderillas, which form the frontier with Chile, and the eastern Corderillas, towering over Lake Titicaca, itself 3820 meters above sea level. These are the highest mountains in the country, some of them over 6000 meters.

A long boat trip on Lake Titicaca was the touristic peak of our visit. The infinite loneliness of the landscape was captivating. At this height, there are only limited fauna and flora manifesting themselves in the most curious forms. Probably only a few of the reed huts that line the banks of the lake still form genuine fishing villages, and the reed boats are used more in the service of tourism than in fishing.

Because of the tremendous differences in the living standards, the complex composition of the population, and the unimaginable social gradient, Bolivia has a long history of political unrest. A succession of coups d' état brought a new president practically every recent year. The situation changed only in 1971 when the then Colonel Hugo Banzer Suárez seized power and became the fifty-eighth president since the Republic of Bolivia was established in 1825. Banzer's seven-year tenure set a new record.

He was, however, overthrown in July 1978 by General Juan Pereda Asbún, the disputed winner in the first presidential elections held since 1966. For a short time, the United States withheld economic aid as a sign of displeasure over the military coup. General Pereda subsequently scheduled presidential and congressional elections for July 1979.

The country obviously needs foreign capital in order to exploit its rich mineral wealth and oil and natural gas deposits. There is as yet no chemical industry in the true sense of the word. It can, however, be assumed that some of the existing projects, such as additional refineries and particularly fertilizer plants, may be realized in the foreseeable future.

Hoechst gained early access to the country through an import company founded by some German immigrants named Zetsche. Hoechst Boliviana, which grew out of the Zetzsche company, is located in a small building on one of the main streets in the city center. The building also serves as a warehouse for less bulky products, such as pharmaceuticals and cosmetics. Mrs. Zetzsche, widow of the founder, is successfully devoting most of her attention to this latter business. We were delighted to find that the small European team had succeeded, together with nationals working for our company, in building up a flourishing business in spite of many climatic problems. If the country retains its political stability—the border clashes with Chile and Paraguay seem at long last to be over—healthy future development is likely. Bolivia has no direct access to either the Atlantic or the Pacific. This circumstance, which became almost a national trauma, was the cause of many earlier wars. Argentina has now granted a freeport to Bolivia since the American oil companies, which help with opening up the oil deposits, need access to the sea. Since Bolivia is a member of the Andes Pact, foreign participations are limited to minority holdings.

A bracelet made of small gold nuggets which the Indians still sieve out of the river sand reminds us of this impressive visit. There is not much gold in Bolivia, but there are ample deposits of tin, zinc, and silver.

22

Between Yesterday and Tomorrow

In this story of my trips around the globe, not much has been said about West Germany itself, though I crisscrossed the country often enough during those years. I had many opportunities to play more than an observer's role in some of the major events of the chemical industry during this era.

I think I can safely ignore the years up to Erhard's currency reform. As a result of the dismemberment of the I. G. Farbenindustrie and Allied control, the three major successor companies to I.G. were not established until 1951. As it happened, the new managements of these companies were headed by three men who had spent their decisive development years in positions of senior responsibility, regarded, as it were, as "crown princes" of the concern: Ulrich Haberland at Bayer, Carl Wurster at BASF, and Karl Winnacker at Hoechst. To simplify the story, I proposed to leave aside for the moment other successors to the former I.G., such as Chemische Werke Hüls, Cassella, and Anorgana.

In industry no less than in politics, the fate of a company and its people is decisively influenced by the people at the top, as the history of the three I.G. successors clearly shows. Ulrich Haberland's thinking and actions were shaped by the world renown which Bayer had gained during the I.G. era with pharmaceuticals, agrochemicals, and dye-stuffs. Haberland, the chemist who quickly evolved into an entrepreneur of considerable élan, surrounded himself with people who had chosen the international arena even before the war. He wanted to establish Bayer as the leader of the German chemical industry. Since fusion of the successor companies within Germany was out of the question, Haberland, as the first and often lone "globe-trotter in chemistry," tried unsuccessfully to induce their growing foreign outposts to opt for reunification abroad. Winnacker had learned from Hoechst's past that to abandon commercial independence also means to lose control over industrial affairs.

Although our colleagues from Bayer (Leverkusen) were able, with the aid of the British authorities, to involve themselves in the surviving foreign activities of I.G., they inherited at the same time the expensive duplication of the various divisions abroad.

But in Germany too, Haberland tried to assure Bayer's primacy. On its behalf, he acquired one-third of the Cassella shares through the banks and the stock exchange. BASF and Hoechst followed swiftly, thus preventing any one of the "big three" from becoming the sole landlord in Frankfurt-Fechenheim, the home of Cassella. The company was thus neutralized for quite some time.

Soon after regaining its independence, Bayer incorporated the Agfa photographic division. It is typical of the entrepreneurial vision of Kurt Hansen, Haberland's successor, that he forged the link between Agfa and Gevaert. Hansen had realized that otherwise it would be extremely difficult to compete with Kodak in the long run.

In contrast with Hoechst, Bayer decided to secure its petrochemical base through the establishment of Erdölchemie in partnership with British Petroleum.

The large field of rubber additives was complemented by a 50 percent share in Bunawerke Hüls GmbH. The next logical step was to acquire a 33 percent interest in Hüls itself through its holding company, Chemie-Verwaltungs AG.

Carl Wurster's commercial policy in Ludwigshafen was more concerned with internal consolidation and with fully exploiting the great potential of organic chemistry. Through partnership with Shell in Wesseling, he created the raw material base that enabled him to realize the enormous expansion in high-pressure polyethylene capacities. With an output of 1 million tons a year, they were easily the largest in Western Europe. In my view, the establishment of BASF's foreign organization was probably pursued with greater reticence than was displayed by either Bayer or Hoechst. BASF did not have a pharmaceuticals business at that time and, unlike Bayer and Hoechst, was not concerned with the recovery of trademarks.

The appointments of Wurster as chairman of the supervisory board of BASF in 1965, and of Bernhard Thimm as head of the board of management, ushered in an impressive phase of expansion. The Phrix acquisition proved a serious mistake, but Thimm quickly liquidated the matter. He turned his attention to improving the company's position in fiber raw materials. Moreover, with the acquisition of Siegle GmbH in Stuttgart, a family concern on friendly terms with the former I.G., BASF succeeded in firmly consolidating its stake in the pigment field, thus gaining a notable position in the paint and printing ink industry.

In the mid-sixties, the first postwar generation of private industrialists began to think about preserving its "heritage." In many cases, as with Siegle (in which, incidentally, Bayer and Hoechst had also been interested), the management came to the conclusion that the best way to secure the future of their companies and of their employees was to seek association with a large-scale concern. In this way, Glasurit, the well-known paint company, joined BASF, followed by Herbol, another paint company acquired in fair competition with Bayer.

These actions abolished a taboo which had survived in Germany far longer than in many neighboring European countries, namely, the unwritten law never to enter fields that were properly the province of customers. As a result, the association of many family concerns with the large-scale chemical companies became almost inevitable. (Du Pont and Imperial Chemical Industries, of course, had been leaders in the paint sector for many decades.) The two other large German paint companies therefore followed the same path: Wiederholdt joined ICI and Herberts was taken over by Hoechst.

Wintershall Joins BASF

BASF must have long regretted its lack of a pharmaceutical division. However, through its own research, the company was able to establish sizable capacities for the production of vitamins A and E. The acquisition of the Nordmark works provided the first entrée into producing pharmaceutical specialties. The acquisition of the renowned and research-intensive Knoll GmbH, with which BASF had been friendly for many years, consolidated this process.

At the end of the sixties it became clear that Wintershall AG, a successful business converted from family to foundation ownership, was seeking association with a large-scale chemical company. Wintershall had been developed between the the World Wars by the brilliant entrepreneur, Auguste Rotsberg, from modest beginnings into the largest German potash company. In 1931, it entered the field of oil production by gaining an interest in Gewerkschaft Elwerath. The company also held a strong position in the field of natural gas production in northern Germany, and it acquired a number of other interests, especially in oil distribution.

The Rosberg heirs were not interested in continuing the company. Dr. Josef Rust, chairman of the Wintershall board and one of the few executors, was for many years a member of the Hoechst supervisory board. He often drew our attention to impending changes at Wintershall. Hoechst felt, however, that the Wintershall product line did not offer an attractive addition to its own range. In consequence, Wintershall was acquired by BASF through an exchange of shares.

This meant an enormous addition to BASF's activities. As is the case with most acquisitions of any size, BASF was confronted with problems and compelled to effect a degree of rationalization. On the whole, however, the Wintershall takeover was a successful rounding off.

During the same period, BASF also increased its holdings abroad. The acquisition of Wyandotte in Michigan, in the United States, in 1969 proved a particularly successful move. The versatile production program of that company has been greatly enlarged and made into a profitable operation.

BASF branched into consumer products through electronics for entertainment. Starting with tapes and cassettes, the company soon became the leader in this field in Germany and also established powerful positions abroad. Matthias Seefelder, who succeeded Bernard Thimm in 1974, has faithfully continued the policy of his predecessor in regard to consumer products and expansion in the United States.

Among the three successor companies, Hoechst clearly had the most meager inheritance. Expansion of the site had been neglected ever since the I.G. era. When I spent my first I.G. year in Hoechst in 1937, I perceived the company as a sleeping giant.

After the war, dismemberment was more thorough in the American zone of occupation than elsewhere. When Karl Winnacker returned to

Hoechst in 1951 to prepare the reconstitution of the company, he encountered a host of human and technical problems. He has described this fully in his memoirs *Challenging Years.* But Winnacker remained optimistic even in the face of the most difficult and unusual situations.

The former Knapsack, Griesheim, Offenbach, Gersthofen, and Kalle works were operated as independent units by the trustees. Although these companies had developed a remarkable degree of autonomy, they were integrated in a surprisingly short time. Gendorf in Bavaria, treated like a stepchild during this period, eventually also joined the new grouping.

The takeover of the Bobingen perlon factory was a courageous step in those days. Production was limited and economical. But the acquisition provided Hoechst, with its slender product line, with a launching site into an entirely new field: man-made fibers. With Bobingen as its initial facility, Hoechst was able to acquire from ICI a license for polyester yarns and filaments, thus ushering in the golden years of the Trevira era.

Acquisition of Spinnstoffabrik Zehlendorf AG and of Süddeutsche Chemiefaser AG complemented and enlarged the range of products with rayon staple and acrylic fibers. To shorten the links with customers, Hoechst also took over two texturizing concerns. Once we have overcome the present depression in the man-made fiber industry—and there is no doubt in my mind that the day is not too far off—we shall again be able to derive satisfaction from this vital field.

The Hoechst pharmaceuticals side, almost 100 years old and always linked with the names of outstanding scientists, had to start entirely afresh. Since 1934, its products had been known to doctors and patients only under the Bayer cross. To establish the new tower-and-bridge symbol was by no means easy, even among our own staff. The establishment of a selling organization was even more difficult. A small management team literally had to fight its way through to success.

It was a long haul before Hoechst was able to catch up with the leaders in this field. Although Behringwerke in Marburg, in accordance with the historical facts, had been returned to Hoechst, Bayer had insisted on the formation of a joint selling organization. Its later dissolution in order to realize integration between Hoechst and Behring involved the most difficult negotiations in which I participated in those years.

First Steps into Petrochemistry

Dyestuffs and pharmaceuticals soon became strong pillars of Hoechst. Acetylene chemistry, based on carbide from Knapsack, was less firmly founded. Technologically, it had been outpaced by petrochemistry in which, at first, we had practically no experience. There was not even a refinery in the neighborhood. We therefore decided to obtain a license for the Ziegler patents for the production of high-density polyethylene, although initially only on a laboratory scale. The process was developed for large-scale production in a remarkably short period of time.

Production of polypropylene was also begun, although the patents for this process were held by Montecatini. It took years of complex negotiations before a satisfactory agreement was reached.

To obtain the appropriate raw materials, Hoechst, in conjunction with the American engineering company, Lumnus, constructed a coker in which crude oil was cracked into hydrocarbons at high temperatures. Then Hoechst developed a high-temperature pyrolysis process in which crude oil was again cracked to produce ethylene and propylene.

Hoechst was faced by the basic question of whether to follow the path of backward integration and to take a direct interest in raw material and energy sources, as had Bayer, BASF, and many others. After much thought, we decided instead to collaborate with the oil companies. In 1961, we concluded a long-term supply contract with the American Caltex company, a joint venture of Standard Oil of California and the Texas Company, which had erected a refinery with a cracker at Kelsterbach, not far from Hoechst. During the oil crisis of 1973, this arrangement proved at least as effective as the direct refinery participations of other companies.

Wacker, a 50 percent Hoechst subsidiary, pursued a similar course with Marathon Oil. At first, Marathon constructed a technologically not entirely successful Wulff plant and eventually a cracker to supply the Burghausen works in upper Bavaria. Hoechst took a one-third interest in Ruhrchemie AG, established during the war by the German coal industry. As a diversification into chemicals, it helped to put the company on a sound footing once more. Ruhrchemie decided to participate in the olefin pipelines in the Rhine/Ruhr area which are fed by Esso, Union Rheinische Kraftstoffwerke (UK) Wesseling and others.

The oil crisis of 1973 and its consequent political pressures led us to look anew at our concept of long-term supply contracts with oil companies. We ultimately concluded that our approach was right in principle. Nevertheless, we thought it appropriate to acquire a 25 percent interest in UK Wesseling with whom long-term supply contracts, for example, for ammonia and methanol, had existed for many years. A viable solution of the energy cost problem—and energy is a raw material of the chemical industry—will be decisive for the future competitiveness of German chemical industry in general and of Hoechst in particular.

Involvement in consumer-oriented fields has proved of considerable value for Hoechst, especially in times of varying economic fortunes. Good examples are reprographics as developed by Kalle and cosmetics which we took up in the seventies. The Marbert and Jade lines, recently supplemented by those of Mouson and Balenciaga, are providing welcome additions in this borderland of chemistry. Roussel-Uclaf has similar feelings regarding the presence of Parfums Rochas S.A. in its product mix.

A Major Realignment

Over the years, Hoechst had come to regard the neutralization of Cassella by the big three I.G. successor companies, each holding equal shares, as detrimental to the progress of that company. After all, Cassella had for many years cooperated with Hoechst, finally as part of I.G.

In the 1950s, behind-the-scenes dealings prevented Cassella from joining Hoechst when the successor companies were established. In fact, it was not until January 1, 1970, that Hoechst was able to increase its Cassella interest to 75.59 percent. The agreement came at the end of protracted negotiations during which both the Bayer and BASF holdings in Cassella were made over to Hoechst.

In return, we traded off our 51 percent interest in Chemie-Verwaltungs AG which, in turn, had a 50 percent interest in Chemische Werke Hüls. We had acquired these shares by a lucky coincidence some years before and now transferred them to Bayer and Hibernia, each getting 50 percent. We also yielded our 50 percent holding in Synthesekautschuk-Beteiligungsgessellschaft GmbH, a company in which Bayer was interested. In addition to the Cassella stock, Hoechst also received a substantial cash payment in the transaction.

A long historical tradition was thus reestablished. The dyestuffs, surfactants, auxiliaries, and synthetic resin lines were enlarged with important products. The Cassella pharmaceutical business had been successfully rebuilt and was of considerable importance. In particular, the cardiac and circulatory drug Intensain had achieved renown in medicine. With the merger, Cassella gained access to the worldwide Hoechst marketing organization.

This realignment opened the way for new developments. Herbert Grünewald, who had succeeded Kurt Hansen in 1974, soon realized that half-ownership of Chemische Werke Hüls offered few prospects for entrepreneurial activity. With the blessing of the West German cartel office, therefore, Leverkusen sold its Hüls interest to Veba, which now has a majority interest in the company. Since Veba has built up its chemical operations from the raw materials side, its Hüls holding offers great promise.

Liberality and the European Idea

My work for Hoechst has been so demanding that, apart from my limited leisure time, I have had little opportunity to occupy myself with general economic problems. However, economic and commercial policy has engaged my attention since the end of the war. Whether addressing the Association of the German Chemical Industry or the public at large, and irrespective of all the ups and downs in politics and trade, I have always held to two principles.

The first is the liberal economic concept which ensured the rapid postwar rise not only of West Germany but of all highly industrialized countries. This applies especially to the European Economic Community, which has been extraordinarily successful in industrialization. There must be serious doubts, however, whether its common agricultural policy will succeed in the face of national egoism.

Unfortunately, nationalism and protectionism are once more gaining ground around the world, making it all the more important for us to cling tenaciously to the basically liberal character of our economic policy. Obviously, a liberal trading policy cannot be a one-way affair: it needs partners with the same liberal attitude. Therefore, where we engage in commercial dealings with countries that operate a state economy or follow the principles of self-sufficiency, and that have no intention of accepting the rules of a market economy, we must, of course, use a different set of rules ourselves.

Western Europe must solidly defend its liberal economic policies.

Clearly, in case of serious structural crises, excessive liberality must be moderated, especially where national funds or those of the EEC, instruments that are often misused, have falsified the true competitive positions.

My other, closely related, principle is a firm belief in a Europe that is unified politically as well as economically. Although there were ironical smiles when Roy Jenkins, current president of the EEC Commission, proposed a short-term currency policy, he was certainly not illogical. Of course, a European monetary system presupposes common economic and financial policies. Indeed, these policies are a prerequisite for true monetary union.

But why should we not try to go the other way around? This route is now being attempted through Giscard's and Schmidt's bold initiative in proposing the European Currency System.

At present, Europe is unfortunately a long way from political union. There is scarcely a statesman in the EEC with the courage and prestige to go beyond a mere rhetorical profession of faith to initiate workable, farsighted plans for political unity.

Nevertheless, I remain convinced that Europe will be unified one day. It seems to me the only way in which the Western world can survive politically or economically. Whenever I see the uninhibited manner with which the young people of the West meet one another, how they regard national frontiers as troublesome formalities and indeed ignore them, my deep belief in a united Europe is strengthened.

Liberality and the Third World

Many of my fellow citizens overlook the revolutionary changes taking place outside the comparatively stable European area. Much has altered in the world since the oil crisis of 1973. We cannot speak merely of a Third World anymore. There are the Fourth and Fifth Worlds. The differences in living standards in the rich and the poor developing countries have sharpened alarmingly.

Some of the developing countries have such enormous oil incomes that they can invest only a part in their own economies. The rest they place, not surprisingly, in countries that offer an attractive and reliable rate of interest. To date, the world has not yet succeeded in controlling these vast funds or in warding off the dangers to individual currencies, let alone in arresting inflation. In their own countries, the rich developing nations are taking giant strides in an attempt to catch up with the

West both industrially and, to a degree, socially. But they tend to underestimate the time needed for the expansion, if not establishment, of the required infrastructure, and for creating the technological and social conditions of growth. And, as events in Iran have recently demonstrated, there is a strong contrast between Islamic traditions and Western-style industrialization.

In the long term, industrialization of the Arab oil countries will be successfully accomplished and they will establish their own petrochemical industries. Cheap raw materials, some of them (like the natural gas currently discharged into the atmosphere) costing nothing at all, location advantages, and low wages and salaries will favor the mass production of a variety of goods. In five to ten years, they may well compete in Western markets, even if only to a limited extent. In view of these prospects, the chemical industry must seriously consider which standard mass-production capacities in Western Europe should be expanded. The oil-producing countries will be able to supply these products at prices with which we shall hardly be able to compete.

One advantage remains with the Western nations. The oil countries have no tradition in research or development. They are unlikely to produce new research discoveries or innovations for a long time to come. Research offers one chance for survival of the Western chemical industry. The prime aim, therefore, must be to intensify research and to manufacture more highly developed products to maintain our industrial leadership.

We shall be successful in this endeavor only if, together with the developing countries, we can establish an economic and political climate that promotes successful cooperation. The new wave of nationalism, which we all thought had receded after World War II, does not favor these efforts. Nationalism is rising not just locally but worldwide. The Andes Pact in Latin America is characterized by nationalism just as much as are certain measures in India, Malaysia, the Middle East, or some African states. These nationalistic currents are usually to the detriment of foreign investors, ranging from disregard of patent rights to the limitation of foreign holdings and to many other fields.

Are Foreign Investments Risky?

Hoechst's investment decisions have generally been governed by two criteria: market and partner. These will also apply in the future to overseas investments in which the national partner holds the majority of shares. The political and economic climate in which foreign investors operate must also be taken into account. Are they being offered a fair chance? Are they welcome as investors, and are they being offered conditions that preclude political, economic, and financial discrimination? These are the decisive questions today in developing countries. It is becoming increasingly clear that in past years people generally preferred to place their investments in countries promising a favorable and stable investment climate. This is true primarily for the United States. Although there is sharp competition in the American chemical industry, West European investments escalated between 1973 and 1979. They are maintaining a high level and will probably continue to grow, thus creating a counterbalance to United States investment in Western Europe.

These developments continue the danger that the gap between the developing and the industrialized nations is not likely to be narrowed. It must be hoped that the governments of the world will realize that the living standards of human beings will be improved in the long term only if there is fair play on all sides and if the free exchange of goods and capital is ensured.

A few words about the multinationals. Who could better fulfill the worldwide tasks of the economy than companies to whom international activities are as natural as they are vital? Research into, and development of, new products is becoming more and more expensive and complex. Very often it takes more than ten years from the first synthesis to the marketing of a new product. Companies are likely to engage in that kind of expenditure only if there is a reasonable prospect of utilizing products like new pharmaceuticals or agrochemicals on a worldwide basis.

Anyone who has experienced the many positive aspects of an international presence and of international cooperation simply cannot understand the politically and ideologically motivated campaigns mounted against the multinationals. They are much better than their reputation. In the Western world, it is the multinational corporations among the chemical companies that are making decisive contributions to the improvement of the quality of life. Pharmaceutical products have played an essential role in ensuring a better and longer life. And in the developing countries, it is the same multinationals that are providing the enterprise and the means for creating new industries, new jobs, and new training facilities—that are leading the fight against human misery.

The Future of Chemistry

It looks as though the years of impetuous economic growth are over. Scientists take the view that the invention of new, revolutionary groups of products is unlikely in the near future. I doubt whether anyone can make accurate forecasts of this kind. Quite apart from the fact that I am not a great believer in such forecasts, I do not share this frequently voiced skepticism. Achievements and breakthroughs in scientific research are not the fruits only of systematic planning by large research groups; Dame Fortune has always been an important catalyst in the acquisition of scientific knowledge. And we should never forget that individual genius can still change the world.

Many aspects of human life urgently need scientific and technological progress. Chemistry and physics can help to achieve revolutionary breakthroughs. I am convinced that our industry is one of the most lively and interesting. In the future, too, it will achieve a higher growth rate than many other fields that appear to have reached the limits of their development potentialities.

Also, in the chemical industry the creative resources of scientists and managers will determine whether their companies are reduced to mediocrity or whether they will hold the leading positions that some of them gained rather cozily during the postwar boom years.

Who Forms a Company's Image?

In these postwar years, many industrialists have deliberately kept out of political life. They occasionally may even have looked with disdain upon the "inferior" machinations of politics. Many leading industrialists, at any rate in Germany, had cause to remember the difficulties to which they exposed themselves through political involvement.

However understandable and humanly motivated this political abstinence may have been, it did industry little good. The mass media delighted in denigrating industrialists, but they didn't really know about the work, responsibility, and life of these leaders. The public had to base its ideas on the opulent stereotypes of the television screen.

Fortunately, today's generation of industrialists is better prepared to involve itself in society's development and to discuss its own role and work with a broad public. Modern management no longer has any reason to shy away from political engagement. It is a firm believer in free enterprise and this attitude should help to break down many ideological barriers that have formed from time to time between the discussion partners.

A New Approach to Social Problems

What are the obligations of modern management? I feel they are, first, to employees. Next, to the shareholders, the owners of the company. And finally, to society at large, which is reaping many social, cultural, and financial benefits despite an excessive fiscal burden which siphons off more than half the profits to "redistribute" them expensively and without sufficiently democratic guidance.

If the German economy were heading toward a socialist state dominated by trade unions, I would advise my children to seek their fortunes elsewhere. However, I passionately believe that our system of a liberal, socially committed economy stands a very good chance of weathering the present storms and of proving its superiority in the future. But it would be foolhardy to push the unjustly labeled capitalist system to the limits of its endurance.

Unemployment of over a million, in either Germany or another country, is an unacceptable state of affairs. At the beginning of the 1930s, my father was unemployed for more than a year. I was old enough at the time to appreciate what it must mean to a man in the prime of life to be thrown out of work, especially when there was no comprehensive social insurance. Trade unions which endanger jobs through excessive wage demands are acting irresponsibly and threaten our democratic society.

In Germany, it would help if the labor situation were not presented quite so dramatically and if public opinion were not frequently misled by inadequately evaluated statistics. At present, quite a few of the jobless individuals have a limited interest, or none at all, in doing a job. It is also too often forgotten that the German economic system has succeeded in creating more jobs than its working population can fill. We continue to employ a large foreign labor force whose jobs could be filled with German workers only with difficulty.

The Importance of Public Relations

Almost every day, public opinion expresses itself in attitudes, campaigns, and political measures. Industrialists are in a minority. We must, therefore, convince other people and other groups that there is no system which could be more successful and which would safeguard personal freedom and welfare. That is why public relations is so important to us.

From the beginning, I have regarded full support of our public relations and publicity activities as an essential task. Our publicity has grown rapidly in line with the growth of the various product divisions. No one today doubts its importance in achieving success in highly competitive markets. But in times of temporary recession, it is difficult to ignore demands for curtailment of the advertising budget, although a very good argument can be made for doing precisely the opposite. However, as is so often necessary, a compromise has to be found between the desirable and the practical.

A successful example of new avenues in public relations is our neighborhood newspaper *Blick auf Hoechst* ("Look at Hoechst"). Despite initial skepticism within the organization, this publication has become a highly successful, and as yet unique, cornerstone of our communications with the outside world.

Public confidence is indispensable for the economic success of a company. Confidence is gained through information, through continuing dialogue. It cannot be engineered. It must be carefully tended and constantly nurtured. It is totally wrong to use public relations only to achieve certain commercial objectives and to hide behind the all-too-frequent "No comment" the rest of the time.

Intelligent public relations help develop and mold the image of a company, thus strengthening its fundamental freedoms. Those who wish to change the system, whatever their creed, have little chance of success when ignorance is replaced by information and ideological claims by factual discussion. Hoechst will therefore communicate with its public even more fully in the future.

Much also needs to be done for the chemical industry on the European level. As cofounder and current president of CEFIC (Conseil Européen des Fédérations de l'Industrie Chimique), I have set myself the task of strengthening cooperation between the European chemical industry associations, chemical company managements, and the EEC authorities in Brussels.

I hope CEFIC will help to crystallize a European viewpoint which can provide vital assistance for the Brussels Commission, especially in harmonizing solutions that might be limited by national laws and legislation. An obvious field is environment protection, where every European state now has its own set of rules. We are not only concerned with presenting Hoechst fairly and honestly to the public. We also want to instill in our employees a positive attitude toward the company. It is not enough to proclaim a management philosophy and to develop principles for the support and guidance of our youth. They must be infused with life.

The selection of suitable staff is surely one of the most important management functions in a company. Remembering my own experience, I not only took an interest for many years in new appointments to senior positions in the sales organization but also made sure that Hoechst educated commercial trainees in both practice and theory. This system has been much refined over the years and ensures that the most able young people have the opportunity to study business economics. I am always delighted to find how many of our people, both in Germany and abroad, have entered management by this avenue.

A New Form of Living Together

We shall have to find a new form of living together in practice, a form that makes it clear to people that they are involved in making decisions and that they possess reasonable freedom of action and personal responsibility for implementing them. I believe it is decisive that the young executives, engineers, chemists, physicists—whatever their particular field—have the feeling of working in a company that offers equality of opportunity in the true sense. Each one of our people should have an opportunity to achieve a position in the company, through training, ability to engage in teamwork, and application to the job, that makes identifying with the company's tasks and targets seem worthwhile.

It is the task of the chemical industry to translate the results of scientific research into products and then to make them available to the consumer. However, I believe this description does not do full justice to the entrepreneurial side of the task. Our industry, with its myriad of constantly changing products, demands many different qualities in its people. This certainly applies to company management, which in the chemical industry is composed mainly of people who have made their careers in the company, or at any rate in that particular branch.

In West Germany, in contrast with other industrialized countries, one often hears that chemical companies must be headed by scientists. In my view, this view will change. As in all management functions, intelligence, character, diligence, and the ability to fit into a team are decisive characteristics that carry as much weight as technical knowl-edge.

There can be hardly another branch of industry in which worldwide activity is so obvious and vital. Research into, and development of, new products are becoming increasingly expensive. Companies can afford innovation only if they can market the results. For example, the glamorous aura of the large international chemical companies attracts active young people today just as it inspired me four decades ago to seek the way into the I.G. Farbenindustrie.

Index of Persons

403

Subject Index

Bibliography

Bäumler, E.: Ein Jahrhundert Chemie, Düsseldorf, 1963.
Carr, William H. A.: Die du Ponts, Stuttgart, 1965.
Chemistry in the Economy: American Chemical Society Study, National Science Foundation, Washington, D. C., 1973.
Faber, Gustav: Brasilien, Weltmacht von morgen, Erdmann-Verlag, 1977.
ICI: Auszug aus einer Sonderbeilage "The Times", London, 1962.
Lorei, Madlen—Kirn, Richard: Frankfurt und die drei wilden Jahre, Frankfurt, 1962.
Lilge, Herbert: Deutschland 1954–1963, Hannover, 1967.
Mahoney, John Thomas: Vom Heftpflaster bis zum Antihistamin, Düsseldorf, 1961.
Reader W. J.: Imperial Chemical Industries, A History, London, 1970.
Roeder, O. G.: Reiseführer Indonesien, Frankfurt, 1976.
Whitehead, Don: The Dow Story, New York, 1968.